CW01180409

Monte Carlo Simulation of Semiconductor Devices

Monte Carlo Simulation of Semiconductor Devices

C. Moglestue

Fraunhofer Institute of Applied Solid State Physics
Freiburg im Breisgau, Germany

CHAPMAN & HALL
London · Glasgow · New York · Tokyo · Melbourne · Madras

To my wife, Dela, and children Andrew and Helen

Co-published by James & James (Science Publishers) Ltd, 5 Castle Road, London NW1 8PR and Chapman & Hall, 2-6 Boundary Row, London SE1 8HN

Chapman & Hall, 2-6 Boundary Row, London SE1 8HN, UK

Blackie Academic & Professional, Wester Cleddens Road, Bishopbriggs, Glasgow G64 2NZ, UK

Chapman & Hall Inc., 29 West 35th Street, New York NY10001, USA

Chapman & Hall Japan, Thomson Publishing Japan, Hirakawacho Nemoto Building, 6F, 1-7-11 Hirakawa-cho, Chiyoda-ku, Tokyo 102, Japan

Chapman & Hall Australia, Thomas Nelson Australia, 102 Dodds Street, South Melbourne, Victoria 3205, Australia

Chapman & Hall India, R. Seshadri, 32 Second Main Road, CIT East, Madras 600 035, India

---

First edition 1993

Typeset in 10/11 pt Times Roman
Printed in Great Britain at the University Press, Cambridge

ISBN 0 412 47770-X

A catalogue record for this book is available from the British Library

Library of Congress Cataloging-in-Publication data available

Contents

Preface

Particle simulation of semiconductor devices is a rather new field which has started to catch the interest of the world's scientific community. It represents a time-continuous solution of Boltzmann's transport equation, or its quantum mechanical equivalent, and the field equation, without encountering the usual numerical problems associated with the direct solution. The technique is based on first physical principles by following in detail the transport histories of individual particles and gives a profound insight into the physics of semiconductor devices. The method can be applied to devices of any geometrical complexity and material composition. It yields an accurate description of the device, which is not limited by the assumptions made behind the alternative drift diffusion and hydrodynamic models, which represent approximate solutions to the transport equation. While the development of the particle modelling technique has been hampered in the past by the cost of computer time, today this should not be held against using a method which gives a profound physical insight into individual devices and can be used to predict the properties of devices not yet manufactured. Employed in this way it can save the developer much time and large sums of money, both important considerations for the laboratory which wants to keep abreast of the field of device research. Applying it to already existing electronic components may lead to novel ideas for their improvement. The Monte Carlo particle simulation technique is applicable to microelectronic components of any arbitrary shape and complexity.

Since the purpose of this book is to explain the method, the reader does not need to have any previous knowledge of particle simulation. The book is intended for the research physicist, the electrical engineer and the graduate student who wishes to learn more about this new modelling technique. The method will first be explained in general terms, then in the form of illustrative examples. Readers will also draw from the author's experience in modelling so that they can write their own model software.

To make it easier for the reader to appreciate the method, the book opens with a few chapters reviewing the essence of solid state physics. Here we discuss crystal vibrations, the dynamics of electrons and the interaction between electrons and the lattice, an important ingredient in transport theory. The material should be sufficient to furnish the reader with the necessary physical background, but no rigorous proof of the theory will be given as this can already be found in the existing literature. The main topic, *i.e.* the Monte Carlo particle model, will then be explained in elementary terms and then enlarged upon in the remainder of the book. The book also contains elements of electrical engineering but it is neither intended to be a textbook of this nor of transport theory. Most of the simulated results which are included as examples to illustrate the method have been published elsewhere, but there are also some results that have not yet appeared in the literature. Some examples have also been selected to illustrate the power and the potential of the method.

The author's simulated results have been obtained from the Science and Engineering Research Council's IBM 360/195 at Appleton-Rutherford Laboratory; the VAX machines at GEC Hirst Research Centre at Wembley; the VAX installations at the Fraunhofer Institute of Applied Solid State Physics at Freiburg im Breisgau (Germany); and at the CRAY XMP at the Naval Research Laboratory in Washington DC. All these organisations are herewith acknowledged for their generosity in donating computer time.

The symbols adorning the heads of each chapter represent Mayan numerals, and originate from buildings at Quiriguá and Palenque (Ifrah, 1981).

I would also like to express my gratitude towards Dr Feenstra *et al.* of IBM, Yorktown Heights, and Dr J. Rosenzweig for permission to reproduce photographs. I would also like to thank my colleagues Dr A. Axmann, Dr M. Berroth, Dr J. Braunstein, Dr U. Nowotny and Dr J. Rosenzweig for refereeing the manuscript and Dr Prof. H. Rupprecht for his encouragement to produce this book. Finally I greatly acknowledge Mrs C. Hindrichs for typing the manuscript.

Freiburg im Breisgau, 1992

1

The Foundation of Modelling

1.1 INTRODUCTION

Solid state electronics has had an enormous impact on our lives and now modern civilisation is almost unimaginable without it. Almost all our machines make use of electronic controls; computers enable calculations of complexity which were previously impracticable; administration is supported by information processing machinery. Financial institutions like banks, insurance companies and the government rely increasingly on computers. Medium size and even small firms utilise electronic book-keeping and stock control as this promises to be more reliable and accurate than the traditional methods.

The amount of data and the speed with which it is communicated between humans and machines increases steadily. Modern electronic entertainment is also a way of transmitting information, although it is a one-way process. Future development will probably see this changing into a dialogue.

The speed of communication and the complexity of information processing will undoubtedly increase in future years. This makes it necessary to employ faster circuits, and therefore faster components. There seems to be an unwritten law stating that the capacity of whatever we build will be exploited to the full and that we are then going to require more. This spurs us on to develop increasingly sophisticated systems. Further applications of electronics are envisaged such as video telephones, high definition television, anti-collision and guiding equipment for road vehicles etc. The quest for new markets drives many institutions to invent alternative uses for electronic equipment.

This development has been enabled by the steady miniaturisation of individual components, and by making their manufacture cheaper. The miniaturisation started by making the first transistors smaller; today, quantum mechanical effects such as electron tunnelling through barriers and conduction through narrow channels with quantised transport are being taken advantage of when developing better devices. This process will only stop when we reach the limits set by the laws of physics and it is almost certain that we will come up against them.

Such a development puts new requirements on the device physicist and engineer. Unfortunately, there is a great gap between these two groups of researchers because of their different training. In our opinion this gap should be bridged and we hope this book goes some way towards achieving this.

Considering a single electronic component like a transistor or a diode, the usual approach is to consider the current as a fluid. Although this is only a figurative description, it does give results that can be verified experimentally. In reality, the current consists of individual particles that move through the device. Their motion consists of a sequence of free flight terminated by scattering. The paths of these particles are rather erratic. This is why, in spite of its success, the fluid description is wrong. The correct theoretical description of any semiconductor device can only be obtained by solving Boltzmann's transport equation, or the quantum mechanical equivalent of it, and Maxwell's field equations. The self-consistent solution of these equations is a daunting task which has challenged many prominent minds. Several simplifications had to be made in order to find a solution and this has resulted in the development of e.g. the drift-diffusion and the hydrodynamic models treating the mobile charges as a fluid. The validity of these models is limited and cannot be applied to many of the devices we make today. Fortunately there exists an indirect way to solve the problem without having to make any of the restrictions in validity imposed by these models. The main purpose of this book is to introduce the reader to a technique which is so powerful that it can describe devices of any shape and complexity.

It is important that the reader becomes familiarised with the idea of the current consisting of individual particles. Each particle has its own transport history. Device description consists of following the transport histories of these particles. Both the duration of the free flights and the cause of their termination are distributed stochastically. Choosing random numbers of the same respective distribution, it is possible to calculate transport histories of individual particles and hence simulate the characteristics of the device. This method, known as the Monte Carlo particle method, represents a space–time continuous solution of the field and transport equations. The technique is suitable for studying both the steady state and the dynamic characteristics of devices. Because we are adapting such a first-principles approach, the user gains a profound physical insight into the function of the device. The method can also be considered as a super microscope revealing everything to the most minute detail. It also generates a 'slow motion' picture of what takes place. The method actually mimics nature itself. The simulation technique consists alternatively of following a large number of particles one by one for a set time, then determining the field with the particles in their actual position at the end of that time.

Monte Carlo particle modelling is now well established. The name 'Monte Carlo' for this method was coined by the Italian physicist Fermi when he worked at the Manhattan Project to develop the first nuclear bomb. He used random numbers to calculate the direction of neutrons generated by fission of uranium in a pile. Results of self-consistent Monte Carlo calculations were published as early as the 1970s (Moglestue and Beard, 1979 and Warriner, 1977). Only recently has the method gained in popularity and now several groups around the world concern themselves with Monte Carlo modelling of devices, especially in the USA e.g. Kizilyalli and Hess (1987), Kizilyalli *et al.* (1987), Wang and Hess (1985), Fischetti and Laux (1988) and Park and Brennan (1990). The foremost modellers in Japan are Througnumchai *et al.* (1986) and Tomisawa *et al.* (1988). Here in Europe we can mention Jacoboni and Reggiani (1979 and 1983), Fauquembergue *et al.* (1980) and Nedjalkov and Vitanov (1989). This list is far from exhaustive: references to earlier and other works can be found in the cited papers.

The cause of the initial reluctance towards Monte Carlo simulation is that the method is rather intensive in computer time. This has frightened off many potential users who see the cost of computer time standing in their way. Today this cost is falling and the trend is expected to continue. Adequate work stations large enough for Monte Carlo modelling can today be purchased for less than £20 000.

The amount of computer time needed should no longer inhibit the researcher. An accurate model pays for itself by its predictive power and by the profound understanding it can give. Accurate modelling is just as important as chip manufacture and a model prediction can save the developers considerable sums of money.

The book is aimed at the researcher who wishes to carry out particle modelling of microelectronic devices of any kind or who just wants to become familiarised with the method. It is suitable as a reference text but can also be used as a course book at undergraduate level, or for self study. It is assumed that the reader has some knowledge of elementary quantum mechanics, linear algebra, electromagnetism and elementary electronics. The experienced research engineer, the physicist and the beginner – all will benefit from this book.

We shall start our presentation by stating Maxwell's field and Boltzmann's transport equations. Then a few of the conventional approaches to device modelling will be reviewed, mainly for the purpose of demonstrating their limitations.

To appreciate the Monte Carlo particle model, the reader should be familiar with the physics governing crystal vibrations (phonons), transport of electrons and the interaction between them and the crystal lattice. While the reader may obtain the necessary knowledge elsewhere, the necessary physics will be presented in the following chapters: crystal dynamics and structures in Chapter 2; band structure and electronic states in Chapter 3; carrier scattering in Chapter 4. The physics of quantum transport will be introduced in later chapters.

Chapter 5 introduces the essential part of the Monte Carlo model itself. We shall expand on this in subsequent chapters in the form of heuristic examples. The material will be presented in such a way that readers should be able to write their own simulator. In Chapter 6 the method will be applied to calculate bulk properties of semiconductor materials such as the drift velocity of the carriers versus the applied field. We shall also see how macroscopic concepts such as mobility and relaxation time can be extracted from model calculations. When we are satisfied that the simulator reproduces measured bulk data it will be used to calculate devices.

The method of solving Poisson's field equation in two and three dimensions by means of fast Fourier transforms will be given in Chapter 7. Then, in Chapter 8, we shall embark on device modelling, the best approach to this being to go through particular examples of modelling field-effect transistors. Our choice of examples is limited to what is necessary to understand the Monte Carlo particle model. New aspects, such as contacts, reflection from the surface and the local field will be described extensively. The transistor characteristics, the transconductance, the drain admittance and the negative differential resistivity will also be treated here. Note that no fitting parameters enter device modelling, only fundamental material properties will be made use of.

Chapter 8 is restricted to steady state properties. In Chapter 9 we discuss dynamic properties such as transients, microwave characteristics, the equivalent circuit and gain. These can be calculated by Fourier analysis of the response to a small change in bias. Emphasis will be laid both on the physical and electrical engineering interpretation. The problem of comparison with experimental data will be addressed in Chapter 10.

In Chapter 10 we progress to composite semiconductor devices like MODFETs and others. Quantum effects such as tunnelling and division of the conduction or valence band into subbands will be discussed. The necessary additional quantum physics will be introduced here.

The subject of Chapter 11 will be devices with two types of carriers i.e. those with ambipolar transport. We shall concentrate on bipolar transistors, photodetectors and lasers. Physical effects such as bremsstrahlung and avalanche breakdown also belong to this chapter.

Monte Carlo particle modelling represents a statistical first principles approach. The modelled fluctuations in the particle distribution reflect the true noise, therefore this type of modelling is also well suited for noise studies. A theory of noise will be discussed in Chapter 12, where concepts like the intrinsic minimum noise figure of the transistor will be introduced. This theory is the only one describing noise by considering its origin. Noise consists of three components: noise due to fluctuations in the distribution of the velocities of the particles; collective motion of the opposite charges (plasma waves) and turbulent flow which is analogous to that of gases and liquids.

The book closes with a consideration of computer implications of the model. Some aspects of coding will also be discussed. This section will also go through the development of the model and the origin of its name. Some of the material presented in this book has not been published previously. The reader may find the book biased towards gallium arsenide, because most of the examples are from devices made from this material. The examples are chosen to emphasise the salient point of device modelling, but the choice of certain geometries has been determined by past problems the author has solved. The method is equally applicable to any semiconductor material, and also to *silicon*.

Throughout the book consistent use is made of units complying with the Système International. The only exception to this is the use of the electronvolt to express energies of particles and excitations. The most commonly used symbols have been listed at the end of the book. To avoid an unnecessary proliferation of symbols, infrequently used characters are given a different meaning in different chapters. Within each chapter however, use is consistent throughout. Symbols will be defined the first time they are used.

The cited literature does not pretend to be complete, serving only to support the topics we are going to treat here.

1.2 THE FOUNDATION OF ELECTRONIC TRANSPORT

Maxwell formulated the four fundamental equations describing electromagnetism in 1864. Although these equations are familiar, we shall formulate them here:

1) $$\nabla \bullet \mathbf{D} = \rho_e \tag{1.1}$$

is the most important one in device modelling. **D** represents the *electric displacement* which radiates from the source and ρ_e the space-dependent charge density which is built up from stationary and mobile charges.

2) $$\nabla \bullet \mathbf{B} = 0 \tag{1.2}$$

states that the *magnetic flux*, **B** has no sources.

3) $$\nabla \times \mathbf{F} = -\partial \mathbf{B}/\partial t \tag{1.3}$$

says that when the magnetic flux changes in time an *electric field* **F** is being induced. The relationship between the electric field and the displacement reads

$$\mathbf{D} = \boldsymbol{\varepsilon} \bullet \varepsilon_0 \mathbf{F} \tag{1.4}$$

where ε_0 represents the *permittivity of vacuum*, a fundamental constant, the value of which has been given in the Appendix. $\boldsymbol{\varepsilon}$ represents the *dielectric tensor* which is of order 3 × 3. For homogeneous material this tensor becomes diagonal with all diagonal elements equal, in which case $\boldsymbol{\varepsilon}$ can be replaced by a scalar, *the dielectric constant*, which we symbolise by ε. In vacuum ε still has a meaning, it takes the scalar value 1.

This dielectric tensor describes how the material reacts to the electric field: the field attempts to separate the negative and the positive charges; this results in the formation of dipoles, the material polarises. This can be expressed mathematically as

$$\mathbf{D} = \varepsilon_0 (\mathsf{I} + \mathsf{X}) \bullet \mathbf{F} \tag{1.5}$$

where I and X denote the unit and *susceptibility tensor*, respectively. The polarisation is expressed by

$$\mathbf{P} = \varepsilon_0 \mathsf{X} \bullet \mathbf{F} = \varepsilon_0 (\boldsymbol{\varepsilon} - \mathsf{I}) \bullet \mathbf{F}. \tag{1.6}$$

The efficiency of the polarisation depends on the frequency of the electric field and on the ability of the charges to follow it. Generally, the dielectric tensor is complex, its imaginary part represents the *spectral function*, which is related to the Green's function which describes the interaction between the many charges composing the material.

4)
$$\nabla \times \mathbf{H} = \mathbf{J} + \partial \mathbf{D}/\partial t \tag{1.7}$$

states that the current, **J**, flowing through a unit area perpendicular to it generates a *magnetic field*, **H**. The evolution of the displacement also induces a magnetic field. The relationship between the magnetic flux and field is

$$\mathbf{B} = \boldsymbol{\mu} \mu_0 \bullet \mathbf{H} \tag{1.8}$$

where μ_0 represents the *permeability of vacuum*, the value of which has been quoted in the Appendix. $\boldsymbol{\mu}$ represents the *relative permeability* of the medium, which for most practical cases is a scalar μ. To be accurate, it, too, is a 3 × 3 tensor. The permeability tensor describes how the orbiting electrons and, to a weaker degree, the nuclear spins attempt to align in the magnetic field.

The reader, if used to the cgs, or the Gaussian system of units, may find these equations somewhat unfamiliar. Equations (1.1–8) have been formulated in the Système International (SI) of units, which will be used throughout the book. This system uses the ampere, the metre, the second and the kilogram as the basic units, all the other ones are derived from these four.

The *electrostatic potential*, Φ, is defined through

$$\mathbf{F} = -\nabla \Phi. \tag{1.9}$$

Differentiating this, making use of Equations (1.1) and (1.4) yields

$$\nabla^2\Phi = -\boldsymbol{\varepsilon}^{-1}\rho_e/\varepsilon_0 - (\nabla\boldsymbol{\varepsilon})\bullet\nabla\Phi/\varepsilon_0 \tag{1.10}$$

which is *Poisson's field equation*, a fundamental equation of device modelling. It relates the electrostatic potential to the electric charge.

The magnetic flux can be derived from the *magnetic vector potential* **A**:

$$\mathbf{B} = \nabla \times \mathbf{A}. \tag{1.11}$$

Inserting this into Equation (1.3) gives the relation

$$\nabla \times (\mathbf{F} + \partial\mathbf{A}/\partial t) = 0 \tag{1.12}$$

which implies that

$$\mathbf{F} = -\partial\mathbf{A}/\partial t - \nabla\Phi \tag{1.13}$$

which is Equation (1.9) with the additional term in the time differential in **A**. Equation (1.13) can be verified by forming the curl on both sides. The vector potential is defined by an arbitrary additional vector which can be chosen by convention. The *Coulomb gauge*, defined by choosing

$$\nabla \times \mathbf{A} = 0 \tag{1.14}$$

represents one of them. We shall make use of the vector potential when discussing interaction between electrons and photons.

The magnetic field is caused by currents i.e. moving charges; in a frame of reference where the charge is at rest, the induced magnetic field also vanishes.

1.3 THE TRANSPORT EQUATION

Boltzmann's transport equation, or the quantum mechanical equivalent of it, represents the second leg on which transport theory stands. It describes a system of many particles statistically. A *semiconductor device* is a system of electrically charged particles that interact with an externally applied electric and magnetic field and with each other. The particles are distributed both in geometrical and velocity space. We prefer, however, to work with the wave vector rather than the velocity which is related to it in a simple manner described in Section 3.5.

The particle distribution function is symbolised by

$$f = f(\mathbf{r}, \mathbf{k}, t) \tag{1.15}$$

where **r** and **k** represent geometrical position and wave vector, respectively. The dynamics of f is governed by the law:

$$\partial f/\partial t + \mathbf{v}\bullet\nabla f + (1/\hbar)\partial\mathbf{p}/\partial t\bullet\nabla_k f = \partial f/\partial t|_{\text{coll}}, \tag{1.16}$$

Boltzmann's transport equation, which describes the evolution of the distribution through the motion of the particles, the forces acting on them and their scattering.

The term $\partial f/\partial t$ represents a change in the distribution due to particles being added or subtracted by scattering, by trapping or release of trapped particles, or by generation or recombination.

The second term on the left represents the product of the group velocity, $\mathbf{v}$, of the particles and the spatial gradient of the distribution. The exact meaning of the former will be made clear in Section 3.5 – for the time being the group velocity can be taken as corresponding to the classical velocity. The gradient in the distribution expresses temperature differences within the distribution.

The third term, $1/\hbar\ \partial \mathbf{p}/\partial t$, represents the acceleration driven by the Lorentz force:

$$\partial \mathbf{p}/\partial t = e(\mathbf{F} + \mathbf{v} \times \mathbf{B}). \tag{1.17}$$

This term expresses the influence of the electric and magnetic fields on the distribution. The fields are local, i.e. consisting of external fields and those generated by the other particles of the system.

The right hand side $\partial f/\partial t|_{\text{coll}}$ expresses the rapid changes due to scattering:

$$\partial f/\partial t|_{\text{coll}} = \sum_{k'} [f(\mathbf{r}, \mathbf{k}', t)\ W(\mathbf{k}', \mathbf{k}) - f(\mathbf{r}, \mathbf{k}, t)\ W(\mathbf{k}, \mathbf{k}')] \tag{1.18}$$

where $W(\mathbf{k}, \mathbf{k}')$ represents the rate at which a particle scatters from state $\mathbf{k}$ to $\mathbf{k}'$. The first term expresses the flow of particles into the state $\mathbf{k}$, the last term the loss by scattering into another state. The scattering can be considered instantaneous. In Equation (1.16) the scattering is instantaneous – this approximation has so far not produced any serious error in device simulation. Finite duration of scattering has to be considered in the corresponding quantum mechanical transport equation.

1.4 CONVENTIONAL MODELS

Most of the semiconductor device models in use are based on approximate solutions of Boltzmann's transport equation and, of course, Maxwell's field equations. The most popular models will be briefly examined here – a more rigorous treatment has been given by Snowden (1989). Boltzmann's transport equation is solved by moment expansion: the lowest moment is equivalent to the *drift-diffusion model* which has been reviewed by Selberherr (1984) and Engl *et al.* (1983).

The drift diffusion model maintains *current continuity*:

$$\frac{\partial n_e}{\partial t} = \frac{1}{e} \nabla \bullet \mathbf{J}_n - \mathbf{R} \tag{1.19a}$$

for electrons and

$$\frac{\partial p_h}{\partial t} = \frac{1}{e} \nabla \bullet \mathbf{J}_p - \mathbf{R} \tag{1.19b}$$

for holes. Here $\mathbf{J}_n$ and $\mathbf{J}_p$ represent the electron and hole current densities respectively; R the generation-recombination rate of charge per unit volume; e the magnitude of the elementary electronic charge; n_e and p_h the number density of electrons and holes respectively. The current densities are given by

$$\mathbf{J}_n = en_e\mu_n\mathbf{F} + eD_n\nabla n \tag{1.20a}$$

for electrons and

$$\mathbf{J}_p = ep_h\mu_p\mathbf{F} - eD_p\nabla p \tag{1.20b}$$

for holes. Here μ_p and μ_n represent hole and electron *mobilities* respectively: these are macroscopic quantities independent of the field in the Ohmic regime. D_p and D_n denote, respectively, the hole and electron diffusion constants, which are given by *Einstein's relation*:

$$D_n = k_B T_{Lt}\mu_n/e \tag{1.21a}$$

for electrons and

$$D_p = k_B T_{Lt}\mu_p/e \tag{1.21b}$$

for holes. Furthermore, k_B denotes Boltzmann's constant, T_{Lt} the absolute lattice temperature and $\mathbf{F}$ the electric field.

This model implies that the velocity of the carriers adjusts itself instantaneously to the local electric field and that the velocity is determined uniquely by it. The electrons have mass, however: they therefore possess inertia which implies that it takes time to adjust to the local electric field. An electron may travel about half a micrometre in gallium arsenide before it has 'adjusted': this figure is nearer 0.2 μm in silicon because of the lower mobility of the electrons. This model will therefore become too crude for devices with geometrical dimensions of this order of magnitude.

For large devices this is not a serious problem because the areas of such non-stationary transport are small in comparison with the other parts. This model has enjoyed great popularity among device engineers because of its simplicity and ease of use. Many attempts have therefore been made to adapt it to cases beyond its validity. Some researchers have even gone so far as to fit various parameters to experimental data obtained for definite devices in order to use it to extrapolate to devices of nearly the same design. However, this is not a valid argument for keeping to it in favour of more accurate models. Furthermore, this model is unable to describe transients because the model assumes $\partial f/\partial t = 0$ and Equations (1.20) and (1.21) do not involve time.

The lack of spread in particle velocity for a given field makes the model a quiet one. It is therefore not able to predict noise properties or other phenomena based on the spread in the velocity distribution of the particles.

The *hydrodynamic model*, also known as the energy transport model, presents a great improvement over the drift-diffusion approach, e.g. Constant (1980). It requires greater computational effort than the simple model, and makes use of momentum and energy relaxation times which are macroscopic parameters too. Equation (1.19), describing current continuity, still applies. It also expresses particle continuity when $\mathbf{J}_n$ and $\mathbf{J}_p$ are replaced by $en\mathbf{v}_n$ and $-ep\mathbf{v}_p$, respectively, where $\mathbf{v}_n$ and $\mathbf{v}_p$ denote the electron and hole drift velocities respectively. Momentum conservation is expressed through

$$\frac{\partial \mathbf{v}}{\partial t} + \mathbf{v}\nabla\mathbf{v} + \frac{e\mathbf{F}}{m^*} + \frac{1}{m^* n_e}\nabla(nk_B T_e) + \frac{\mathbf{v}(E)}{\tau_p(E)} = 0 \tag{1.22}$$

and energy conservation through

$$\frac{\partial E}{\partial t} + e\mathbf{v}\bullet\mathbf{F} + \mathbf{v}\bullet\nabla E + \frac{1}{n}\nabla\bullet(n\mathbf{v}\,k_B T_e) + \frac{E - E_0}{\tau_E(E)} = 0 \tag{1.23}$$

with

$$E = \frac{1}{2}m^*\mathbf{v}^2 + \frac{3}{2}k_B T_e \tag{1.24}$$

expressing the average electron energy. Here T_e represents the electron temperature, and τ_p and τ_E the *momentum* and *energy relaxation times* respectively, which themselves are functions of the average electron energy. These can be measured, but during recent years they have also been extracted from Monte Carlo simulation. This is especially the case for devices of complex geometry where they cannot be measured easily and where a quantum mechanical treatment of the transport is expected. E_0 denotes the equilibrium energy given by Equation (1.24) with the lattice temperature T_{Lt} replaced by the electron temperature T_e. Furthermore, m^* represents the effective mass of the carriers and $\mathbf{v}$ the carrier drift velocity.

This concept is based on the idea that the distribution of the electrons is given by

$$f(E) \propto \exp[-(E - E_0)/(k_B T_e)], \tag{1.25}$$

the shifted Maxwellian distribution. We shall see in Chapter 6 that this applies well to silicon, but not in polar semiconductors like e.g. gallium arsenide because of polar optical phonon scattering. Curtice and Yun (1981) have developed the temperature model on the basis of this equation and have been successful in applying it.

Equations (1.22) to (1.24) also apply to holes where the various parameters entering them represent the corresponding ones for holes. The hydrodynamic model has the same weakness as the drift-diffusion one, namely it does not yield any distribution function for the velocities: this model is therefore also a quiet one. Various refinements have been made to this model to apply it to heterojunction devices, but these are questionable.

The only correct approach to semiconductor device modelling is to solve the field and the transport equations in space and time. However, since the direct solution is unfortunately infested with numerical problems, an analytic solution can only be obtained in trivial cases.

The Monte Carlo particle model does not suffer from any of these problems. It represents an indirect solution of the transport and the field equations. This has been proved for a homogeneous electric field by Fawcett *et al.* (1970), Price (1973 and 1981) and more recently by Nedjalkov and Vitanov (1989, 1990 and 1991). In the case of non-homogeneous fields the proof can be obtained by verifying that the carrier distribution calculated from the Monte Carlo particle simulator satisfies the transport equation.

The particle model is also known as the ensemble model, and has been used by several groups around the world as mentioned in Section 1.1. The model involves following the detailed transport histories of a large number of particles for a set time. The field is then determined at the end of this time from the instantaneous position of the particles, the background ionised dopants, the position of the trapped charges and the applied bias. The alternating particle pushing and solution of the field equation is continued as long as necessary. The field is determined often enough that the solution can be considered time continuous, just as the numerical solution of the field equation over a mesh can be considered continuous in space.

The transport consists of sequences of free flights interrupted by scattering. We follow the particles one by one: they move without any consideration for other particles the presence of which is only felt through the local field they set up. The interaction between them is taken care of when solving the field equation because the field charges they carry contribute to the local fields.

Such a scheme is considerably more computer intensive than any of the other models described here, but with the advent of fast computers and work stations there should not be any objection to using the Monte Carlo particle model. A possible saving of computer time can perhaps be obtained combining it with any of the simpler models. Higman *et al.* (1989) and Lebwohl and Price (1971a) suggested a scheme using the particle model for hot electrons and the drift-diffusion approach for the other electrons. Cheng *et al.* (1988) use the Monte Carlo model in certain areas of the device, the drift-diffusion model for the rest of it. Bandyopadhyay *et al.* (1987) have developed a more refined way of drawing up criteria for which areas should be described by the particle model. A very recent discussion on how to combine the drift-diffusion with the Monte Carlo model was published by Kosina *et al.* (1991).

2

Essential Crystallography and Crystal Dynamics

2.1 INTRODUCTION

In the next three chapters we shall review the essential semiconductor physics which describes the transport of electrical charge. A general view will be given, sufficient to be able to appreciate Monte Carlo particle modelling of transport. Students who wish to go deeper into the subject should consult the many excellent textbooks which are available. The physicist who already specialises in crystal dynamics should use these three chapters to become familiarised with what is required and to get a view of the required terminology.

This chapter opens by exposing the principles of crystal formation and how semiconductors, which in their purest form make perfect insulators, acquire their conductive properties. The most commonly applied semiconductors have a crystal structure like that of diamond or zinc blende. Readers will familiarise themselves with this structure in the next section. The Hamiltonian, or the total energy of the entire crystal, consisting of the nuclei and the electrons surrounding them, will then be formulated in Section 2.3. This Hamiltonian consists of three parts: one describing the interaction between the nuclei; one discussing the electron; and one studying the interaction between the nuclei and the electrons, which is the centrepiece of electronic transport. The study of electron dynamics will be carried out in Chapter 3, while the interaction between the electrons and the lattice will be addressed in Chapter 4.

The description of the crystal lattice dynamics presented here closely follows that of Ziman (1960). It is not meant to be fully rigorous, as this would involve too much mathematical detail. The presentation is aimed at demonstrating how the phonon spectrum can be calculated from first principles. A rigorous mathematical quantum mechanical treatment can be found elsewhere. This philosophy of outlining the matter without rigorous proofs also applies to the subjects of the next two chapters.

Chapter 2 also introduces concepts like the reciprocal lattice, and the Brillouin zone, which may loosely be said to be the vibrating atom's or the electron's view of the lattice. A formalism treating the creation and disappearance of quanta of vibrational energy, phonons, will be introduced in Section 2.6. This formalism, which is very popular with theoretical physicists, proves very useful in the description of interaction between different forms of energy. We shall make use of Dirac's notation which will be introduced in Section 2.6.

Phonons have particle properties. Their distribution is governed by Bose-Einstein statistics which will be described in the last section.

2.2 THE ATOMIC ARRANGEMENT IN SEMICONDUCTORS

All solid matter consists of atoms arranged in a regular lattice. The nuclei possess positive charge which is entirely compensated for by that of the electrons surrounding them. These electrons form *closed shells*. The number of electrons within each shell is governed by the laws of quantum physics which will be explained in Section 3.3. In most types of atoms some electrons are left over after the closed shells have been formed. The only way they can obtain closed shells is to get electrons from other atoms. This means that they have to share electrons and therefore arrange themselves to form complexes such as molecules to form liquids or gases or periodic arrays as crystals. Only the latter is of interest to the semiconductor physicist. The only atoms that have the right number of electrons to form only closed shells are those of the eighth column of the periodic system, the inert gases such as helium, neon etc., elements which do not partake in compounds.

Compounds are formed by atoms sharing, usually, eight electrons, known as *valence electrons*. Atoms of the first column of the periodic system, e.g. lithium, sodium etc. have only one electron left when the complete shells have been formed. Fluorine, chlorine etc. of the seventh column lack one electron to complete the last shell. The latter type of atoms form compounds with the former one by annexing the extra electron of the former so that the deprived atom becomes bound to the halogen by electrostatic attraction. An *ionic bond* has been formed. Ordinary table salt, NaCl, is an example of such a bond.

Atoms of the fourth column of the periodic table of elements, e.g. carbon, silicon, germanium etc. have four surplus electrons or lack four electrons to complete the last shell. They share their outermost electrons equally, each valence electron belonging equally to each of the atoms, thus forming a *covalent bond*. NaCl and Si are extreme examples of ionic and covalent bonds: most compounds are formed by a mixed degree of ionicity and covalence (e.g. Harrison, 1980).

Considering the energy of the electrons of the solid, we find that it forms bands with gaps between them. Each band corresponds to one or more complete shells. In silicon and many other materials a gap starts just when the last electron completes the construction of the crystal. An externally applied electric field cannot cause electrons to move because electrical conduction requires that electrons can move into free states which are usually not accessible in this case. The material is a perfect *insulator*. All semiconductors belong to this category of materials. In a metal there is no such gap above the last electron, the required free states for electronic motion are available, therefore the metal becomes a *conductor*.

Returning to our insulator, replacing some of the atoms with ones that have more electrons, we observe that these extra electrons start to fill the band above the gap – they populate the *conduction band*. These electrons are allowed to move in an external electric field, whereby the material has become a weak conductor or an *n-type semiconductor*. On the contrary, replacing some atoms with ones that have less electrons, the valence band is no longer full, there is room for the electrons to move in the valence band; the electrical conduction takes place by motion of vacancies, or *holes* in the valence band. We are now dealing with a *p-type semiconductor*. The band structure will be discussed in more detail in the next chapter where several rigorous mathematical concepts will be introduced.

Both silicon and germanium possess an energy gap between the valence and the conduction band. Other semiconductors consisting of atoms with four valence electrons are C, C_xSi_{1-x}, Si_xGe_{1-x} where x represents the mole fraction of the relevant atom. Also atoms with respectively three and five valence atoms combine to form *III-V semiconductors*. Examples are GaAs, GaP, GaSb, InSb, InAs, InP, AlP, AlAs, and AlSb. Ternary alloys such as $Al_xGa_{1-x}As$, $In_xGa_{1-x}As$ or quaternary alloys, e.g. $Al_xIn_yGa_{1-x-y}As$ will also serve as semiconductors. (Here x and y represent mole fractions.) It is also possible to form *II-VI semiconductors* from atoms with respectively two and six valence electrons: CdS, InSe, PbTe, CdTe, etc.

The great majority of semiconductors finding practical application in electronics have a crystal structure like that of *diamond* or *zinc blende* (ZnS), which is *face-centred cubic*: the perfect crystal can be considered as being built up from cubes with one type of atom (e.g. Zn) in each of the corners and in the middle of each face – the other species of atom are found displaced one quarter of the length along the diagonal, Figs 2.1 and 2.2. It will be useful to become familiarised with this structure as it will be referred to several times. We shall use GaAs as an example, although any other face-centred cubic structure would serve the purpose equally well. Each gallium (arsenic) atom lies in the centre of a tetrahedron with arsenic (gallium) atoms at the corners. The side of the cube is a, the *lattice constant*.

In a Cartesian coordinate system with axes along three orthogonal cube edges, Fig. 2.1, the basic vectors of the cubic unit cell are:

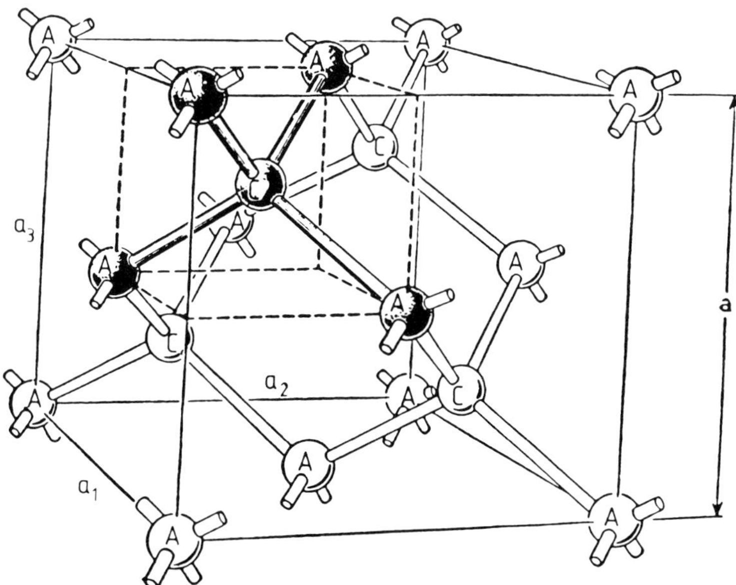

Figure 2.1 *The atomic arrangement of the zinc blende structure in the cubic unit cell of GaAs showing the tetrahedral bond arrangement. A and C represent a- and c-sites respectively.* $\mathbf{a}_1$, $\mathbf{a}_2$ *and* $\mathbf{a}_3$ *indicate the three basic vectors of the cubic unit cell. From Kittel (1961)*

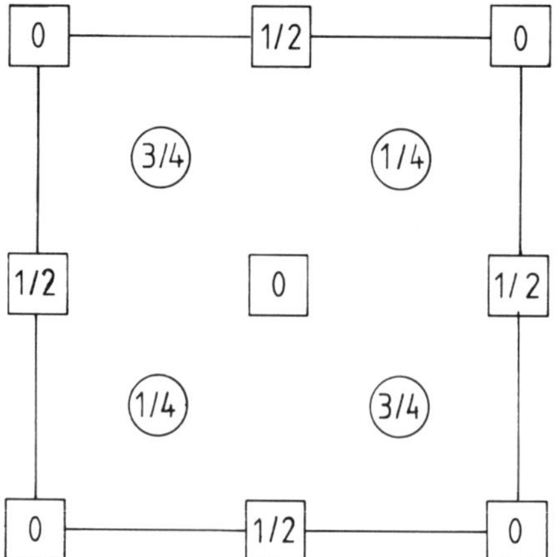

Figure 2.2 *Planar view of the cubic unit cell of GaAs. Squares represent gallium, circles arsenic. The numbers indicate the height of the atoms above the (x, y) plane in units of the lattice constant a. From Kittel (1961)*

$$\mathbf{a}_1 = \begin{pmatrix} 1 \\ 0 \\ 0 \end{pmatrix} \quad \mathbf{a}_2 = \begin{pmatrix} 0 \\ 1 \\ 0 \end{pmatrix} \quad \mathbf{a}_3 = \begin{pmatrix} 0 \\ 0 \\ 1 \end{pmatrix} \tag{2.1}$$

in units of a. Figure 2.1 shows the orientation of these vectors in the cubic cell. The gallium (arsenic) atoms have coordinates

$$\begin{pmatrix} 0 \\ 0 \\ 0 \end{pmatrix}, \begin{pmatrix} 0 \\ \frac{1}{2} \\ \frac{1}{2} \end{pmatrix}, \begin{pmatrix} \frac{1}{2} \\ 0 \\ \frac{1}{2} \end{pmatrix} \text{ and } \begin{pmatrix} \frac{1}{2} \\ \frac{1}{2} \\ 0 \end{pmatrix}, \tag{2.2a}$$

and the other species occupy the positions

$$\begin{pmatrix} \frac{1}{4} \\ \frac{1}{4} \\ \frac{1}{4} \end{pmatrix}, \begin{pmatrix} \frac{1}{4} \\ \frac{3}{4} \\ \frac{3}{4} \end{pmatrix}, \begin{pmatrix} \frac{3}{4} \\ \frac{1}{4} \\ \frac{3}{4} \end{pmatrix} \text{ and } \begin{pmatrix} \frac{3}{4} \\ \frac{3}{4} \\ \frac{1}{4} \end{pmatrix}, \tag{2.2b}$$

also in units of a, Fig. 2.2. The zinc blende structure has the property that the two species of atoms form alternating layers along the (110), or face diagonal, direction. Figure 2.3 shows a scanning tunnelling microscope picture of the (110) surface of gallium arsenide, taken by Feenstra *et al.* in 1987. The photograph clearly demonstrates the layered arrangement of the individual atoms.

This definition of the cubic unit cell is not unique but it is the conventional one. The definition of the *primitive unit cell*, the *Bravais cell*, Fig. 2.4, however, which is spanned by the basis

$$\mathbf{b}_1 = \tfrac{1}{2}a(\mathbf{a}_2 + \mathbf{a}_3) \tag{2.3a}$$

Figure 2.3 *Combined colour scanning tunnelling microscopic image of the GaAs (110) surface, showing the gallium atoms in blue and the arsenic atoms in red. The blue image was acquired by tunnelling into empty states which are localised around the gallium atoms, and the red one by tunnelling out of filled states localised around the surface arsenic atoms. Reproduced with permission of Feenstra* et al. *(1987). Note, this figure is reproduced in colour as Plate 1*

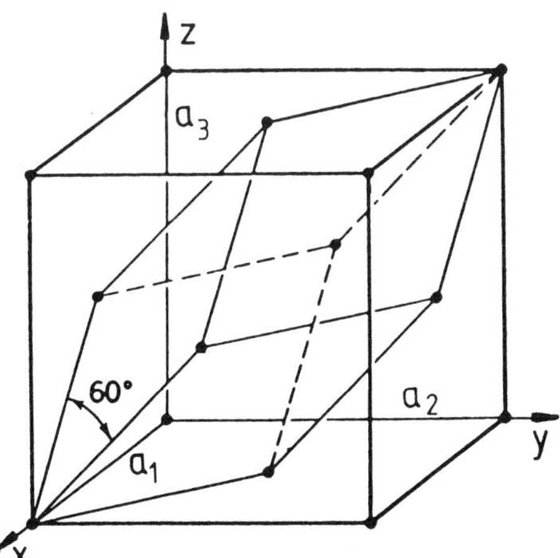

Figure 2.4 *The relationship between the Bravais cell and the cubic unit cell*

$$\mathbf{b}_2 = \tfrac{1}{2}a(\mathbf{a}_1 + \mathbf{a}_3) \tag{2.3b}$$

$$\mathbf{b}_3 = \tfrac{1}{2}a(\mathbf{a}_1 + \mathbf{a}_2) \tag{2.3c}$$

is unique. It contains only two atoms, namely at

$$\mathbf{r}_a = \begin{pmatrix} 0 \\ 0 \\ 0 \end{pmatrix} \tag{2.4a}$$

and

$$\mathbf{r}_c = \frac{a}{4} \begin{pmatrix} 1 \\ 1 \\ 1 \end{pmatrix} \tag{2.4b}$$

which, for convenience, we shall refer to as the anion (a) and the cation (c) site respectively, although the chemical bond between them has little to do with ionicity.

The basis $\{\mathbf{b}_1, \mathbf{b}_2, \mathbf{b}_3\}$ spans the *direct*, or *covariant space*. When we consider momenta or wave vectors of particles, it is convenient to work in the *reciprocal*, or *contravariant, space*. The space vectors are, apart from a factor 2π, vectors of the reciprocal space. The basis vectors in this space are

$$\left.\begin{aligned} \mathbf{b}_1^* &= \mathbf{b}_2 \times \mathbf{b}_3 / V_u \\ \mathbf{b}_2^* &= \mathbf{b}_3 \times \mathbf{b}_1 / V_u \\ \mathbf{b}_3^* &= \mathbf{b}_1 \times \mathbf{b}_2 / V_u \end{aligned}\right\} \tag{2.5}$$

where V_u is the volume of the primitive unit cell;

$$V_u = \mathbf{b}_1 \times \mathbf{b}_2 \bullet \mathbf{b}_3 = \tfrac{1}{4} a^3. \tag{2.6}$$

For the zinc blende or diamond structure the basis of the reciprocal primitive unit cell works out, in terms of the basis vectors of the orthogonal unit cell, Equation (2.1), to be

$$\mathbf{b}_1^* = \frac{1}{a}(-\mathbf{a}_1 + \mathbf{a}_2 + \mathbf{a}_3) \tag{2.7a}$$

$$\mathbf{b}_2^* = \frac{1}{a}(\mathbf{a}_1 - \mathbf{a}_2 + \mathbf{a}_3) \tag{2.7b}$$

$$\mathbf{b}_3^* = \frac{1}{a}(\mathbf{a}_1 + \mathbf{a}_2 - \mathbf{a}_3) \tag{2.7c}$$

which also defines the basis vectors of the primitive unit cell of the *body-centred cubic lattice*. The triplet $\{\mathrm{b}_1^*, \mathrm{b}_2^*, \mathrm{b}_3^*\}$ spans the *contravariant lattice*. Many authors would call this the reciprocal lattice space, but we shall reserve this term for the space spanned by the vectors $2\pi\ \mathbf{b}_i^*$. We shall see below that the momentum and the wave vectors of the electrons are related to the contravariant space by a factor 2π; thus when we study the motion of the current carriers we work in the contravariant space.

A lattice point in the contravariant lattice is given by

$$\mathbf{g} = \frac{1}{a}(l_1\mathbf{b}_1^* + l_2\mathbf{b}_2^* + l_3\mathbf{b}_3^*) \tag{2.8}$$

where l_1, l_2 and l_3 represent integers. The eight nearest lattice points to the origin are the eight vectors

$$\mathbf{g}_3 = \frac{1}{a}(\pm\mathbf{b}_1^* \pm \mathbf{b}_2^* \pm \mathbf{b}_3^*) \tag{2.9a}$$

the six next-nearest ones are

$$\mathbf{g}_4 = \pm\frac{2\mathbf{b}_1^*}{a}; \pm\frac{2\mathbf{b}_2^*}{a}; \pm\frac{2\mathbf{b}_3^*}{a}. \tag{2.9b}$$

(The subscript indicates the squared length of **g**, *the contravariant lattice vector*, in units of $1/a$.)

Figure 2.5 shows the polyhedron defined by the perpendicular planes bisecting the straight lines from the origin to the eight points $\{\mathbf{g}_3\}$ and the six points $\{\mathbf{g}_4\}$. Any planes bisecting the lines connecting more distant reciprocal lattice points with the origin will be too far away to intersect this polyhedron. It has a shape like an octahedron with truncated corners and has six square and eight hexagonal

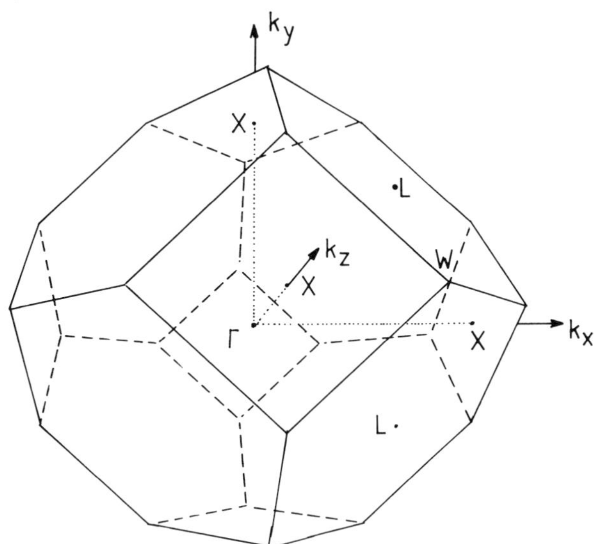

Figure 2.5 *The Brillouin zone for the body-centred cubic (diamond or zinc blende) lattice. It is formed by the planes bisecting the line segments from the origin to the eight points $x = \pm 2$, $y = \pm 2$, $z = \pm 2$, and the eight points of coordinates $(\pm 1, \pm 1, \pm 1)$. The length of each edge of the polyhedron is $= \sqrt{2}$. Γ represents the centre, the other symbols identify points at the surface and main directions of symmetry*

faces, all the edges are equally long; $\sqrt{2}/a$. The entire space can be stacked with polyhedra in such a way that no gaps will be left between them.

We should note the rotational symmetry of this polyhedron: the matrix representation of the four-fold symmetry around the vector $\mathbf{b}_1^*$, the $\langle 100\rangle$ direction, is

$$\mathbf{S}_x = \begin{pmatrix} 1 & 0 & 0 \\ 0 & 0 & 1 \\ 0 & -1 & 0 \end{pmatrix}. \tag{2.10a}$$

$\mathbf{S}_x$ is the transformation matrix representing a rotation of $\pi/2$ around the x axis in the reciprocal space. It rotates a vector 90° about the x axis. A four-fold application of $\mathbf{S}_x$ brings us back to the point where we started, i.e. $\mathbf{S}_x^4 = \mathbf{I}$ (the identity matrix). The four-fold symmetry around $\mathbf{b}_2^*$ is

$$\mathbf{S}_y = \begin{pmatrix} 0 & 0 & -1 \\ 0 & 1 & 0 \\ 1 & 0 & 0 \end{pmatrix}, \tag{2.10b}$$

the four-fold symmetry around $\mathbf{b}_3^*$ is

$$\mathbf{S}_z = \begin{pmatrix} 0 & 1 & 0 \\ -1 & 0 & 0 \\ 0 & 0 & 1 \end{pmatrix} \tag{2.10c}$$

and the six-fold symmetry around the [111] direction is

$$\mathbf{S}_1 = \begin{pmatrix} \frac{2}{3} & \frac{2}{3} & -\frac{1}{3} \\ -\frac{1}{3} & \frac{2}{3} & \frac{2}{3} \\ \frac{2}{3} & -\frac{1}{3} & \frac{2}{3} \end{pmatrix}. \tag{2.11}$$

A rotation around any other seven of the equivalent $\langle 111\rangle$ directions can be obtained by combining the relation given by Equation (2.11) with any of the rotations Equation (2.10). This makes it sufficient to describe only that 1/48 of the polyhedron, i.e. that part lying in the first octant limited by the planes $x = y$, $y = z$ and $x = z$.

Later we shall need the scalar product of contravariant and covariant vectors as argument of the exponential function. The *wave vector* is defined as 2π times the contravariant vector. In the wave vector space the above-mentioned polyhedron also grows linearly by this factor in all directions to become the *Brillouin zone* of the face-centred cubic lattice – we shall meet it again in Section 2.7.

2.3 THE CRYSTAL HAMILTONIAN

As already said, the crystal is built up of cells with atoms, each cell being identical for perfect structures. The arrangement of the atoms looks the same whatever cell it is viewed from. This is an expression of the *cosmological principle.* Relative to a chosen cell corner, the corresponding corner of the other Bravais cells occupy the positions

$$\mathbf{r}_\mathbf{l} = l_1\mathbf{b}_1 + l_2\mathbf{b}_2 + l_3\mathbf{b}_3 \tag{2.12}$$

where $\mathbf{b}_1$, $\mathbf{b}_2$ and $\mathbf{b}_3$ represent vectors along the three edges of the primitive unit cell, and $\mathbf{l}$ is short for the triplet of the integers l_1, l_2 and l_3. Each Bravais cell contains $\mathbf{N}_u$ atoms. In particular $\mathbf{N}_u = 2$ for the zinc blende or the diamond structure.

Relative to the chosen origin atom $N°\ b$ of the unit cell at $\mathbf{r}_\mathbf{l}$ is situated at

$$\mathbf{r}_{\mathbf{l}b} = \mathbf{r}_\mathbf{l} + \boldsymbol{\alpha}_{1b}\mathbf{b}_1 + \boldsymbol{\alpha}_{2b}\mathbf{b}_2 + \boldsymbol{\alpha}_{3b}\mathbf{b}_3 \tag{2.13}$$

where $\boldsymbol{\alpha}_{1b}$, $\boldsymbol{\alpha}_{2b}$ and $\boldsymbol{\alpha}_{3b}$ represent non-negative numbers smaller than or equal to one. The index b runs from l to $\mathbf{N}_u$. This atom vibrates around its equilibrium position: its instantaneous position is

$$\mathbf{r}_{\mathbf{l}b} = \mathbf{r}^0_{\mathbf{l}b} + \mathbf{S}_{\mathbf{l}b} \tag{2.14}$$

where $\mathbf{S}_{\mathbf{l}b}$ represents its displacement.

The atom, which has mass M_b, vibrates with kinetic energy

$$T_{\mathbf{l}b} = \frac{\mathbf{P}^2_{\mathbf{l}b}}{2M_b} \tag{2.15}$$

where $\mathbf{P}_{\mathbf{l}b}$ represents its momentum which is a quantum mechanical operator:

$$\mathbf{P}_{\mathbf{l}b} = -i\hbar\frac{\partial}{\partial\mathbf{S}_{\mathbf{l}b}}\,. \tag{2.16a}$$

We prefer to describe the motion in a coordinate system with origin at the corner of the chosen primitive cell, and with axes along the cube edges of the cubic cell enclosing it, as indicated in Fig. 2.1. In this system $\mathbf{P}_{\mathbf{l}b}$ has components

$$-i\hbar\begin{pmatrix}\dfrac{\partial}{\partial x_{\mathbf{l}b}}\\[2mm] \dfrac{\partial}{\partial y_{\mathbf{l}b}}\\[2mm] \dfrac{\partial}{\partial z_{\mathbf{l}b}}\end{pmatrix} \tag{2.16b}$$

when

$$\mathbf{S}_{\mathbf{l}b} = \begin{pmatrix}x_{\mathbf{l}b}\\ y_{\mathbf{l}b}\\ z_{\mathbf{l}b}\end{pmatrix} \tag{2.17}$$

(x_{lb}, y_{lb} and z_{lb} represent Cartesian components of $\mathbf{S}_{lb}$). The forces that act on the atom tend to restore it to its equilibrium position. The potential energy of a displaced atom is

$$U_{lb} = \sum_{l'b'} U_{lb;l'b'}\left(\mathbf{r}_{l'b'} - \mathbf{r}_{lb}\right) \tag{2.18}$$

so that its total energy is $T_{lb} + U_{lb}$.

Summing the kinetic and potential energies over the entire crystal yields

$$\mathcal{H}_{Lt} = \sum_{l,b} \frac{\mathbf{P}_{lb}^2}{2M_b} + \frac{1}{2}\sum_{l,l',b,b'} U_{lb;l'b'}\left(\mathbf{r}_{l'b'} - \mathbf{r}_{lb}\right). \tag{2.19}$$

The factor $\frac{1}{2}$ has been introduced into the potential term to compensate for the double counting of them. The sums run over the cells from $\mathbf{l} = \mathbf{0}$ to $\mathbf{N} = \{N_1, N_2, N_3\}$, where $N_i (i = 1, 2, 3)$ represents the number of cells in the direction of basis vector $N°\ i$ of the Bravais cell, and over all the atoms, labelled b, within the cell.

An electron of mass m_0 has kinetic energy $\mathbf{p}_i^2/(2m_0)$ and momentum $\mathbf{p}_i$ and does not need to be bound to a particular atom. The total energy of the electrons, consisting of their kinetic and potential one is

$$\mathcal{H}_e = \sum_i \frac{\mathbf{p}_i^2}{2m_0} + \frac{1}{2}\sum_{i,j} \frac{e^2}{4\pi\varepsilon_0 |\mathbf{r}_i - \mathbf{r}_j|} \tag{2.20}$$

where $\mathbf{r}_i$ represents the instantaneous position of electron $N°\ i$, i and j run over all electrons, but $i \neq j$ in the second sum. (Notice that the equilibrium position of the nuclei bear double indices in contrast to the instantaneous position of the electrons, which only get one scalar index because they are not bound to a particular nucleus. The reference corner of the primitive unit cell, however, has one vector index, which is shorthand for three indices.)

The first sum runs over all the electrons, $\mathbf{p}_i$ denotes the operator:

$$\mathbf{p}_i = -i\hbar \begin{pmatrix} \dfrac{\partial}{\partial x_i} \\ \dfrac{\partial}{\partial y_i} \\ \dfrac{\partial}{\partial z_i} \end{pmatrix} \tag{2.21}$$

in our Cartesian coordinates. The elementary electronic charge is represented by e, and ε_0 stands for the permittivity of vacuum. The factor $\frac{1}{2}$ of the second sum has been introduced because the terms are counted twice.

Furthermore, the electron at position $\mathbf{r}_i$ is attracted by the nucleus at $\mathbf{r}_{lb}$ which has a charge $-Ze$. The total potential energy of all the electrons due to this interaction, the *exchange energy*, is

$$\mathcal{H}_{ex} = -\frac{1}{2}\sum_{i,\mathbf{l},b} \frac{Z_b e^2}{4\pi\varepsilon_0 |\mathbf{r}_i - \mathbf{r}_{\mathbf{l}b}|} . \tag{2.22}$$

i runs over all electrons, $\mathbf{l}$ over all Bravais cells and b over all atoms in the Bravais cell.

This rather innocent-looking expression contains a wealth of information: e.g. how do the electrons follow the vibrating nuclei? This question, which has occupied the mind of several prominent physicists, will be addressed in Chapter 4.

The Schrödinger equation for the entire crystal is thus

$$(\mathcal{H}_{Lt} + \mathcal{H}_e + \mathcal{H}_{ex})\,\Psi_T = E_T\Psi_T \tag{2.23}$$

with E_T representing the total energy, consisting of the vibrational one of the nuclei in the lattice, the electronic one and the exchange one due to Coulombic interaction between the electrons and the vibrating lattice; and Ψ_T the wave function of the entire crystal, which depends on all electronic and nuclear coordinates.

The task of solving Equation (2.23) analytically is impossible. However, we can come a long way towards a solution by making use of symmetry and perturbation theory. If the term H_{ex} was absent, the wave function could be written in the form

$$\Psi_T = \Psi_{Lt}(\ldots, \mathbf{r}_{\mathbf{l}b}, \ldots)\,\Psi_e(\ldots, \mathbf{r}_i, \ldots) \tag{2.24}$$

where, as above, $\mathbf{l}$ runs through all the nuclei, and i through the electrons. The wave function Ψ_{Lt} depends on the nuclear coordinates, Ψ_e on the electronic ones only. Substituting Equation (2.24) in (2.23), (taking $H_{ex} = 0$) and dividing by $\Psi_T = \Psi_{Lt}\Psi_e$ yields

$$\frac{\mathcal{H}_{Lt}\Psi_{Lt}}{\Psi_{Lt}} + \frac{\mathcal{H}_e\Psi_e}{\Psi_e} = E_T, \tag{2.25}$$

allowing us to write

$$\mathcal{H}_{Lt}\Psi_{Lt} = E_{Lt}\Psi_{Lt}, \tag{2.26}$$

the *lattice Hamiltonian*, and

$$\mathcal{H}_e\Psi_e = E_e\Psi_e, \tag{2.27}$$

the *electronic Hamiltonian*, where

$$E_{Lt} + E_e = E_T. \tag{2.28}$$

The exchange term will be discussed in Chapter 4.

2.4 CRYSTAL VIBRATIONS, PHONONS

We now turn to the nuclear part of the crystal Hamiltonian. The derivation presented below follows mainly Ziman (1960).

Equation (2.26) reads in full:

$$\mathcal{H}_{Lt} = \sum_{\mathbf{l},b} \frac{\mathbf{P}^2_{\mathbf{l}b}}{2M_b} + \frac{1}{2}\sum_{\substack{\mathbf{l},b\\ \mathbf{l}',b'}} U_{\mathbf{l}b;\mathbf{l}'b'}(\mathbf{r}_{\mathbf{l}'b'} - \mathbf{r}_{\mathbf{l}\mathbf{b}}). \tag{2.19}$$

The expression for the potential energy can be expanded around the equilibrium position:

$$\begin{aligned} U_{\mathbf{l}b;\mathbf{l}'b'}(\mathbf{r}_{\mathbf{l}'b'} - \mathbf{r}_{\mathbf{l}b}) = {} & U_{\mathbf{l}b;\mathbf{l}'b'}(\mathbf{r}^0_{\mathbf{l}'b'} - \mathbf{r}^0_{\mathbf{l}b}) \\ & + \mathbf{S}_{\mathbf{l}b} \bullet \frac{\partial}{\partial \mathbf{S}_{\mathbf{l}b}} U_{\mathbf{l}b;\mathbf{l}'b'}(\mathbf{r}^0_{\mathbf{l}'b'} - \mathbf{r}^0_{\mathbf{l}b}) \\ & + \frac{1}{2}\mathbf{S}_{\mathbf{l}b} \bullet \frac{\partial^2}{\partial \mathbf{S}_{\mathbf{l}b}\partial \mathbf{S}_{\mathbf{l}'b'}} U_{\mathbf{l}b;\mathbf{l}'b'}(\mathbf{r}^0_{\mathbf{l}'b'} - \mathbf{r}^0_{\mathbf{l}b}) \bullet \mathbf{S}_{\mathbf{l}'b'} \\ & + \text{higher order terms in } \mathbf{S}_{\mathbf{l}b}. \end{aligned} \tag{2.29}$$

Here $\mathbf{r}^0_{\mathbf{l}b}$ represents the equilibrium position of the atom at $\mathbf{r}_{\mathbf{l}b}$. These 'higher order terms in $\mathbf{S}_{\mathbf{l}b}$' will be discarded: they represent interaction between phonons. The term $\mathbf{U}_{\mathbf{l}b;\mathbf{l}'b'}(\mathbf{r}^0_{\mathbf{l}'b'} - \mathbf{r}^0_{\mathbf{l}b})$ is just a constant, the potential energy of the system in its equilibrium position, and can be incorporated in E_{Lt} of Equation (2.26). Below we assume that this has been done. The term $\frac{\partial}{\partial \mathbf{S}_{\mathbf{l}b}}$ $U_{\mathbf{l}b;\mathbf{l}'b'}(\mathbf{r}^0_{\mathbf{l}'b'} - \mathbf{r}^0_{\mathbf{l}b}) = 0$ because the potential energy passes through a minimum in the equilibrium position. This may not be true in strained lattices, however. The lattice Hamiltonian reads

$$\mathcal{H}_{Lt} = \sum_{\mathbf{l},b} \frac{\mathbf{P}^2_{\mathbf{l}b}}{2M_b} + \frac{1}{2}\sum_{\substack{\mathbf{l},b\\ \mathbf{l}',b'}} \mathbf{S}_{\mathbf{l}b} \bullet \mathbf{G}_{\mathbf{l}b;\mathbf{l}'b'} \bullet \mathbf{S}_{\mathbf{l}'b'} \tag{2.30}$$

where the *interatomic force tensor*, defined as

$$\mathbf{G}_{\mathbf{l}b;\mathbf{l}'b'} \equiv \frac{\partial^2}{\partial \mathbf{S}_{\mathbf{l}b}\partial \mathbf{S}_{\mathbf{l}'b'}} \mathrm{U}_{\mathbf{l}b;\mathbf{l}'b'}(\mathbf{r}^0_{\mathbf{l}'} - \mathbf{r}^0_{\mathbf{l}}), \tag{2.31}$$

represents a tensor of rank 3. The subscript b has vanished from the arguments of $U_{\mathbf{l}b;\mathbf{l}'b'}$. This is justified from the cosmological principle saying that there is a symmetry in the interaction between atoms at the a and the c sites.

As the crystal contains millions of atoms, the error committed by letting it extend indefinitely in all directions is negligible. Assuming that the original crystal is of the same shape as the Bravais cell and has its edges parallel to the three basic vectors of it, the extension to infinity is carried out by periodic repetition along the three edges. We shall make use of this idea of periodic extension again in Chapter 7 when discussing the solution of Poisson's equation by means of fast Fourier transforms. The future may, however, bring microelectronic devices containing only a few thousand atoms, or less, e.g. the quantum dot, in which case such an approach would no longer be valid; a rigorous solution of the original crystal wave equation is then required. Such an approach may also be necessary when describing boundaries between different materials.

The periodic extension allows $\mathbf{P}_{\mathbf{l}b}$ and $\mathbf{S}_{\mathbf{l}b}$ to be Fourier transformed:

$$\mathbf{P}_{\mathbf{l}b} = \frac{1}{\sqrt{N_{cr}}} \sum_{\mathbf{q}} \mathbf{P}_{\mathbf{q}b} \exp(i\,\mathbf{q}\bullet\mathbf{r}_{\mathbf{l}}) \tag{2.32a}$$

$$\mathbf{S}_{\mathbf{l}b} = \frac{1}{\sqrt{N_{cr}}} \sum_{\mathbf{q}} \mathbf{S}_{\mathbf{q}b} \exp(-i\,\mathbf{q}\bullet\mathbf{r}_{\mathbf{l}}) \tag{2.32b}$$

with the wave vector

$$\mathbf{q} = \frac{2\pi l_1}{N_1}\mathbf{b}_1^* + \frac{2\pi l_2}{N_2}\mathbf{b}_2^* + \frac{2\pi l_3}{N_3}\mathbf{b}_3^*. \tag{2.33}$$

As before, the crystal measures respectively $N_1\ N_2\ N_3$ primitive cells along the three directions, i.e. it contains $N = N_1\ N_2\ N_3$ Bravais cells. l_j $(j = 1,2,3)$ can take the values 0, 1, . . ., $N_j - 1$ in the exponential of Equation (2.32) before the value of the exponential repeats. The vector $\mathbf{q}$ therefore spans a grid of points in the *reciprocal lattice space*.

The inverse transforms read

$$\mathbf{P}_{\mathbf{q}b} = \frac{1}{\sqrt{N_{cr}}} \sum_{\mathbf{l}} \mathbf{P}_{\mathbf{l}b} \exp(-i\,\mathbf{q}\bullet\mathbf{r}_{\mathbf{l}}) \tag{2.34a}$$

and

$$\mathbf{S}_{\mathbf{q}b} = \frac{1}{\sqrt{N_{cr}}} \sum_{\mathbf{l}} \mathbf{S}_{\mathbf{l}b} \exp(i\,\mathbf{q}\bullet\mathbf{r}_{\mathbf{l}}), \tag{2.34b}$$

$\mathbf{P}_{\mathbf{q}b}$ and $\mathbf{S}_{\mathbf{q}b}$ are known as *normal coordinates*.

Inserting Equations (2.32) in (2.30) yields

$$\begin{aligned}
\mathcal{H}_{Lt} &= \frac{1}{2N_{cr}} \sum_{\substack{\mathbf{q},\mathbf{q}' \\ \mathbf{l},b}} \frac{1}{M_b} \mathbf{P}_{\mathbf{q}b} \bullet \mathbf{P}_{\mathbf{q}'b} \exp[i(\mathbf{q}+\mathbf{q}')\bullet\mathbf{r}_{\mathbf{l}}] \\
&+ \frac{1}{2N_{cr}} \sum_{\substack{\mathbf{q},\mathbf{q}' \\ \mathbf{l},\mathbf{l}',b,b'}} \mathbf{S}_{\mathbf{q}b} \exp(-i\mathbf{q}\bullet\mathbf{r}_{\mathbf{l}})\bullet\mathbf{G}_{\mathbf{l}b;\mathbf{l}'b'}\bullet\mathbf{S}_{\mathbf{q}'b'} \exp(-i\mathbf{q}'\bullet\mathbf{r}_{\mathbf{l}'}) \\
&= \frac{1}{2}\sum_{\mathbf{q},b} \frac{\mathbf{P}_{\mathbf{q}b}\bullet\mathbf{P}_{-\mathbf{q}b}}{2M_b N_{cr}} + \frac{1}{\mathbf{q},\mathbf{q}'} \sum_{\substack{\mathbf{l},\mathbf{l}' \\ b,b'}} \mathbf{S}_{\mathbf{q}b}\bullet \sum \mathbf{G}_{bb'} \exp(i\,\mathbf{q}'\bullet\mathbf{h})
\end{aligned}$$

$$\times \exp[-i(\mathbf{q}+\mathbf{q}')\bullet\mathbf{r}_\mathbf{l}]\bullet\mathbf{S}_{\mathbf{q}'b'}. \tag{2.35}$$

The first term after the last equality sign has been arrived at by means of the identity

$$\sum_\mathbf{l}\exp[i(\mathbf{q}+\mathbf{q}')\bullet\mathbf{r}_\mathbf{l}] = \begin{cases} N_{cr} \text{ when} & \mathbf{q}+\mathbf{q}'=\mathbf{0} \\ 0 \text{ otherwise} \end{cases} \tag{2.36}$$

which is strictly valid only when **l** runs to infinity. This is the case when the periodic extension of the crystal discussed above can be applied, i.e. for most bulk crystals met with in practical life. In the second term we have made use of the cosmological principle which says that $\mathbf{G}_{\mathbf{l}b;\mathbf{l}'b'}$ only depends on $\mathbf{h}=\mathbf{r}_\mathbf{l}^0-\mathbf{r}_\mathbf{l}^0$ so that we may write

$$\mathbf{G}_{bb'}(\mathbf{h}) = \mathbf{G}_{\mathbf{l},b;\mathbf{l}',b'}. \tag{2.37}$$

The summation over $\mathbf{l}'$ therefore turns into one over $\mathbf{h}$. The sum

$$\mathbf{G}_{bb'\mathbf{q}} \equiv \sum_\mathbf{h}\mathbf{G}_{bb'}\exp(i\mathbf{q}\bullet\mathbf{h}) \tag{2.38}$$

represents the Fourier transform of the force constant. Summing over **l** as well, making use of the relation (2.36),

$$\mathcal{H}_{Lt} = \frac{1}{2}\sum_{\mathbf{q},b}\frac{\mathbf{P}_{qb}\bullet\mathbf{P}_{-qb}}{M_b} + \frac{1}{2}\sum_{\mathbf{q},b,b'}\mathbf{S}_{\mathbf{q}b}\bullet\mathbf{G}_{bb'\mathbf{q}}\bullet\mathbf{S}_{\mathbf{q}b'}. \tag{2.39}$$

With the further convention that

$$\mathbf{S}^*_{\mathbf{q}b} = \mathbf{S}_{-\mathbf{q}b} \tag{2.40a}$$

and

$$\mathbf{P}^*_{\mathbf{q}b} = \mathbf{P}_{-\mathbf{q}b} \tag{2.40b}$$

the crystal Hamiltonian can be recast into

$$\mathcal{H}_{Lt} = \sum_\mathbf{q}\mathcal{H}_\mathbf{q} \tag{2.41}$$

with

$$\mathcal{H}_\mathbf{q} = \frac{1}{2}\left\{\sum_b\frac{1}{M_b}\mathbf{P}_{\mathbf{q}b}\bullet\mathbf{P}^*_{\mathbf{q}b} + \sum_{b,b'}\mathbf{S}_{\mathbf{q}b}\bullet\mathbf{G}_{bb'\mathbf{q}}\bullet\mathbf{S}^*_{\mathbf{q}b'}\right\}. \tag{2.42}$$

The asterisk denotes the Hermitian conjugate.

By means of Fourier's theorem be crystal has been decoupled into individual oscillators. The wavefunction Ψ_{Lt}, Equation (2.26), can now be factorised:

$$\Psi_{Lt} = \prod_{\mathbf{q}} \Psi_{\mathbf{q}}(\mathbf{S}_{\mathbf{q}b}) \tag{2.43}$$

where $\Psi_{\mathbf{q}}$ only depends on $\mathbf{S}_{\mathbf{q}b}$, $b = 1, 2, \ldots, N_u$. The Schrödinger equation reads

$$\mathcal{H}_{\mathbf{q}} \Psi_{\mathbf{q}} = E_{\mathbf{q}} \Psi_{\mathbf{q}} \tag{2.44}$$

with

$$\sum_{\mathbf{q}} E_{\mathbf{q}} = E_{Lt}. \tag{2.45}$$

2.5 THE PHONON SPECTRUM

The Schrödinger equation for a single oscillator reads

$$\mathcal{H}_{\mathbf{q}} \Psi_{\mathbf{q}} = E_{\mathbf{q}} \Psi_{\mathbf{q}}. \tag{2.44}$$

To solve it, i.e. find the relationship between the energy E_q of the individual oscillators and the wave vector $\mathbf{q}$ we start from the corresponding classical equation of motion

$$M_b \frac{d^2 \mathbf{S}_{\mathbf{q}b}}{dt^2} + \sum_{b'} \mathbf{G}_{bb'\mathbf{q}} \bullet \mathbf{S}_{\mathbf{q}b'} = \mathbf{0}. \tag{2.46}$$

This equation can be derived from Equation (2.44) by means of the Lagrangian and making use of the relation between pairs of conjugate variables. These concepts we shall not need any further so we limit ourselves to write down the result. The interested reader can find the rigorous derivation from Equation (2.44) in the literature, e.g. Kubo and Nagamiya (1969).

The solution of Equation (2.46) is of the form

$$\mathbf{S}_{\mathbf{q}b} = \frac{s^p_{\mathbf{q}b}}{\sqrt{M_b}} \xi^p_{\mathbf{q}b} \exp(i\hbar\omega^p_{\mathbf{q}}) \tag{2.47}$$

where $\xi^p_{\mathbf{q}b}$ represents a dimensionless unit vector in the direction of the vibration of the oscillator, *the polarisation vector*, and $s^p_{\mathbf{q}b}$ an amplitude factor to give $\mathbf{S}_{\mathbf{q}b}$ the correct dimension. Substituting this into Equation (2.46) yields a set of $3N_u$ equations:

$$\sum_{b',j} \left[\frac{1}{\sqrt{M_b M_{b'}}} \mathbf{G}^{ij}_{bb'\mathbf{q}} - \omega^{p2}_{\mathbf{q}} \boldsymbol{\delta}_{bb'} \boldsymbol{\delta}_{ij} \right] \xi_{\mathbf{q}b'}{}^p_j = 0 \tag{2.48}$$

where i and j represent Cartesian coordinate components and $\hbar\omega^p_{\mathbf{q}}$ the energy associated with the polarisation, p, of $\xi^p_{\mathbf{q}b}$. There are $3N_u$ such equations, they have a non-trivial solution (i.e. non-zero $\xi^p_{\mathbf{q}b}$) only when their determinant vanishes. The vanishing determinant defines a secular equation in $\omega^p_{\mathbf{q}b}$ of degree $3N_u$ with generally $3N_u$ roots. For each root there is a corresponding set of values of $\xi_{\mathbf{q}bj}{}^p$ which have been distinguished by the additional label p, the polarisation. We may choose the magnitude of the vectors $\xi^p_{\mathbf{q}b}$ of which $\xi_{\mathbf{q}bj}{}^p$ is a Cartesian component in such a way that

$$\sum_b \xi^{*p}_{\mathbf{q}b} \bullet \xi^{p'}_{\mathbf{q}b} = \delta^{pp'}. \tag{2.49}$$

The polarisation does not necessarily coincide with the coordinate axes, this is the reason why $\xi^p_{\mathbf{q}b}$ has been given the additional index p. The $3N_u$ solutions of the secular equation (2.48) form triplets, one for each atom in the Bravais cell, corresponding to the three directions of polarisation.

For cubic semiconductors Equation (2.48) becomes, written out in full:

$$\begin{pmatrix}
G^{xx}_{aa}/M_a-\omega^2, & G^{xy}_{aa}/M_a, & G^{xz}_{aa}/M_a, & G^{xx}_{ac}/\sqrt{M_aM_c}, & G^{xy}_{ac}/\sqrt{M_aM_c}, & G^{xz}_{ac}/\sqrt{M_aM_c}, \\
G^{yx}_{aa}/M_a, & G^{yy}_{aa}/M_a-\omega^2, & G^{yz}_{aa}/M_a, & G^{yx}_{ac}/\sqrt{M_aM_c}, & G^{yy}_{ac}/\sqrt{M_aM_c}, & G^{yz}_{ac}/\sqrt{M_aM_c}, \\
G^{zx}_{aa}/M_a, & G^{zy}_{aa}/M_a, & G^{zz}_{aa}/M_a-\omega^2, & G^{zx}_{ac}/\sqrt{M_aM_c}, & G^{zy}_{ac}/\sqrt{M_aM_c}, & G^{zz}_{ac}/\sqrt{M_aM_c}, \\
G^{xx}_{ac}/\sqrt{M_aM_c}, & G^{xy}_{ac}/\sqrt{M_aM_c}, & G^{xz}_{ac}/\sqrt{M_aM_c}, & G^{xx}_{cc}/M_c-\omega^2, & G^{xy}_{cc}/M_c, & G^{xz}_{cc}/M_c, \\
G^{yx}_{ac}/\sqrt{M_aM_c}, & G^{yy}_{ac}/\sqrt{M_aM_c}, & G^{yz}_{ac}/\sqrt{M_aM_c}, & G^{yx}_{cc}/M_c, & G^{yy}_{cc}/M_c-\omega^2, & G^{yz}_{cc}/M_c, \\
G^{zx}_{ac}/\sqrt{M_aM_c}, & G^{zy}_{ac}/\sqrt{M_aM_c}, & G^{zz}_{ac}/\sqrt{M_aM_c}, & G^{zx}_{cc}/M_c, & G^{zy}_{cc}/M_c, & G^{zz}_{cc}/M_c-\omega^2,
\end{pmatrix}
\begin{pmatrix} \xi^x_{\mathbf{q}a} \\ \xi^y_{\mathbf{q}a} \\ \xi^z_{\mathbf{q}a} \\ \xi^x_{\mathbf{q}c} \\ \xi^y_{\mathbf{q}c} \\ \xi^z_{\mathbf{q}c} \end{pmatrix} = 0. \tag{2.50}$$

For brevity we have dropped the labels $\mathbf{q}$ from $\mathbf{G}^{jj'}_{bb'\mathbf{q}}$, p and $\mathbf{q}$ from $\omega^p_{\mathbf{q}}$ and p from $\xi_{\mathbf{q}bi}{}^p$. The subscripts a and c refer to anion and cation sites respectively, as defined in Section 2.2.

From Equation (2.38) and the comments preceding it we observe that $\mathbf{G}_{bb'\mathbf{q}}$ represents an interaction between pairs of atoms, and depends only on the distance between them. It is reasonable to assume that the strength of the interaction reduces with increasing distance. On writing down the force constant $G_{bb,\mathbf{q}}$ we start with the four nearest neighbours, then the 12 next nearest ones etc., truncating at those pairs sufficiently distant that their contribution to Equation (2.48) becomes insignificant. The force constants and tables of the nearest neighbours out to the sixth have been given by Herman (1959). The position of the lattice atoms out to the sixth nearest neighbours have been listed in Table 2.1. Symmetry yields the following simplifications

$$G^{ij}_{ac,\mathbf{q}} = G^{*\,ji}_{ac,\mathbf{q}} \tag{2.51a}$$

$$G^{ij}_{ac,\mathbf{q}} = G^{*\,ji}_{ca,\mathbf{q}}. \tag{2.51b}$$

Out to the sixth neighbour

$$G^{i,i}_{bb,\mathbf{q}} = 4\nu_{11} + 4\nu_{21}\left(1 - \cos\tfrac{1}{2}aq_{i+1}\cos\tfrac{1}{2}aq_{i+2}\right)$$

Table 2.1 Neighbours of an atom in a crystal lattice. (The central atom is considered to be at (0, 0, 0), an *a*-site. The nearest neighbours to the atom at the *c*-site at (1/4, 1/4, 1/4) can be found, relative to this site, by interchanging the roles of *c* and *a*)

Neighbour	Position in units of $a/4$			Neighbour	Position in units of $a/4$		
1 (*c*-site)	1	1	1	4 (*a*-site)	0	0	4
	1	−1	−1		0	4	0
	−1	1	−1		4	0	0
	−1	−1	1		0	0	−4
					0	−4	0
2 (*a*-site)	2	2	0		−4	0	0
	2	0	2				
	0	2	2	5 (*c*-site)	3	3	1
	−2	2	0		3	1	3
	−2	0	2		1	3	3
	0	−2	2		−3	3	−1
	2	−2	0		−3	−1	3
	2	0	−2		−1	−3	3
	0	2	−2		3	−3	−1
	−2	−2	0		3	−1	−3
	−2	0	−2		−1	3	−3
	0	−2	−2		−3	−3	1
					−3	1	−3
3 (*c*-site)	−1	−1	−3		1	−3	−3
	−1	−3	−1				
	−3	−1	−1	6 (*a*-site)	2	2	4
	1	−1	3		2	4	2
	1	3	−1		4	2	2
	3	1	−1		−2	2	−4
	−1	1	3		−2	−4	2
	−1	3	1		−4	−2	2
	3	−1	1		2	−2	−4
	1	1	−3		2	−4	−2
	1	−3	1		−4	2	−2
	−3	1	1		−2	−2	4
					−2	4	−2
					4	−2	−2

$$
\begin{aligned}
&+ 4\nu_{31}\left(2 - \cos\tfrac{1}{2}aq_{i+2}\cos\tfrac{1}{2}aq_i - \cos\tfrac{1}{2}aq_i\cos\tfrac{1}{2}aq_{i+1}\right)\\
&+ 8\nu_{41} + 4\nu_{51} + 2\nu_{61}\left(1 - \cos aq_i\right)\\
&+ 2\nu_{71}\left(2 - \cos aq_{i+1} - \cos aq_{i+2}\right)\\
&+ 8\nu_{81} + 4\nu_{91} + 8\nu_{101}\left(1 - \cos aq_i\cos\tfrac{1}{2}aq_{i+1}\cos\tfrac{1}{2}aq_{i+2}\right)\\
&+ 8\nu_{111}\left(2 - \cos aq_{i+1}\cos\tfrac{1}{2}aq_{i+2}\cos\tfrac{1}{2}aq_i\right.\\
&\left. - \cos aq_{i+2}\cos\tfrac{1}{2}aq_i\cos\tfrac{1}{2}aq_{i+1}\right) \qquad (2.52\text{a})
\end{aligned}
$$

with b representing a or c,

$$
\begin{aligned}
G^{i,i+1}_{aa,\mathbf{q}} &= 4\nu_{12}\sin\tfrac{1}{2}aq_i\sin\tfrac{1}{2}aq_{i+1} + 8\nu_{32}\sin\tfrac{1}{2}aq_i\sin\tfrac{1}{2}aq_{i+1}\cos aq_{i+2}\\
&\quad - 4i\nu_{22}\sin\tfrac{1}{2}aq_{i+2}\left(\cos\tfrac{1}{2}aq_i - \cos\tfrac{1}{2}aq_{i+1}\right)
\end{aligned}
$$

$$+ 8\nu_{42}\cos\tfrac{1}{2}aq_{i+2}(\sin\tfrac{1}{2}aq_i\sin aq_{i+1} - \sin\tfrac{1}{2}aq_{i+1}\sin aq_i)$$
$$- 8i\nu_{52}\sin\tfrac{1}{2}aq_{i+2}(\cos\tfrac{1}{2}aq_i\cos aq_{i+1} - \cos\tfrac{1}{2}aq_{i+1}\cos aq_i), \tag{2.52b}$$

$$\begin{aligned}
G_{ac,\mathbf{q}}^{i,i+1} = &-\nu_{11}[1 + \exp(\tfrac{1}{2}iaq_i)\exp(\tfrac{1}{2}iaq_{i+1})\\
&+ \exp(-\tfrac{1}{2}iaq_{i+1})\exp(-\tfrac{1}{2}iaq_{i+2})\\
&+ \exp(-\tfrac{1}{2}iaq_{i+2})\exp(-\tfrac{1}{2}iaq_i)]\\
&- \nu_{51}\{[\exp(-\tfrac{1}{2}iaq_{i+1})\exp(-\tfrac{1}{2}iaq_{i+2}) + 1]\exp(-iaq_i)\\
&+ [\exp(-\tfrac{1}{2}iaq_{i+1}) + \exp(-\tfrac{1}{2}iaq_{i+2})]\exp(-\tfrac{1}{2}iaq_i)\}\\
&- \nu_{41}\{[\exp(-\tfrac{1}{2}iaq_{i+2})\exp(-\tfrac{1}{2}iaq_i) + 1]\exp(-iaq_{i+1})\\
&+ [\exp(-\tfrac{1}{2}iaq_{i+2}) + \exp(-\tfrac{1}{2}iaq_i)]\exp(-\tfrac{1}{2}iaq_{i+1})\}\\
&- \nu_{41}\{(\exp(-\tfrac{1}{2}iaq_i)\exp(-\tfrac{1}{2}iaq_{i+1}) + 1]\exp(-iaq_{i+2})\\
&+ [\exp(-\tfrac{1}{2}iaq_i) + \exp(-\tfrac{1}{2}iaq_{i+1})]\exp(-\tfrac{1}{2}iaq_{i+2})\}\\
&- \nu_{91}\{[\exp(-iaq_{i+1})\exp(\tfrac{1}{2}iaq_{i+2})\\
&+ \exp(-iaq_{i+2})\exp(\tfrac{1}{2}iaq_{i+1})]\exp(-\tfrac{1}{2}iaq_i)\\
&+ \exp(\tfrac{1}{2}iaq_{i+1})\exp(\tfrac{1}{2}iaq_{i+2}) + \exp(-iaq_{i+1})\exp(-\tfrac{1}{2}iaq_{i+2})\}\\
&- \nu_{81}\{[\exp(-iaq_{i+2})\exp(\tfrac{1}{2}iaq_i)\\
&+ \exp(-iaq_i)\exp(\tfrac{1}{2}iaq_{i+2})]\exp(-\tfrac{1}{2}iaq_{i+1})\\
&+ \exp(\tfrac{1}{2}iaq_{i+2})\exp(\tfrac{1}{2}iaq_i) + \exp(-iaq_{i+2})\exp(-iaq_i)\}\\
&- \nu_{81}\{[\exp(-iaq_i)\exp(\tfrac{1}{2}iaq_{i+1})\\
&+ \exp(-iaq_{i+1})\exp(\tfrac{1}{2}iaq_i)]]\exp(-\tfrac{1}{2}iaq_{i+2})\\
&+ \exp(\tfrac{1}{2}iaq_i)\exp(\tfrac{1}{2}iaq_{i+1}) + \exp(-iaq_i)\exp(-iaq_{i+1})\},
\end{aligned} \tag{2.52c}$$

$$\begin{aligned}
G_{ac,\mathbf{q}}^{i,i+2} = &-\nu_{14}[1 + \exp(-\tfrac{1}{2}iaq_i)\exp(-\tfrac{1}{2}iaq_{i+1})\\
&- \exp(-\tfrac{1}{2}iaq_{i+1})\exp(-\tfrac{1}{2}iaq_{i+2})\\
&- \exp(-\tfrac{1}{2}iaq_{i+2})\exp(-\tfrac{1}{2}iaq_i)]\\
&- \nu_{24}\{[\exp(-\tfrac{1}{2}iaq_{i+1})\exp(-\tfrac{1}{2}iaq_{i+2}) - 1]\exp(-iaq_i)\\
&- [\exp(-\tfrac{1}{2}iaq_{i+1}) - \exp(-\tfrac{1}{2}iaq_{i+2})]\exp(\tfrac{1}{2}iaq_i)\}\\
&- \nu_{24}\{[\exp(-\tfrac{1}{2}iaq_{i+2})\exp(-\tfrac{1}{2}iaq_i) - 1]\exp(-iaq_{i+1})\\
&+ [\exp(-\tfrac{1}{2}iaq_{i+2}) - \exp(\tfrac{1}{2}iaq_i)]\exp(-\tfrac{1}{2}iaq_{i+1})\}\\
&- \nu_{34}\{[\exp(-\tfrac{1}{2}iaq_i)\exp(-\tfrac{1}{2}iaq_{i+1}) + 1]\exp(-iaq_{i+2})\\
&- [\exp(-\tfrac{1}{2}iaq_i) + \exp(-\tfrac{1}{2}iaq_{i+1})]\exp(\tfrac{1}{2}iaq_{i+2})\}\\
&- \nu_{44}\{[\exp(-iaq_{i+1})\exp(\tfrac{1}{2}iaq_{i+2})
\end{aligned}$$

$$- \exp(-iaq_{i+2}) \exp(\tfrac{1}{2}iaq_{i+1})] \exp(-\tfrac{1}{2}iaq_i)$$
$$+ \exp(\tfrac{1}{2}iaq_{i+1}) \exp(\tfrac{1}{2}iaq_{i+2}) - \exp(-iaq_{i+1}) \exp(-iaq_{i+2})\}$$
$$+ \nu_{44}\{[\exp(-iaq_{i+2}) \exp(\tfrac{1}{2}iaq_i)$$
$$- \exp(-iaq_i) \exp(\tfrac{1}{2}iaq_{i+2})] \exp(-\tfrac{1}{2}iaq_{i+1})$$
$$- \exp(\tfrac{1}{2}iaq_{i+2}) \exp(\tfrac{1}{2}iaq_i) + \exp(-iaq_{i+2}) \exp(-iaq_i)\}$$
$$+ \nu_{54}\{[\exp(-iaq_i) \exp(\tfrac{1}{2}iaq_{i+1})$$
$$+ \exp(-iaq_{i+1}) \exp(\tfrac{1}{2}iaq_i)] \exp(-\tfrac{1}{2}iaq_{i+2})$$
$$- \exp(-\tfrac{1}{2}iaq_i) \exp(-\tfrac{1}{2}iaq_{i+1}) - \exp(-iaq_i) \exp(-iaq_{i+1})\}. \quad (2.52d)$$

The indices of the Cartesian components of the wave vector $\mathbf{q}$, $i + 1$ and $i + 2$, should be read modulo 3, i.e. they can only take the values 1, 2 and 3, representing the x, y and z Cartesian components.

Here a large range of new constants, ν_{11}, ν_{12} etc., have been introduced, their numerical values are unfortunately not found in standard semiconductor tables. Recently Strauch and Dorner (1990) have calculated the phonon spectrum of GaAs by a least square fit to experimental data, Fig. 2.6. Although the fit is good it does not give a unique set of values for the constants, and the fit does not improve significantly by extending to more distant neighbour atoms. Another source of uncertainty is due to the distribution of isotopes which causes a broadening of the phonon energies for a given wave number.

One general feature of all crystals is that the secular determinant has three degenerate roots $\omega_{\mathbf{q}}^p$ for $\mathbf{q} = 0$. For other values of $\mathbf{q}$ there are generally $3N_c$ distinct roots: in some directions of symmetry several of these may coincide. Each of the roots form generally distinct continuous functions of $\mathbf{q}$, which are referred to as *branches of the phonon spectrum* or *phonon branches*.

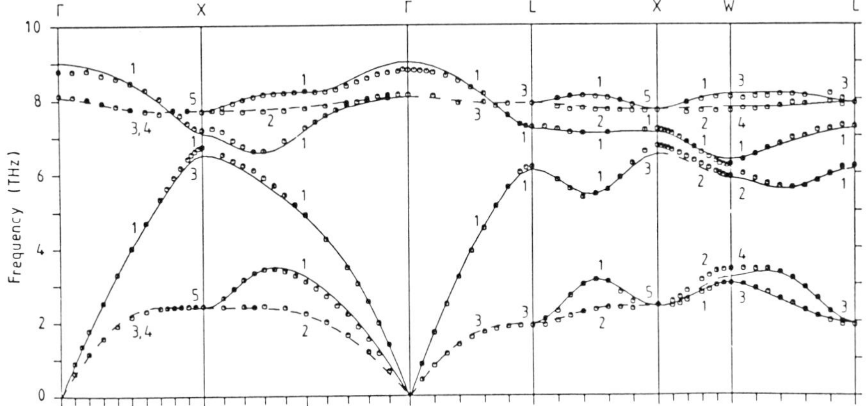

Figure 2.6 *The phonon spectrum of GaAs along the main crystallographic directions. The labelling is the same as in Fig. 2.5. This type of spectrum is typical for crystals of zinc blende or diamond structure. From Strauch and Dorner (1990)*

The three branches that approach zero as $\mathbf{q} \to 0$ represent the *acoustic branches* of the phonon spectrum, since they correspond in the limit of long wavelength to the ordinary elastic waves of a continuum. The other branches are known as *optical modes* of vibrations: neighbouring atoms vibrate against each other, as opposed to the acoustic modes, where they move in unison. The frequencies of the optical modes are about the same as those of infra-red electromagnetic radiation, hence their name. The acoustic modes and the *optical branches* can be grouped in threes: two with motion perpendicular to the direction of propagation, the *transversal mode*; and one along it, the *longitudinal mode*. The phonon branches are perpendicular to the surface of the Brillouin zone. A graphical representation of the complete spectrum is impracticable. It is customary to show the phonon spectra along selected directions in the Brillouin zone, as seen in the different panels in Fig. 2.6.

2.6 THE BRA AND KET NOTATION. THE BOSON CREATION AND ANNIHILATION OPERATORS

Below, the following shorthand notation will be used: in quantum physics, generally, the state of a system or particle is described by a wave probability distribution function, Ψ_n, being the solution of the equation

$$\mathcal{O}\Psi_n = \Omega_n\Psi_n \tag{2.53}$$

where $\mathcal{O}$ represents a quantum mechanical operator, Ω_n the eigenvalue and Ψ_n the corresponding eigenfunction. In many cases only discrete values of Ω_n will be found, which are identified by integer subscripts. When Ω forms a continuum, the variables on which Ψ depends serve as a label. The expectation value of the operator $\mathcal{O}$ is

$$\langle\mathcal{O}\rangle = \int \Psi_n \mathcal{O}\Psi_n d^3\mathbf{r} \tag{2.54}$$

where $\mathbf{r}$ denotes the space coordinates, or domain, over which Ψ_n has been defined. The integral runs over the entire domain where Ψ_n exists. In the case of discrete eigenvalues the integral is replaced by a sum.

Following the shorthand convention introduced by Dirac, we write

$$|n\rangle = \Psi_n \tag{2.55a}$$

the *bra*, whereby Equation (2.53) takes the form

$$\mathcal{O}|n\rangle = \Omega_n|n\rangle. \tag{2.53$'$}$$

The complex conjugate of Ψ_n, Ψ_n^*, is written

$$\langle n| = \Psi_n^* \tag{2.55b}$$

the *ket*. This convention is extended to the expectation value of $\mathcal{O}$:

$$\langle\mathcal{O}\rangle = \langle n|\mathcal{O}|n\rangle. \tag{2.54$'$}$$

Every time this symbol is met, we should understand it as the integral form (2.54). The orthogonality relation of the eigenfunctions is written

$$\langle n|m\rangle = \boldsymbol{\delta}_{nm} \tag{2.56a}$$

for discrete eigenvalues and

$$\langle \mathbf{q}|\mathbf{q}'\rangle = \boldsymbol{\delta}(\mathbf{q}-\mathbf{q}') \tag{2.56b}$$

for continuous ones when the eigenfunctions are functions of $\mathbf{q}$. This representation is short for

$$\int \Psi_n^* \Psi_m d^3\mathbf{r} = \boldsymbol{\delta}_{nm} \tag{2.56a$'$}$$

and

$$\int \Psi_{\mathbf{q}}^* \Psi_{\mathbf{q}'}\, d^3\mathbf{r} = \boldsymbol{\delta}(\mathbf{q}-\mathbf{q}'), \tag{2.56b$'$}$$

respectively; $\boldsymbol{\delta}_{nm}$ represents the *Kronecker symbol*

$$\boldsymbol{\delta}_{nm} = \begin{cases} 1 \text{ for } n = m \\ 0 \text{ otherwise} \end{cases} \tag{2.57a}$$

and δ the delta function

$$\boldsymbol{\delta}(\mathbf{q}-\mathbf{q}') \begin{cases} \rightarrow \infty \text{ as } \mathbf{q} \rightarrow \mathbf{q}' \\ = 0 \text{ otherwise,} \end{cases} \tag{2.57b}$$

but such that

$$\int_{-\infty}^{\infty} \boldsymbol{\delta}(\mathbf{q}-\mathbf{q}')d^3\mathbf{q} = 1. \tag{2.58}$$

The wave equation for the oscillator, Equation (2.44) with $H_{\mathbf{q}}$ given by Equation (2.42), can be solved for each mode of polarisation. The eigenvalues are

$$E_n = \hbar\omega_{\mathbf{q}}^p(n + \tfrac{1}{2}) \tag{2.59}$$

and the corresponding eigenfunction is a Hermitian polynomial of degree n in q^p, the component of $\mathbf{q}$ along the polarisation p, $H_n(q^p)$.

The operator

$$a_{\mathbf{q}}^p = \frac{1}{\sqrt{2\hbar\omega_{\mathbf{q}}^p}} \sum_b \frac{1}{\sqrt{M_b}} \{\boldsymbol{\xi}_{\mathbf{q}b}^{*p} \bullet (\mathbf{P}_{\mathbf{q}b} - i\omega_{\mathbf{q}}^p M_b \mathbf{S}_{\mathbf{q}b}^*)\} \tag{2.60a}$$

and its complex conjugate

$$a_{\mathbf{q}}^{p+} = \frac{1}{\sqrt{2\hbar\omega_{\mathbf{q}}^{p}}}\sum_{b}\frac{1}{\sqrt{M_b}}\{\boldsymbol{\xi}_{\mathbf{q}b}^{p}\bullet(\mathbf{P}_{\mathbf{q}b}^{*} + i\omega_{\mathbf{q}}^{p}M_b\mathbf{S}_{\mathbf{q}b})\} \quad (2.60b)$$

are defined. They are known respectively as annihilation and creation operators for phonons. Conversely

$$\mathbf{S}_{lb} = -i\sum_{\mathbf{q},p}\left(\frac{\hbar}{2N_{cr}M_b\omega_{\mathbf{q}}^{p}}\right)^{\frac{1}{2}}\exp(-i\,\mathbf{q}\bullet\mathbf{r}_l)\boldsymbol{\xi}_{\mathbf{q}b}^{p}(a_{\mathbf{q}}^{+p} - a_{-\mathbf{q}}^{p}) \quad (2.61a)$$

and

$$\mathbf{P}_{lb} = \sum_{\mathbf{q},p}\left(\frac{\hbar\omega_{\mathbf{q}}^{p}M_b}{2N_{cr}}\right)^{\frac{1}{2}}\exp(i\,\mathbf{q}\bullet\mathbf{r}_l)\boldsymbol{\xi}_{\mathbf{q}b}^{p}(a_{\mathbf{q}}^{+p} - a_{-\mathbf{q}}^{p}). \quad (2.61b)$$

In terms of these new operators the Hamiltonian (2.42) reads

$$\mathcal{H}_{\mathbf{q}} = \frac{1}{2}\sum_{p}\hbar\omega_{\mathbf{q}}^{p}(a_{\mathbf{q}}^{+p}a_{\mathbf{q}}^{p} + a_{\mathbf{q}}^{p}a_{\mathbf{q}}^{+p}). \quad (2.62)$$

The operators $\mathbf{a}_q^p$ and $\mathbf{a}_q^{+p}$ obey the following commutation relation:

$$[a_{\mathbf{q}}^{p}, a_{\mathbf{q}'}^{+p'}] \equiv a_{\mathbf{q}}^{p}a_{\mathbf{q}'}^{+p'} - a_{\mathbf{q}'}^{+p'}a_{\mathbf{q}}^{p} = \delta_{pp'}\,\delta(\mathbf{q} - \mathbf{q}'). \quad (2.63)$$

Making use of this, Equation (2.62) can be written

$$\mathcal{H}_{\mathbf{q}} = \hbar\omega_{\mathbf{q}}^{p}\sum_{p}(a_{\mathbf{q}}^{+p}a_{\mathbf{q}}^{p} + \tfrac{1}{2}). \quad (2.64)$$

Operating $a_{\mathbf{q}}^{p}$ upon the eigenfunction H_n yields a Hermitian polynomial which is one degree lower:

$$a_{\mathbf{q}}^{p}H_n(q^p) = \sqrt{n}\,H_{n-1}(q^p). \quad (2.65)$$

In the bra and ket notation this is written

$$a_{\mathbf{q}}^{p}|n_{\mathbf{q}}\rangle = \sqrt{n_{\mathbf{q}}}|n_{\mathbf{q}}-1\rangle. \quad (2.66a)$$

The subscript $\mathbf{q}$ has been added to state that the wave function $|n_{\mathbf{q}}\rangle$ refers to an oscillation of wave vector $\mathbf{q}$. When $a_{\mathbf{q}}^{+p}$ operates on $H_n(q)$, where q^p represents a Cartesian component of $\mathbf{q}$, we get a polynomial of degree $n + 1$, i.e.

$$a_{\mathbf{q}}^{+p}|n_{\mathbf{q}}\rangle = \sqrt{n_{\mathbf{q}} + 1}\,|n_{\mathbf{q}} + 1\rangle. \quad (2.66b)$$

The Schrödinger equation associated with Equation (2.64) is

$$\mathcal{H}_{\mathbf{q}}|n_{\mathbf{q}}\rangle = \hbar\omega_{\mathbf{q}}^{p}(a_{\mathbf{q}}^{+p}a_{\mathbf{q}}^{p} + \tfrac{1}{2})|n_{\mathbf{q}}\rangle = \hbar\omega_{\mathbf{q}}^{p}(n_{\mathbf{q}} + \tfrac{1}{2})|n_{\mathbf{q}}\rangle. \quad (2.67)$$

The last part has been obtained using Equations (2.66a,b). This result tells that the eigenvalues are placed equidistantly $\hbar\omega_{\mathbf{q}}^{p}$ apart. Each time $a_{\mathbf{q}p}$ or $a_{\mathbf{q}p}^{+}$ is applied to H_n, the corresponding eigenvalue lowers or rises by this amount of energy. Differently formulated: such an operator destroys or creates a quantum, $\hbar\omega_{\mathbf{q}}^{p}$ of energy – they are therefore referred to as *annihilation* and *creation operators*, respectively.

This quantum of energy is known as a *phonon*. An oscillator attains an integral number of phonons above its ground state energy, $\frac{1}{2}\ \hbar\omega_{\mathbf{q}}^{p}$. The phonons behave like particles i.e. they propagate through the crystal. Interaction between them is possible through the higher order terms of Equation (2.29) which we have neglected. The phonons are known as *quasiparticles* because of their nature. Heat conduction can be described as the transport of phonons. In this picture $|n_{\mathbf{q}}\rangle$ represents the wave function of $n_{\mathbf{q}}$ phonons each of wave number $\mathbf{q}$.

Annihilation and creation operators have found widespread use in modern theoretical physics. Phonons have no spin; a phonon interacting with an electron does not cause its spin to flip. Particles and quasiparticles of zero or integral spins (in units of $\hbar$) have one common property: they obey Bose-Einstein statistics. This means that more particles can coexist in the same quantum state. The spinless elementary particles like pions and kaons, and the photon with spin $\hbar$ are examples of such particles. The elementary particle physicist often goes as far as postulating particles in terms of their annihilation and creation operators. The reader should not be deterred by this formalism, nor overestimate its value. It only serves to clear the thought, and does not represent any saving in computational effort to solve Schrödinger's equation. Very often a description in terms of these operators is sufficient to state a fact.

The displacement from the equilibrium position and the momentum is described in terms of annihilation and creation operators through Equations (2.61). The use of them is often referred to as *second quantisation*.

In the light of this new concept, Equation (2.64) says that an energy state is first destroyed and then recreated. This formalism is most useful when studying interaction between phonons and electrons – use of it will be made in Chapter 4. Many authors prefer this formalism, e.g. Ting and Nee (1986). An electron of wave vector $\mathbf{k}$ absorbs a phonon of wave vector $\mathbf{q}$ to attain a wave vector $\mathbf{k} + \mathbf{q}$. In terms of these operators the electron of wave vector $\mathbf{k}$ and the phonon are both annihilated and replaced by an electron of wave vector $\mathbf{k} + \mathbf{q}$. In the next chapter we shall meet creation and annihilation operators for electrons, which have somewhat different commutation relations from those we have seen here.

2.7 THE BRILLOUIN ZONE AND THE DENSITY OF STATES

In a periodic lattice it is sufficient to consider phonons inside a limited volume of the wave vector space of the Brillouin zone. Any process referring to a wave vector outside it can be reduced to one inside it by translational symmetry.

The density of states describes how many energy states there are within a small volume of the wave vector space.

2.7.1 The Brillouin zone

The amplitude with which an atom vibrates is given by Equation (2.32b) which demonstrates that the vibration is composed of a number of harmonic oscillations. Considering one particular phonon, its amplitude at the atom in position $\mathbf{r}_{lb}$ is $\mathbf{S}_{\mathbf{q}b}\exp(-i\,\mathbf{q}\bullet\mathbf{r}_l)$. The cosmological principle says that the amplitude is the same at an atom at $\mathbf{r}_l + \mathbf{R}_N$ where

$$\mathbf{R}_N = \sum_{j=1}^{3} N_j \mathbf{b}_j \tag{2.68}$$

where $\mathbf{b}_j$ represents a crystallographic basis vector of length a defined by Equations (2.3) and N_j the number of Bravais cells in the direction of the basis vector $\mathbf{b}_j$; in short, the j direction. This means that

$$\mathbf{S}_{\mathbf{q}b}\exp(-i\mathbf{q}\bullet\mathbf{r}_l) = \mathbf{S}_{\mathbf{q}b}\exp[-i\mathbf{q}\bullet\mathbf{r}_l + i\mathbf{q}\bullet(N_1\mathbf{b}_1 + N_2\mathbf{b}_2 + N_3\mathbf{b}_3)]. \tag{2.69}$$

In this case

$$\exp[i\mathbf{q}\bullet(N_1\mathbf{b}_1 + N_2\mathbf{b}_2 + N_3\mathbf{b}_3)] = 1 \tag{2.70}$$

which is fulfilled when

$$q_j = \frac{2\pi l_j}{N_j a}, l_j = 1, 2, \ldots, N_j \tag{2.71}$$

for $j = 1, 2, 3$. The possible values of $\mathbf{q}_j$ are thus the same as in Equation (2.33). There are thus $N_{cr} = N_1 N_2 N_3$ different triplets $\{l_1, l_2, l_3\}$ and therefore N_{cr} different phonon wave vectors. These lie inside the volume in $\mathbf{q}$-space limited by the vectors $\{\mathbf{G}_i = 2\pi\mathbf{g}_i\}$ where $\mathbf{g}_i$ is given by the expressions $\mathbf{g}_3$ and $\mathbf{g}_4$ defined by Equation (2.9). The shape of the Brillouin zone has been described in Section 2.2, and shown in Fig. 2.5.

2.7.2 The density of states

Equation (2.71) states that $\mathbf{q}$ forms a lattice in the phonon wave vector space. The distance between the lattice points in the j direction is

$$\Delta q_j = \frac{2\pi}{N_j a} = \frac{2\pi}{L_j} \tag{2.72}$$

where L_j is the length of the crystal in the j direction. When the $\mathbf{q}$-space is divided into cells of volume

$$\Delta^3\mathbf{q} \equiv \Delta q_1 \Delta q_2 \Delta q_3 = (2\pi)^3/V_{cr} \tag{2.73}$$

where V_{cr} represents the volume of the crystal, each cell will contain one lattice point.

The *density of states* is defined as

$$D_S(\mathbf{q}) = \frac{1}{8\pi^3} \,. \tag{2.74}$$

Summing this over a volume of **q**-space gives the number of lattice points inside this volume. When the crystal is sufficiently large, the lattice points lie sufficiently close that they can, without making any appreciable error, be considered forming a continuum. Then we may write

$$d^3\mathbf{q} = dq_x dq_y dq_z \tag{2.75}$$

instead of $\Delta^3\mathbf{q}$, and inside an infinitesimal volume element there are

$$D_S(\mathbf{q})d^3\mathbf{q} = \frac{1}{8\pi^3}\, d^3\mathbf{q} \tag{2.76}$$

states. In polar coordinates

$$d^3\mathbf{q} = q^2 dq \sin\theta \, d\theta \, d\phi \tag{2.76'}$$

where ϕ represents the asimuth and θ the polar angle. See Fig. 2.7.

2.8 BOSE-EINSTEIN STATISTICS

Particles or quasiparticles with zero or integral spin (in units of $\hbar$), known as *bosons*, can coexist in the same quantum state, in contrast to particles of half-

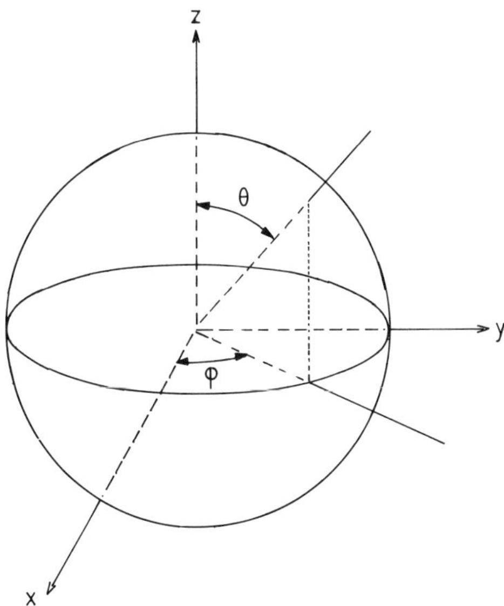

Figure 2.7 *The relationship between polar and Cartesian coordinates. ϕ: azimuth angle, θ: polar angle*

integral spin, like electrons or holes. Bosons occupying the same quantum state cannot be distinguished from each other.

Following Feller (1961) and Landau and Lifshitz (1959) we consider N particles of energy ranging over a range of values. This range can be imagined divided into consequent small subranges: subrange j contains N_j particles and $G_j \geqslant N_j$ energy states. Any of these particles can occupy a definite energy state, so that the total number of ways of distributing the N_j particles amongst the G_j states is

$$\Gamma_j = \frac{(G_j + N_j - 1)!}{(G_j - 1)!N_j!}\,. \tag{2.77}$$

To see this, we may consider the ways in which N_j identical balls (particles) can be distributed over a long, one ball diameter wide box with G_j compartments (energy states). The $G_j - 1$ walls separating the compartments can be moved in order to create all possible arrangements of the balls over the G_j cells. It is possible to arrange the set consisting of the $G_j - 1$ separation and the N_j balls in $(G_j - 1 + N_j)!$ different ways. As the separations and the balls are indistinguishable, like the particles, this number of arrangements of balls and walls has to be divided by $(G_j - 1)!$ and $N_j!$, the number of ways of arranging the $G_j - 1$ separations and the N_j balls respectively.

The total number of arrangements over the entire energy range is the product of the number of arrangements for each energy subrange:

$$\sigma = \prod_{j=1}^{\infty} \Gamma_j. \tag{2.78}$$

The product runs to infinity when the subranges shrink in such a way that their total length remains unchanged.

The *entropy* is the logarithm of σ:

$$S_e = k_B \sum \log\Gamma_j = k_B \sum_{j=1}^{\infty} \{\log(G_j + N_j - 1)! - \log(G_j - 1)!\log N_j!\} \tag{2.79}$$

where the factor k_B, Boltzmann's constant, has been inserted to comply with the conventional definition. If G_j and N_j are large numbers, the logarithms can be approximated by Stirling's formula:

$$\log n! = n\log n - n. \tag{2.80}$$

Figure 2.8 shows that this approximation becomes more accurate with growing n. With this approximation

$$S_e = k_B \sum_{j=1}^{\infty} \{(G_j + N_j)\log(G_j + N_j) - G_j - N_j - G_j\log G_j + G_j - N_j\log N_j + N_j\} \tag{2.81}$$

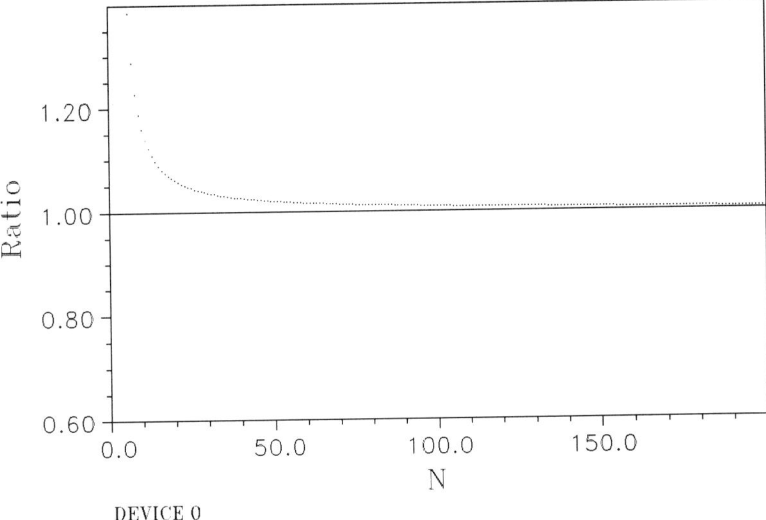

Figure 2.8 *The ratio of log n! and the approximation to it calculated from Sterlings' formula (log n! = n log n − n) against n. The full line represents unit ratio*

where the term 1 has been discarded as $1 \ll G_j$. The most probable distribution is the one which renders S_e largest, subject to the condition

$$E_T = \sum_{j=1}^{\infty} N_j E_j, \tag{2.82}$$

i.e. that the total energy, E_T, of the ensemble is fixed. The requirement that S_e be maximum is another expression for the second law of thermodynamics. Maximum entropy has the physical significance of greatest disorder in the system. The greatest value of S_e, subject to the Condition (2.82) can be obtained from

$$\frac{\partial}{\partial N_j}\left\{S_e + \beta' \sum_{j=1}^{\infty} (E_T - N_j E_j)\right\} = 0 \tag{2.83}$$

yielding

$$f_j \equiv \frac{N_j}{G_j} = \left(\exp\left[\frac{\beta' E_j}{k_B}\right] - 1\right)^{-1}. \tag{2.84}$$

f_j expresses the probability that a state is occupied. It is valid for all the energy intervals, j, so that this subscript may be dropped. The Lagrange multiplier β' has the significance of $1/T_f$, the inverse of the *temperature*. The occupation probability can be formulated as

$$f(E) = \frac{1}{\exp[E_T/(k_B T_f)] - 1}. \tag{2.85}$$

Note that this has been derived without assuming that the quasiparticles, or bosons, need be in thermal equilibrium, a concept that we shall discuss in Section 3.8.

3

Electrons

3.1 INTRODUCTION

Now turning to Equation (2.27) concerning the electrons:

$$\left\{\sum_i \frac{\mathbf{p}_i^2}{2m_0} + \sum_{i,j} \frac{e^2}{4\pi e_0 |\mathbf{r}_i - \mathbf{r}_j|} + U_e(\mathbf{r}_i)\right\}\Psi_e = E_e \Psi_e \tag{3.1}$$

where the first term represents the kinetic energy, the second the mutual Coulombic repulsion between the electrons, and the last the potential energy in the electric field from the ions which, to a first approximation, is given by Equation (2.22) with $\mathbf{r}_{\mathbf{l}b}^0$, the equilibrium position of the nucleus, instead of $\mathbf{r}_{\mathbf{l}b}$. Part of the nuclear charge, Z_b, is screened off by the deeper lying core electrons. The periodicity of the crystal allows us, in analogy with the previous chapter, to write

$$U_e(\mathbf{r}_i) = \sum_{\mathbf{l},b} U_{e,\mathbf{l}b}(\mathbf{r}_i - \mathbf{r}_{\mathbf{l}b}^0). \tag{3.2}$$

Equation (3.1) is even more difficult to solve than the lattice Hamiltonian Equation (2.26). An enormous amount of effort has been put into this problem and describing it will be lengthy, so we shall restrict ourselves to two practical methods of calculating the relationship between the energy and the momentum of the valence and conduction band electrons, the *band structure* in Section 3.5. But before starting, a few widely used concepts have to be introduced.

The purpose of this chapter is to introduce the electronic structure of the crystal. The calculation of the band structure is only one aspect of this. We also ought to develop an understanding of what the electronic orbits look like. A free, electrically neutral atom has just as many electrons as protons in the nucleus. These electrons are arranged in shells, which are complete, or closed, except for the outermost one. (Only the atoms of the inert gases have the outermost shell complete as well.) Each shell contains a fixed number of electrons of nearly the same energy, but electrons from different shells have distinctly different binding energies. In Section 3.3 we shall introduce the electronic orbitals.

The atoms achieve completion of the last shell by sharing valence electrons with other atoms, thus forming compounds. The atoms of the crystal come so close together that electrons, which in perfectly isolated atoms would have the same energy, attain energies spread out over a band. The bands formed by the electrons nearest to the nuclei are the narrowest since these electrons are screened off most by those further away from them. Some energy bands overlap one another, others are separated by gaps.

If the thermal vibrations of the lattice were absent, the electrons would fill up to a certain level in energy, the *Fermi level*, and no electrons would have energy above it.

Apart from spin, no two electrons in a lattice are permitted to have the same state, unlike phonons. Electrons obey a different kind of statistics from bosons, the Fermi–Dirac statistics. The formula for the Fermi–Dirac distribution and the electronic density of states will be derived in Section 3.6.

If the nuclei were absent from the lattice and the electrons did not repel one another, they would behave like free electrons in a confinement defined by the surfaces of the semiconductor crystal. Free electrons exhibit a quadratic relationship between energy and momentum. This, in essence, is Sommerfeld's theory (Section 3.2). The presence of the lattice will still render Sommerfeld's theory valid, provided the electrons are attributed an effective mass different to that of the free ones. A more accurate calculation, based on pseudopotentials, of the relationship between the energy and the momentum of the electrons, the band structure, will be presented in Section 3.5. These energy bands often overlap one another, but there will also be gaps between some of them.

In all semiconductors the last band, the *valence band*, is complete; there is no room for additional electrons in it. There is a gap between it and the next band, the *conduction band*, which is empty in pure material. The energy gap that forms between the valence and the conduction band is an essential characteristic for all semiconductors.

Replacing some of the host atoms by ones with more valence electrons will make the semiconductor *n-type*; the extra electrons will be so loosely bound to the foreign atoms that they easily get excited into the conduction band by thermal agitation. They can only enter this band because there is no room for them in the valence band. On the other hand if the substitute atoms have fewer valence electrons than those of the host material the semiconductor becomes *p-type*. Vacancies in the valence band can easily be created because the foreign atoms take up an extra electron in order to create an electronic environment similar to that of the host. Both types of semiconductors will be discussed in Section 3.8 where we shall introduce the concept of Fermi energy as well.

3.2 FREE ELECTRONS. THE EFFECTIVE MASS APPROXIMATION

The simplest approach to the relationship between the electronic energy and momentum, the band structure, is to consider electrons inside an otherwise empty box. This, in essence, is Sommerfeld's model, which, despite its simplicity, is surprisingly accurate in describing the band structure of semiconductors, provided we use a modified value for the electronic mass.

3.2.1 Sommerfeld's theory

If the potential energy terms of Equation (3.1) were neglected, the wave function, Ψ_e, of Equation (2.27) could be written

$$\Psi_e(\mathbf{r}_1, \mathbf{r}_2, \ldots) = \prod_i \Psi_\mathbf{k}(\mathbf{r}_i) \tag{3.3}$$

with the index i running over the electrons. Below, in Section 3.4, we shall see that the wave function $\Psi_\mathbf{k}$ does not depend on which electron it describes, but on the wave vector of the electron, $\mathbf{k}$, which is related to its energy. This is the reason why the symbol for the wave function has been so labelled. With zero potential energy, Equation (3.1) becomes separable along the same lines as in Section 2.3:

$$\frac{\mathbf{p}_i^2}{2m_0}\psi_\mathbf{k}(\mathbf{r}_i) = E_\mathbf{k}\Psi_\mathbf{k}(\mathbf{r}_i). \tag{3.4}$$

Here $E_\mathbf{k}$ represents the kinetic energy of the electron and $\mathbf{p}_i$ the momentum operator given by Equation (2.21).

This electronic wave function can further be split into its three Cartesian components:

$$\Psi_\mathbf{k}(\mathbf{r}_i) = \Psi_\mathbf{k}^x(x_i)\Psi_\mathbf{k}^y(y_i)\Psi_\mathbf{k}^z(z_i) \tag{3.5}$$

where x_i, y_i and z_i represent the Cartesian components of $\mathbf{r}_i$. The one-dimensional Schrödinger equation reads

$$-\frac{\hbar^2}{2m_0}\frac{d^2\Psi_\mathbf{k}^j}{dx_j^2} = E_\mathbf{k}^j\Psi_\mathbf{k}^j \tag{3.6}$$

where $E_\mathbf{k}^j$ is the energy associated with the motion along the j component $(j = x, y, z)$. $E_\mathbf{k} = E_k^x + E_k^y + E_\mathbf{k}^z$ ($E_\mathbf{k}^j$ is not a vector component of $E_\mathbf{k}$). (We could equally well split $\mathbf{r}_i$ into its polar components.)

The wave function satisfying Equation (3.6) is

$$\Psi_\mathbf{k}^j(x) = \frac{1}{\sqrt{L_j}}\exp(ik_jx_j) \tag{3.7}$$

with L_j and k_j representing the width of the box and the wave number along the j component, respectively. The wave number or wave vector can only take discrete values defined by the requirement that $\Psi_\mathbf{k}^j$ vanishes at the boundaries of the box:

$$\Psi_\mathbf{k}^j(0) = \Psi_\mathbf{k}^j(L_j) = 0. \tag{3.8}$$

This gives

$$k_j = 2\pi l_j/aL_j \tag{3.9}$$

with $l_j = 1, 2, \ldots.$

Similar expressions are obtained for the two other Cartesian components, so that, by Equation (3.5),

$$\Psi_{\mathbf{k}}(\mathbf{r}_i) = \frac{1}{\sqrt{V_{cr}}} \exp[i(k_x x_i + k_y y_i + k_z z_i)]$$

$$= \frac{1}{\sqrt{V_{cr}}} \exp[i\mathbf{k} \bullet \mathbf{r}_i] \tag{3.10}$$

with V_{cr} denoting the volume of the crystal and $\mathbf{k}$ the *wave vector*, which has components k_x, k_y and k_z. The corresponding energy is calculated by inserting Equation (3.10) into (3.4):

$$E_{\mathbf{k}} = \frac{\hbar^2 \mathbf{k}^2}{2m_0}. \tag{3.11}$$

Equation (3.11) states the quadratic relationship between the wave vector and the energy. This expression represents the *band structure* of the electrons in the box. This ensemble of electrons is often referred to as an *electron gas* or an *electron plasma*.

When the volume of the box extends to infinity,

$$\Psi_{\mathbf{k}} = \frac{1}{\sqrt{\pi}} \exp(i\mathbf{k} \bullet \mathbf{r}_i), \tag{3.12}$$

then the electrons are referred to as *free electrons*. The electrons obeying Equation (3.10) are *confined unbound electrons*.

3.2.2 Periodic lattices. The Brillouin zone. The effective mass

The periodic lattice of atoms will now be discussed. Electrons belonging to the same energy band will see an electrostatic potential from the surrounding nuclei and deeper lying core electrons partially shielding off the proton charges. According to the cosmological principle this potential is the same whatever crystallographic cell the electron is in. The Schrödinger equation (3.1), without the mutual electronic repulsion term, is still separable, as above, so that we can derive the one-electron equation

$$\left\{ \sum_i \frac{\mathbf{p}_i^2}{2m_0} + U_e(\mathbf{r}_i) \right\} \Psi_{\mathbf{k}}(\mathbf{r}_i) = E_{\mathbf{k}} \Psi_{\mathbf{k}}(\mathbf{r}_i). \tag{3.13}$$

Bloch (1929) has proved that its solution is

$$\Psi_{\mathbf{k}}(\mathbf{r}_i) = u_{\mathbf{k}}(\mathbf{r}_i) \exp(-i\mathbf{k} \bullet \mathbf{r}_i) \tag{3.14}$$

which means that the periodic potential modulates the confined unbound electron wave function by the factor $u_{\mathbf{k}}(\mathbf{r}) \sqrt{V_{cr}}$, $u_{\mathbf{k}}(\mathbf{r})$. This is known as the *Bloch func-*

tion. At a given site this function is $u_{\mathbf{k}}(\mathbf{r})$: at the equivalent site of the neighbouring crystallographic cell it is $u_{\mathbf{k}}(\mathbf{r})e^{i\mathbf{k}\cdot\mathbf{a}}$ because of the lattice periodicity. Translating the electron the distance across the crystal the wave function is

$$u_{\mathbf{k}}(\mathbf{r})\, e^{i\mathbf{k}\cdot\mathbf{L}} = u_{\mathbf{k}}(\mathbf{r}) \exp(ik_x L_x + ik_y L_y + ik_z L_z). \tag{3.15}$$

As these two sites are identical because of the periodic extension of the lattice we introduced in Section 2.4

$$k_j L_j = 2\pi\, 1_j \tag{3.16}$$

for $j = 1, 2, 3$, or

$$k_j = 2\pi l_j / L_j. \tag{3.17}$$

The wave function of the confined unbound electron, Equation (3.7), is distinct only when $l_j = -N_j/2, -N_j/2 + 1, -N_j/2 + 2, \ldots, -2, -1, 0, 1, 2, \ldots, N_j/2 - 1, N_j/2$. Any other value yields a replica of $\Psi_{\mathbf{k}}$ [Equation (3.14)] for one of the values of l_j already listed.

In three dimensions the functions $\Psi_{\mathbf{k}}$ are distinct only for the different possible values of $\mathbf{k}$ inside the same Brillouin zone we met in connection with the phonons. The phonons and the electrons therefore share the same Brillouin zone.

Treating $U_e(\mathbf{r}_i)$ of Equation (3.1) as a perturbation and neglecting the mutual repulsion between the electrons, first order perturbation calculation yields the wave function

$$\Psi_{\mathbf{k}}^{(1)} = \Psi_{\mathbf{k}} + \sum_{\mathbf{G}}{}' \frac{2m_0 U_{\mathbf{G}} \Psi_{\mathbf{k}+\mathbf{G}}}{\hbar^2\{\mathbf{k}^2 - (\mathbf{k}+\mathbf{G})^2\}}. \tag{3.18}$$

Here $\Psi_{\mathbf{k}}$ represents the unperturbed wave function, Equation (3.10). With second order perturbation theory the corresponding energy becomes

$$E'_{\mathbf{k}} = \frac{\hbar^2\mathbf{k}^2}{2m_0} + U_0 + \sum_{\mathbf{G}}{}' \frac{2m_0 |U_{\mathbf{G}}|^2}{\hbar^2\{\mathbf{k}^2 - (\mathbf{k}+\mathbf{G})^2\}} \tag{3.19}$$

where $U_{\mathbf{G}} = \int \Psi^*_{\mathbf{k}+\mathbf{G}} U_e \Psi_{\mathbf{k}+\mathbf{G}} d^3\mathbf{r}$ and $\mathbf{G}$ is a reciprocal lattice vector which is 2π longer than the contravariant vector introduced in Section 2.2. $U_0 = \int \Psi^*_{\mathbf{k}} U_e \Psi_{\mathbf{k}} d^3\mathbf{r}$ merely represents a shift in the energy. The dash of the summation sign indicates that the summation excludes $\mathbf{G} = \mathbf{0}$. Introducing $E'_{\mathbf{k}} = E'_{\mathbf{k}} - U_0$ the term U_0 can be omitted from Equation (3.19) when $E''_{\mathbf{k}}$ is replaced by $E''_{\mathbf{k}}$. Requiring the quadratic relationship between $E'_{\mathbf{k}}$ and $\mathbf{k}$ of Equation (3.11) to be valid we have to replace the free electron mass m_0 by

$$m^* = \left\{ m_0^{-1} + \sum_{\mathbf{G}}{}' \frac{4\, m_0 |U_{\mathbf{G}}|^2}{\hbar^4[\mathbf{k}^2 - (\mathbf{k}+\mathbf{G})^2]^2} \right\}^{-1} \tag{3.20}$$

m^* is known as the *effective mass*.

In general the effective mass depends weakly on $\mathbf{k}$ when $\mathbf{k}$ is far from $\mathbf{k} + \mathbf{G}$ – for most practical purposes it can be considered constant. Values of the effective mass of a variety of materials have been given in the literature, and collected in the reference work of Landolt and Börnstein (1982).

For metals, the effective mass is larger than that of free electrons; in semiconductors it is smaller. In transistors it is desirable to have a small effective mass, because the electrons then have greater mobility.

Electrons in the perfect lattice move as if it were absent, but they behave as if they had a mass different from that of the free ones. Thus Sommerfeld's theory can still be used, provided the right value for the electronic mass is chosen.

A careful study of the top of the valence band reveals that the empty electronic states also behave like particles, but with a negative effective mass. These particles are referred to as *holes*. Instead of a negative mass, we prefer to attribute a positive mass to them, but then their charge has also to be reckoned positive.

We can thus treat electrons and holes as free particles of a distinct effective mass, described as the *effective mass approximation*.

The periodic lattice is also the cause of the energy gaps mentioned in Sections 3.1 and 2.2. To gain an idea of the origin of these gaps we can follow the work presented by Ziman (1960): when $\mathbf{k}^2 = (\mathbf{k} + \mathbf{G})^2$ the result of our perturbation theory calculation is no longer valid. The unperturbed system is degenerate in energy; two states of the same energy are linked by a reciprocal lattice vector, $\mathbf{G}$. In this case we have to resort to degenerate perturbation theory which tells us that we have to find the energy from the following secular equation:

$$\begin{vmatrix} \dfrac{\hbar^2\mathbf{k}^2}{2m_0} - E & U_{\mathbf{G}} \\ U_{\mathbf{G}} & \dfrac{\hbar^2(\mathbf{k}+\mathbf{G})^2}{2m_0} - E \end{vmatrix} = 0 \tag{3.21}$$

giving

$$E = \frac{1}{2}\left\{\frac{\hbar^2\mathbf{k}^2}{2m_0} + \frac{\mathbf{k}^2(\mathbf{k}+\mathbf{G})^2}{2m_0} \pm \left[\left(\frac{\hbar^2\mathbf{k}^2}{2m_0} + \frac{\mathbf{k}^2(\mathbf{k}+\mathbf{G})^2}{2m_0}\right)^2 + 4|U_{\mathbf{G}}|^2\right]^{\frac{1}{2}}\right\}. \tag{3.22}$$

The energy has been split in two levels and no energy states between them can exist. The analysis presented here is only indicative, a more rigorous proof being beyond the scope of this book.

3.3 THE ONE ELECTRON ORBITAL

The fundamental law of physics that no two electrons can occupy the same state should be kept well in mind in order to appreciate semiconductor theory. To describe the electronic composition of a crystal we first turn our attention to a free atom with nuclear charge $-Z_be$ and Z_b protons. Schrödinger's equation for a single electron orbiting the nucleus reads, in polar coordinates,

$$\left\{-\frac{\hbar^2}{2m_0}\left[\frac{1}{r^2}\frac{\partial}{\partial r}\left(r^2\frac{\partial}{\partial r}\right)+\frac{1}{r^2\sin\theta}\frac{\partial}{\partial\theta}\left(\sin\theta\frac{\partial}{\partial\theta}\right)+\frac{1}{r^2\sin^2\theta}\frac{\partial^2}{\partial\phi^2}\right]\right.$$
$$\left.-\frac{Z_b e^2}{4\pi e_0 r}\right\}\Psi_{nlms}=\Psi_{nlms}. \tag{3.23}$$

Here r represents the distance from the nucleus, ϕ the azimuth and θ the polar angle as shown in Fig. 2.7. The dielectric constant of vacuum has been symbolised by ε_0, and the electronic mass by m_0. The reason why, suddenly, the wave function Ψ has four subscripts will become clear below.

The only physically acceptable solution to this wave equation is, in its normalised form:

$$\Psi_{nlms}(r,\theta,\phi)=\left\{\left(\frac{2Z_b}{na_B}\right)^3\frac{(n-l-1)!}{2n[(n+l)!]^3}\right\}^{\frac{1}{2}}\left(\frac{2Z_b r}{na_B}\right)$$
$$\times\exp\left(-\frac{Z_b r}{na_B}\right)L_{n-l-1}^{2l+1}\left(\frac{2Z_b r}{na_B}\right) \tag{3.24}$$
$$\times\, Y_l^{|m|}(\theta,\phi)s$$

where the *principal quantum number*, n, is a positive integer. The *associated Legendre polynomial* is defined by

$$L_{n-l-1}^{2l+1}(\rho)=-\frac{d^{2l+1}}{d\rho^{2l+1}}\left\{e^{\rho}\frac{d^{n+l}}{d\rho^{n+l}}\left(\rho^{n+l}e^{-\rho}\right)\right\} \tag{3.25}$$

where the *orbital angular quantum number*, l, takes the values

$$l=n-1,n-2,\ldots,1,0 \tag{3.26}$$

for a given value of n, the *magnetic quantum number*, m, can, for a given value of l, take the values

$$m=-l,-l+1,\ldots,-l,0,1,\ldots,l-1,l. \tag{3.27}$$

The *spherical harmonics* are given by

$$Y_l^m(\theta,\phi)=(-1)^m\left(\frac{2l+1}{4\pi}\frac{(l-m)!}{(l+m)!}\right)e^{-im\phi}p_l^m(\cos\phi), \tag{3.28}$$

where the *associated Legendre function* of the first kind, $P_l^m(\xi)$, is obtainable from the generic formula

$$P_l^m(\xi)=\frac{(1-\xi)^{m/2}}{2^l l!}\frac{d^{l+m}}{d\xi^{l+m}}(\xi^2-1)^2. \tag{3.29}$$

The *spin quantum number*, s, takes the values

$$s = \{-\tfrac{1}{2}, \tfrac{1}{2}\}, \tag{3.30}$$

s is often represented by arrows ↑ and ↓ or by the *spinors* $\begin{pmatrix}1\\0\end{pmatrix}$ and $\begin{pmatrix}0\\1\end{pmatrix}$, or as up or down respectively. The spin may naively be considered as the electron's rotation about its own axis. The spin is a relativistic effect and can be described rigorously by formulating the relativistic equivalent of Schrödinger's equation, *Klein–Gordon's equation*.

The spin operator $s = \frac{1}{2}\hbar\sigma$ is represented as a vector where σ has the matrix components

$$\sigma_x = \begin{pmatrix}0 & 1\\1 & 0\end{pmatrix}, \sigma_y = \begin{pmatrix}0 & -i\\i & 0\end{pmatrix} \text{ and } \sigma_z = \begin{pmatrix}1 & 0\\0 & -1\end{pmatrix}, \tag{3.31}$$

the *Pauli spin matrices*. The spin operators, **s**, obey the commutation relations

$$[s_x, s_y] = i\hbar s_z, [s_y, s_z] = i\hbar s_x \text{ and } [s_z, s_x] = i\hbar s_y. \tag{3.32}$$

The quantity

$$a_B = \frac{4\pi\varepsilon_0\hbar^2}{m_0 e^2} \tag{3.33}$$

represents *Bohr's radius*: $a_B = 0.529172$ Å. The energy

$$E_n = -\frac{Z_b^2 e^2}{8\pi\varepsilon_0 a_B n^2} \tag{3.34}$$

depends only on the principal quantum number n. It is thus degenerate in the other three quantum numbers. E_n is reckoned downwards towards the nucleus, or $-E_n$ represents the minimum energy required to liberate the electron from it. For hydrogen this ionisation energy from the ground state ($n = 1$) is 13.61 eV, or one *Rydberg*, which is the unit of energy favoured by many spectroscopists. Another frequently ocurring unit representing radiative energy is the cm^{-1}. This is arrived at by combining the relations $c = \lambda\nu$ and $E = h\nu$, where ν represents the frequency, λ the wave length, c the velocity of light, and h (not $\hbar$) is Planck's constant.

Table 3.1 lists various orbitals applicable up to atomic number 36 (krypton); Table 3.2 the corresponding spherical harmonics. Figure 3.1 shows the radial part of Ψ_{nlms}, i.e. the part of Equation (3.24) without the term $Y_l^{|m|}(\theta, \phi)$, up to $n = 4$.

The exclusion principle states that no two electrons belonging to the same isolated atom can have the same set of quantum numbers n, l, m and s. Electrons with the same principal quantum number form a shell. The nth shell has room for $2n^2$ electrons. Provided the atom is not significantly disturbed, its electrons fill up the lowest possible energy states.

Table 3.1 Electronic orbitals of a single electron surrounding a nucleus of charge Ze where e represents the magnitude of the elementary electronic charge

Orbital	$n\ l$	Function
1s	1 0	$2\left(\frac{Z}{a_B}\right)^{3/2}\exp\left(-\frac{Zr}{a_B}\right) Y_0^0 s$
2s	2 0	$\frac{1}{\sqrt{2}}\left(\frac{Z}{a_B}\right)^{3/2}\left[1-\frac{1}{2}\frac{Zr}{a_B}\right]\exp\left(-\frac{Zr}{2a_B}\right) Y_0^0 s$
2p	2 1	$\frac{1}{2\sqrt{6}}\left(\frac{Z}{a_B}\right)^{5/2} r\exp\left(-\frac{Zr}{2a_B}\right) Y_1^m s$
3s	3 0	$\frac{2}{3\sqrt{3}}\left(\frac{Z}{a_B}\right)^{3/2}\left[1-\frac{2}{3}\frac{Zr}{a_B}+\frac{2}{27}\left(\frac{Zr}{a_B}\right)^2\right]\exp\left(-\frac{Zr}{3a_B}\right) Y_0^0 s$
3p	3 1	$\frac{2}{27\sqrt{6}}\left(\frac{Z}{a_B}\right)^{5/2}\left[1-\frac{1}{6}\frac{Zr}{a_B}\right] r\exp\left(-\frac{Zr}{3a_B}\right) Y_1^m s$
3d	3 2	$\frac{1}{81\sqrt{30}}\left(\frac{Z}{a_B}\right)^{7/2} r^2\exp\left(-\frac{Zr}{3a_B}\right) Y_0^0 s$
4s	4 0	$\frac{1}{4}\left(\frac{Z}{a_B}\right)^{3/2}\left(1-\frac{3}{4}\frac{Zr}{a_B}+\frac{1}{8}\left(\frac{Zr}{a_B}\right)^2-\frac{1}{192}\left(\frac{Zr}{a_B}\right)^3\right]\times\exp\left(-\frac{Zr}{4a_B}\right) Y_0^0 s$
4p	4 1	$\frac{\sqrt{5}}{16\sqrt{3}}\left(\frac{Z}{a_B}\right)^{5/2}\left(1-\frac{1}{4}\frac{Zr}{a_B}+\frac{1}{80}\left(\frac{Zr}{4a_B}\right)^2\right] r\times\exp\left(-\frac{Zr}{4a_B}\right) Y_1^m s$
4d	4 2	$\frac{1}{64\sqrt{5}}\left(\frac{Z}{a_B}\right)^{7/2}\left(1-\frac{1}{12}\frac{Zr}{a_B}\right) r^2\exp\left(-\frac{Zr}{4a_B}\right) Y_2^m s$
4f	4 3	$\frac{1}{768\sqrt{35}}\left(\frac{Z}{a_B}\right)^{9/2} r^3\exp\left(-\frac{Zr}{4a_B}\right) Y_3^m s$

The orbitals with $l = 0$, which are known as *s-orbitals*, have spherical symmetry. The *p-orbitals* ($l = 1$) have cubic symmetry, while $l = 2$ and $l = 3$ are referred to as *d-* and *f-orbitals* respectively. The lowest state ($n = 1$) can contain only two electrons, which have to be s-orbitals. The next shell ($n = 2$) can contain two electrons as s-orbitals and six as p-orbitals. The third shell ($n = 3$) can contain two s-orbitals, six p-orbitals and ten d-orbitals etc. Figure 3.2 shows the charge density of the 1s, 2s, 2p and 3d orbitals, as presented by Coulson (1961).

In an atom with more than one electron there will be mutual repulsion between them. Electrons nearer the core will shield off electrons in the higher orbits from the field of the nucleus. This shielding is not quite perfect, however, and this is the reason why the pseudopotential calculation of the band structure (Section

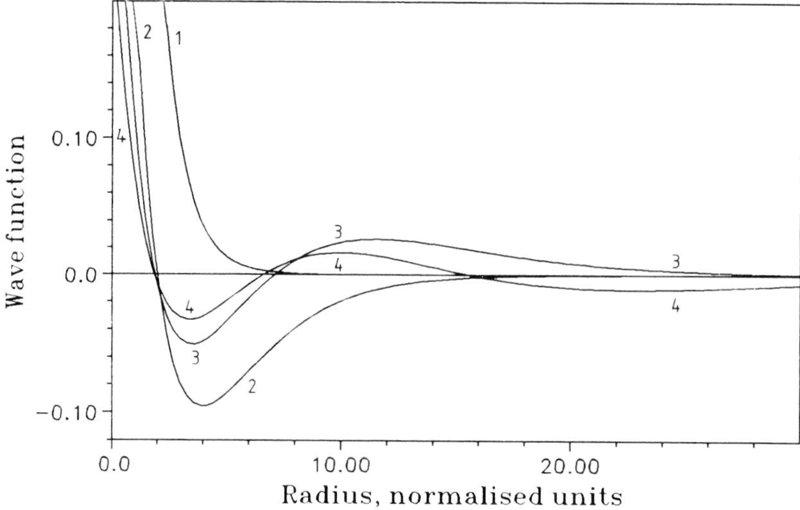

Figure 3.1 *The radial part of the wave function for n = 1, 2, 3 and 4. The parameter indicates the value of n. s = 0*

Table 3.2 Spherical harmonics

s-orbitals	$Y_0^0(\theta, \phi) = \dfrac{1}{2\sqrt{\pi}}$
p-orbitals	$Y_1^0(\theta, \phi) = \dfrac{1}{2}\sqrt{\dfrac{3}{\pi}}\cos\theta$
	$Y_1^{\pm 1}(\theta, \phi) = \dfrac{1}{2}\dfrac{3}{\sqrt{2\pi}}\sin\theta\,(\cos\phi \pm i\sin\phi)$
d-orbitals	$Y_2^0(\theta, \phi) = \dfrac{1}{4}\sqrt{\dfrac{5}{\pi}}\,(3\cos^2\theta - 1)$
	$Y_2^{\pm 1}(\theta, \phi) = \dfrac{1}{2}\sqrt{\dfrac{5}{6\pi}}\,3\cos\theta\sin\theta\,(\cos\phi \pm i\sin\phi)$
	$Y_2^{\pm 2}(\theta, \phi) = \dfrac{3}{4}\sqrt{\dfrac{5}{6\pi}}\sin^2\theta\,(\cos 2\phi \pm \sin 2\phi)$
f-orbitals	$Y_3^0(\theta, \phi) = \dfrac{1}{4}\sqrt{\dfrac{7}{\pi}}\cos\theta\,(5\cos^2\theta - 1)$
	$Y_3^{\pm 1}(\theta, \phi) = \dfrac{3}{8}\sqrt{\dfrac{7}{3\pi}}\sin\theta\,(5\cos^2\theta - 1)\,(\cos\phi \pm i\sin\phi)$
	$Y_3^{\pm 2}(\theta, \phi) = \dfrac{15}{4}\sqrt{\dfrac{7}{30\pi}}\sin^2\theta\cos\theta\,(\cos 2\phi \pm i\sin 2\phi)$
	$Y_3^{\pm 3}(\theta, \phi) = \dfrac{1}{8}\sqrt{\dfrac{35}{\pi}}\sin^3\theta\,(\cos 3\phi \pm i\sin 3\phi)$

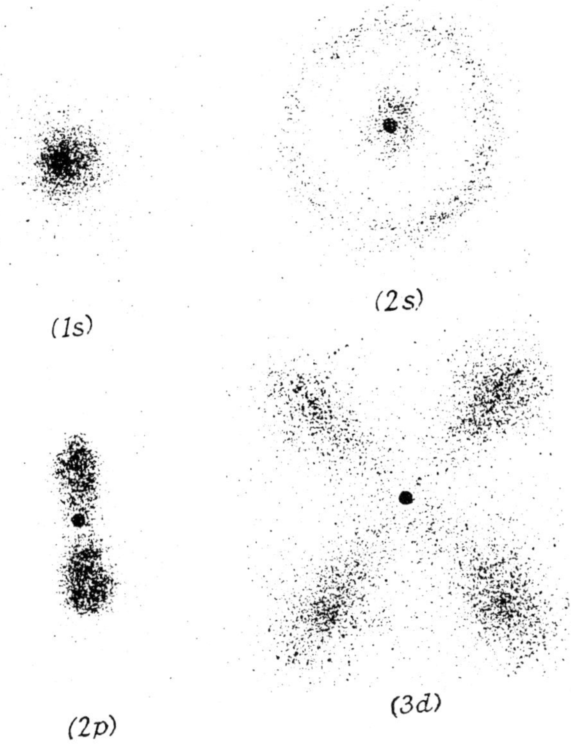

Figure 3.2 *The 1s, 2s, 2p and 3d orbitals. The gray tone indicates the density of the wave function, i.e.* $|\Psi_{nlms}|^2$. *From Coulson (1961)*

3.5.1) is not quite exact although it is rather successful. Both the orbitals, Equation (3.24), and the energy, Equation (3.34) will be modified somewhat: the orbital energy will now also depend on the quantum numbers l and m. Within a shell, the p-orbital electrons tend to have higher energy than that of the s-electrons. The energy of the d-electrons lie near that of the s-electrons of the next shell out. The d-electrons contribute mostly to the top of the valance band of some semiconductors, which is where the conduction by holes occurs. The proximity in energy of the $3d$ electrons to that of the $4s$ electrons is the reason why the ten transition elements, scandium to zinc, lie in the fourth rather than the third row of the periodic table of the elements.

An atom in its ground state will have its lowest electric states occupied: in silicon ($Z_b = 14$) the two innermost shells are full; the third shell contains two s-electrons and two p-electrons. This electron configuration is symbolised by

$$(1s)^2(2s)^2(2p)^6(3s)^2(3p)^2. \tag{3.35}$$

3.4 THE EXCLUSION PRINCIPLE FOR CRYSTALS. THE BRA AND KET NOTATION

In the previous section the exclusion principle for electrons belonging to one atom was discussed. All the electrons were distinguishable through their unique values of the four quantum numbers n, l, m and s. In a system consisting of several atoms being alike a modified version of the exclusion principle applies. To describe it, it is necessary to introduce an additional label.

A change in the quantum state of an electron can formally be described as the destruction of it in its initial state and a re-creation of it in the final state. The necessary formalism, together with a shorthand notation for the electronic wave function, will be introduced.

3.4.1 The exclusion principle. Fermions

A crystal is built up from many atoms. All the atoms of the e.g. silicon crystal have the same electronic configuration (3.35), of which only the two 3s and 3p electrons contribute to the chemical bonds holding the crystal together. Other semiconductors have an analogous electronic structure. How can this be compatible with the exclusion principle which states that no two electrons can be in the same quantum state? The answer is that the mutual interaction between the atoms causes the energy level for each state (n, l, m, s) to broaden into bands of discrete levels. The energy levels of these bands, however, lie so close together that they may be considered to form a continuum in all practical cases of crystals containing more than a few thousand atoms. (Figure 3.3.) To distinguish between these electrons we introduce an additional quantum number, the wave number **k**, which is unique for each electronic state inside the same band. We can, at least in principle, calculate the relationship between **k** and the energy, $E_{nlms\mathbf{k}}$, the band structure. Two possible methods of doing this will be discussed in the next Section. We adopt an *extended exclusion principle for crystals* which states that no two

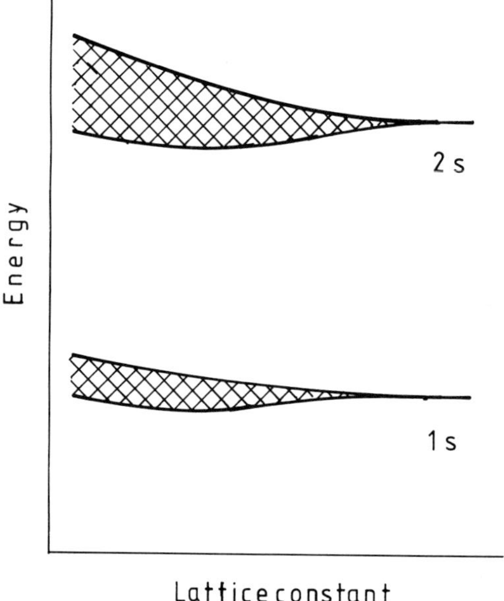

Figure 3.3 *Dependence of energy levels of the 1s and 2s electrons in a crystal against the lattice constant. As the atoms move closer together the coupling between them increases, whereby the energy levels split, as indicated here by the cross-hatched areas*

electrons can share the seven quantum numbers n, l, m, s and $\mathbf{k}$. ($\mathbf{k}$ represents a triplet. Many theorists may probably object to classifying $\mathbf{k}$ as a quantum number, but we consider it as such because it distinguishes between electrons having the same values of n, l, m and s.)

Consider a system of N_e electrons. Let the quantum state of electron i be κ ($= n, l, m, s, \mathbf{k}$) and its wave function Ψ_κ dropping the indices n, l, m and s). In the independent particle approximation the wave function of the entire ensemble of the N_e electrons may be taken to be

$$\Psi_e = \prod_{j=1}^{N_e} \Psi_{\kappa j}(\mathbf{r}_j). \tag{3.36}$$

However, this description is not adequate. The electrons are indistinguishable. The wave function of electron j can equally well be $\Psi_{\kappa i}(\mathbf{r}_j)$ with $i \neq j$. This gives $N_e!$ different ways of arranging the electrons over the N_e given quantum states, and all of them are equally probable. The wave function Ψ_e should therefore be formulated as

$$\Psi_e = \frac{1}{\sqrt{V_{cr} N_e!}} \begin{vmatrix} \Psi_{\kappa 1}(r_1)\Psi_{\kappa 2}(r_1) & \ldots & \Psi_{\kappa N}(r_1) \\ \Psi_{\kappa 1}(r_2)\Psi_{\kappa 2}(r_2) & \ldots & \Psi_{\kappa N}(r_2) \\ \vdots & & \vdots \\ \Psi_{\kappa 1}(r_N)\Psi_{\kappa 2}(r_N) & \ldots & \Psi_{\kappa N}(r_N) \end{vmatrix} \tag{3.37}$$

where V_{cr} represents the volume of the electronic confinement, and the factor $\sqrt{V_{cr} N_e!}$ takes care of the normalisation of Ψ_e, provided the wave function $\Psi_{\kappa j}(\mathbf{r}_j)$ of an individual electron has already been normalised. This determinant is known as the *Slater determinant*.

This way of writing the wave function of the ensemble also reproduces the properties required of such an ensemble. The wave function changes sign when two coordinates are interchanged. In Equation (3.37) this is achieved by interchanging two columns or rows. Also two electrons cannot coexist in the same state: the determinant vanishes when two columns or rows are equal.

It is not practicable, except for small, and therefore perhaps uninteresting ensembles, to calculate the wave function of an electronic system from Equation (3.37). The mutual repulsion between the particles has also been omitted. We have introduced this discussion to illustrate properties of particles or quasiparticles of half integral spin, *fermions*. Examples of fermions are electrons, protons, holes and neutrons. Elementary particles like the Σ and the χ also have spin ½ $\hbar$, the Ω particle has spin $3/2\,\hbar$. All such particles obey the Fermi–Dirac statistics which will be discussed in Section 3.6. Bosons, on the other hand, have integral or no spin. Phonons and pions are spinless, photons have spin $\hbar$. Composite particles consisting of an even number of fermions, like excitons (bound electron–hole pairs), hydrogen atoms (consisting of a proton and an electron) α-particles etc. are bosons, and obey the Bose–Einstein statistics described in Section 2.8.

3.4.2 The bra and the ket notation

As for phonons, and bosons in general, we shall also introduce the bra and the ket notation for fermions. The wave function of an electron in a given quantum state k is represented by

$$\Psi_k(\mathbf{r}) = |1\rangle_k, \tag{3.38a}$$

the *bra*. Complex conjugation gives

$$\Psi_k^*(\mathbf{r}) = {}_k\langle 1|, \tag{3.38b}$$

the *ket*. The state $|0\rangle_k$ represents the unoccupied (empty) state which can only receive a fermion of wave vector **k**. When we do not want to specify whether the state is filled we write $|\mathbf{k}\rangle$ instead.

The *annihilation operator* is defined by

$$c_k|1\rangle_k = |0\rangle_k \tag{3.39a}$$

and the *creation operator*, the complex conjugate of c_k, by

$$c_k^+|0\rangle_k = |1\rangle_k. \tag{3.39b}$$

The operation $c_k^+c_k^+|0\rangle_k$ means that two electrons have been created into the same quantum state. This is impossible, therefore

$$c_k^+c_k^+|0\rangle_k = c_k^+(c_k^+|0\rangle_k) = c_k^+|1\rangle_k = 0. \tag{3.40}$$

In the same way

$$c_k|0\rangle_k = 0. \tag{3.41}$$

Equation (3.39a) has the following interpretation: create an electron in quantum state **k**. In band theory calculation the values for n, l, m and, in the absence of the magnetic field, s are implicitly understood.

At absolute zero temperature, all electron states are filled to a certain energy, the Fermi level; all the states above it are unoccupied. This applies to metals as well as semiconductors, although in the latter the Fermi level lies in the gap between the valence and the conduction band: all valence band states are occupied, the conduction band states are all free. Such a system is known as the *Fermi sea*.

The operator c_k can only destroy an electron which is already there. When c_k destroys an electron below the Fermi sea level, a vacancy or a hole is left behind. The operator c_k is therefore creating a hole below sea level, annihilating an electron above it. Similarly c_k^+ creates an electron above, although creating an electron below the Fermi energy level is the same as filling a hole, thus c_k^+ annihilates the hole of wave vector **k**.

Combining Equations (3.38) and (3.39) we can prove that

$$(c_k c_k^+ + c_k^+ c_k) | 0 \rangle_k = | 0 \rangle_k$$
$$(c_k c_k^+ + c_k^+ c_k) | 1 \rangle_k = | 1 \rangle_k \tag{3.42}$$

which means that

$$c_k c_k^+ + c_k^+ c_k \equiv \{c_k, c_k^+\} = 1. \tag{3.43}$$

This expression is known as the *anticommutator* between c_k and c_k^+. The property of the anticommutator that

$$\{c_k, c_{k'}\} = \{c_k^+, c_k^+\} = 0 \tag{3.44a}$$

and

$$\{c_k, c_{k'}^+\} = \delta(\mathbf{k} - \mathbf{k}') \tag{3.44b}$$

can be proven using the relations given here.

These anticommutation relations, Equations (3.43) and (3.44), express fundamental properties on the coexistence of fermions.

3.5 THE BAND STRUCTURE

An analytical solution of the electronic Schrödinger equation (3.1) cannot be achieved in general. It is possible to solve it numerically for crystals containing only a few hundred electrons using a supercomputer, but most interesting semiconductor devices contain hundreds of thousands of charge carriers or even more, in which case such an approach is impracticable. Several attempts to make a short cut to the band structure have been made; the most interesting ones we shall study are the pseudopotential approach discussed by Cohen and Heine (1970) and the **k•p** perturbation theory approach pioneered by Kane (1956 and 1957). Both methods are based on the electronic wave functions presented in Sections 3.2 and 3.3.

In the former, the nuclei and the core electrons have been substituted by effective charges; in the latter the lattice modulation of the wave functions has been exploited.

A practical simplified analytical band structure for holes and electrons will also be outlined, which is sufficiently accurate for many purposes. Concepts like group velocity of the carriers and the effective mass will be defined in terms of the band structure.

There are other ways to the band structure too, the most interesting being perhaps the eigenvalue method of Chang and Schulman (1982) which is adaptable to pseudopotential, **k•p** and tight binding theory.

3.5.1 The pseudopotential approach

The successful pseudopotential approach consists of dividing the space occupied by the crystal in two parts: the spheres near the nuclei, *A*; and the interstitial space *B*. (Figure 3.4.) The potential in *B* varies only a few electron volts, while

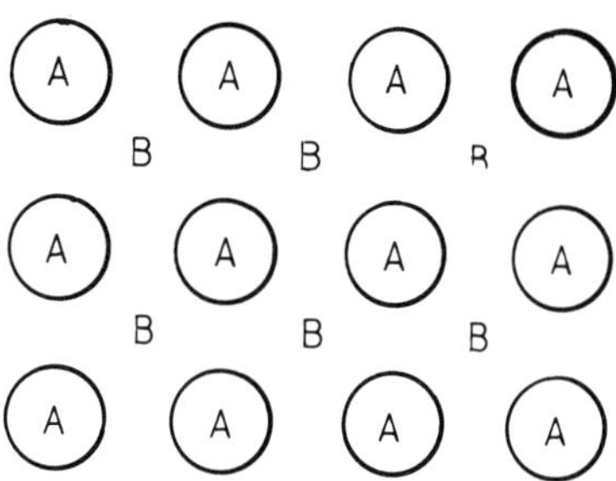

Figure 3.4 *The division of the crystal lattice into two types of regions to calculate the band structure by means of the pseudopotential approach. A represents the areas nearest to the nuclei, in which the actual charge distribution has been replaced by the effective smoothed one. B indicates the interstitial area*

that in A can vary over thousands of electron volts. If Schrödinger's equation can be solved in areas B and A separately, the solutions have to join smoothly at the boundaries between them, i.e.

$$\Psi_A = \Psi_B \tag{3.45a}$$

and

$$\frac{1}{\Psi_A}\frac{d\Psi_A}{d\mathbf{r}} = \frac{1}{\Psi_B}\frac{d\Psi_B}{d\mathbf{r}} \tag{3.45b}$$

where Ψ_A and Ψ_B represent the solution inside and outside the spheres respectively. It does not matter what Ψ_A looks like, as long as Equation (3.45) has been satisfied. The true charge distribution inside the spheres has to satisfy Poisson's equation and Gauss' theorem:

$$\int_{AB} \frac{d}{d\mathbf{r}}\Phi \bullet d\mathbf{A} = -\frac{1}{\varepsilon_0}\int_A \rho_e d^3\mathbf{r}. \tag{3.46}$$

The left-hand side has to be integrated over the interface between A and B, where the electrostatic potential is Φ, and the vector $d\mathbf{A}$ is directed outwards from the nucleus. The right-hand side has to be integrated over the entire interior of the sphere A. Here ρ_e represents the charge density and ε_0 the permittivity of vacuum. The charge distribution inside A will be replaced by a smoother one that still satisfies Equation (3.46) – this substitute charge defines the *pseudopotential*.

This method of calculating the band structure was first used by Phillips and Kleinman (1959). Modifying Löwdin's (1950) perturbation technique Brust (1964) obtained a very accurate band structure for silicon and germanium. He used an empirical pseudopotential method, which involves fitting to data from optical measurements. Soon afterwards Cohen and Bergstresser (1966) calculated the band structure for 14 semiconductors of the zinc blende or diamond structure. An extensive description of the method is given by Cohen and Heine (1970).

The band structure can be calculated by obtaining the eigenvalues of the Hamiltonian

$$\mathcal{H}^{i,j}(\mathbf{k}) = \begin{cases} \dfrac{\hbar^2}{2m_0}\left|\mathbf{k} + \mathrm{G}_i\right|^2 + V_S(0) & \text{for } i = j \\ V_S(\mathbf{G}_i - \mathbf{G}_j)\cos[(\mathbf{G}_i - \mathbf{G}_j)\bullet\boldsymbol{\tau}] & \\ + i\,V_A(\mathbf{G}_i - \mathbf{G}_j)\sin[(\mathbf{G}_i - \mathbf{G}_j)\bullet\boldsymbol{\tau}] & \\ \qquad \text{for } i \neq j & \end{cases} \tag{3.47}$$

with $\mathbf{k}$ representing a wave vector inside the first Brillouin zone, $\mathbf{G}_i$ and $\mathbf{G}_j$ two reciprocal lattice vectors and $\boldsymbol{\tau} = \frac{a}{8}\begin{pmatrix}1\\1\\1\end{pmatrix}$, a being the lattice constant. V_S and V_A denote the symmetric and antisymmetric parts of the pseudopotential form factor respectively:

$$V_S(\mathbf{G}) = \frac{1}{2}\,[v_1(\mathbf{G}) + v_2(\mathbf{G})] \tag{3.48a}$$

$$V_A(\mathbf{G}) = \frac{1}{2}\,[v_1(\mathbf{G}) - v_2(\mathbf{G})] \tag{3.48b}$$

where v_1 and v_2 represent the pseudopotential form factors of the first and the second element of the compound $A^{(\mathrm{III})}B^{(\mathrm{V})}$ or $A^{(\mathrm{II})}B^{(\mathrm{VI})}$ respectively (for GaAs $A^{(\mathrm{III})}$ stands for gallium, $B^{(\mathrm{V})}$ for arsenic). For silicon, germanium and diamond $V_S = v_1 = v_2$ and $V_A = 0$. The indices i and j in the Hamiltonian (3.47) run from 1 up to a value N', the number of reciprocal vectors whose magnitude is less than or equal to G_N, chosen large enough that including further elements into the Hamiltonian makes no significant changes in the eigenvalues. Experience shows that $G_N = 11 \times 2\pi/a$ is sufficient, which gives an $(N_{ps} \times N_{ps})$ Hamiltonian matrix of more than a hundred rows and columns. The various vectors $\mathbf{G}_i$ can be calculated from Equation (2.8).

The eigenvalues represent the band structure. An accelerated computation can be made by truncating the Hamiltonian matrix (3.47) to one with $N'_{ps} \times N'_{ps}$ elements where $N'_{ps} < N_{ps}$. The elements of this matrix are:

$$H^{ij} = \begin{cases} H^{i,i} + \displaystyle\sum_{m=N'_{ps}+1}^{N_{ps}} \frac{H^{i,m^*}H^{m,i}}{H^{i,i} - H^{m,m}} & \text{for } i = \mathrm{j} \\ H^{i,j} + \displaystyle\sum_{m=N'_{ps}+1}^{N_{ps}} \frac{H^{i,m^*}H^{m,j}}{E' - H^{m,m}} & \text{for } i \neq j \end{cases} \tag{3.49}$$

with $1 \leqslant i \leqslant N'_{ps}$ and $1 \leqslant j \leqslant N'_{ps}$. Here

$$E' = \frac{\hbar^2}{2m_0N_V}\sum_{i=1}^{N_V} |\,\mathbf{k} + \mathbf{G}_i|^2 \tag{3.50}$$

where N_V represents the total number of valence electrons.

We find that the pseudopotential approach *ab initio* does not yield any values for the gap between the valence and the conduction band nor of the position and the height of the other minima of the latter. These have to be established by fitting to experimental data. This approach yields a rather accurate description of the band structure. Vogl *et al.* (1983) have developed a theory that makes use of universal constants which are valid for all zinc blende and diamond structure semiconductors. However, although this theory gives a fairly accurate estimate of the density of states, the details of the band structure are not very accurate. We can expect that the future will bring better calculations of the band structure from a more refined approach. Recently pseudopotential calculations have been applied to heterojunction interfaces and surfaces (Zollner *et al.*, 1990).

Figure 3.5 shows the section of the band structure through the (k_x, k_y) Cartesian plane through the Brillouin zone of GaAs, as calculated by Shichijo and Hess (1981). Figures 3.6 and 3.7 show the band structure of silicon and gallium arsenide respectively along the main directions of symmetry. The figures originate from Cohen and Bergstresser (1966). The light and heavy hole valence bands coincide in the centre of the Brillouin zone (the Γ point) where the energy is at a maximum. This is a general property of unstressed semiconductors of the zinc blende or diamond structure. In GaAs the conduction band has its main minimum at the Γ point – it is a *direct semiconductor*. It is well suited as material for photodetectors and lasers. The lowest minima of conduction band of silicon lie near the X points – silicon is an *indirect semiconductor*.

Spin–orbit interaction between the electrons of the valence band has not been considered in obtaining Figs. 3.6 and 3.7. Spin–orbit interaction is a relativistic effect which we shall not discuss further here. The effect of spin-orbit interaction

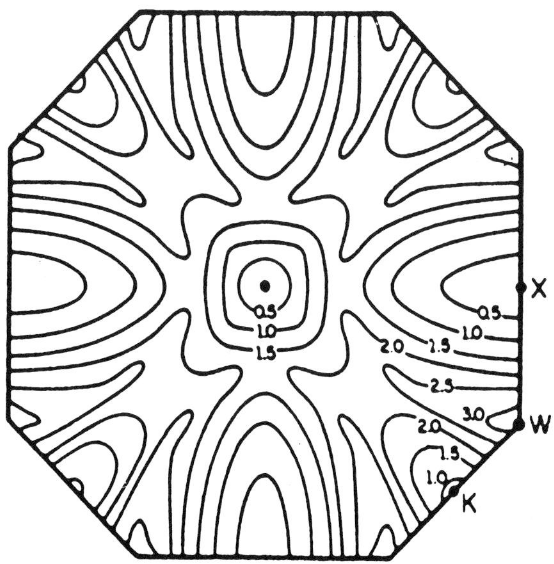

Figure 3.5 *Contours of constant energy of the band structure of GaAs in the (k_x, k_y) Cartesian plane of the Brillouin zone, as calculated from pseudopotential theory. From Shichijo and Hess (1981)*

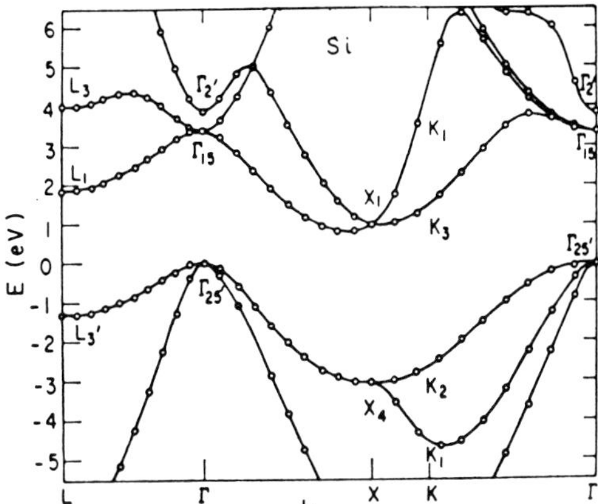

Figure 3.6 *The band structure of Si in the main directions of symmetry calculated from pseudopotential theory. The symbols L, Γ, X and K refer to points and directions in the Brillouin zone as indicated in Fig. 2.5. Si represents an indirect semiconductor in the sense that the lowest minima of the conduction band lie near the (111) faces of the Brillouin zone and the maximum of the valence band lie in the centre. From Cohen and Bergstresser (1966)*

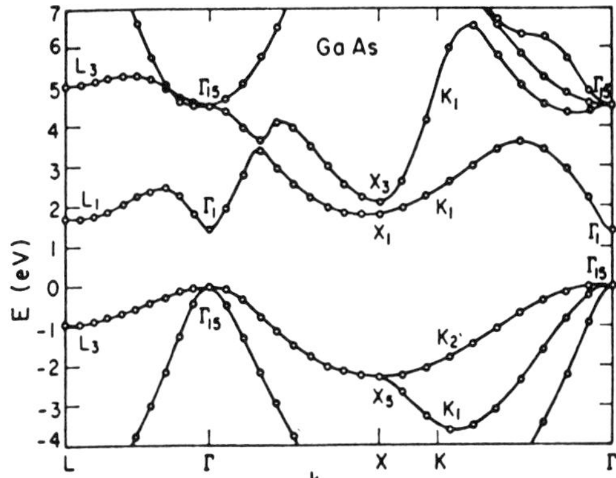

Figure 3.7 *The band structure of GaAs in the main directions of symmetry calculated from pseudopotential theory. The symbols L, Γ, X and K refer to points and directions in the Brillouin zone as indicated in Fig. 2.5. GaAs represents a direct semiconductor in the sense that the maximum of the valence band and the lowest minimum of the conduction band lie in the centre of the Brillouin zone. From Cohen and Bergstresser (1966)*

on the band structure has been summarised by Mašovič and Vukajlović (1983). The result is that an additional valence band arises with maximum at the Γ point and is offset a few decielectron volts further down. A relativistic refinement of this theory has been presented by Milman *et al.* (1990).

3.5.2 $k \bullet p$ theory

The wavefunction of a confined electron is given by Equation (3.14). Substituting this into the Schrödinger equation (3.13) yields

$$\left(\frac{\mathbf{p}^2}{2m_0} + U_e + \frac{\hbar}{m_0}\mathbf{k}\bullet\mathbf{p}\right)\Psi_{\mathbf{k}} = E'_{\mathbf{k}}\Psi_{\mathbf{k}} = \left(E_{\mathbf{k}} - \frac{\hbar^2\mathbf{k}^2}{2m_0}\right)\Psi_{\mathbf{k}} \tag{3.51}$$

which can be solved treating the term containing the vector product **k•p** as a perturbation. Kane (1957) has included spin–orbit, so that Equation (3.51) has been generalised to

$$\left(\frac{\mathbf{p}^2}{2m_0} + U_e + \frac{\hbar}{m_0}\mathbf{k}\bullet\mathbf{p} + \frac{\hbar}{4m_0^2c^2}\left[\frac{d}{d\mathbf{r}}U_e \times \mathbf{p}\right]\bullet\boldsymbol{\sigma} + \frac{\hbar^2}{4m_0^2c^2}\left[\frac{d}{d\mathbf{r}}U_e \times \mathbf{k}\right]\bullet\boldsymbol{\sigma}\right)\Psi_{\mathbf{k}}$$

$$= \left(E_{\mathbf{k}} - \frac{\hbar^2\mathbf{k}^2}{2m_0}\right)\Psi_{\mathbf{k}}. \tag{3.52}$$

The third term represents the **k•p** interaction, the fourth the **k**-independent or atomic-like spin–orbit interaction and the last the **k**-dependent spin–orbit interaction, which turns out to be very small compared to the atomic momentum in the deep interior of the atom where most of the spin–orbit coupling takes place. This term can therefore to the first approximation be neglected. Furthermore, c represents the velocity of light, **p** electron momentum, **k** the wave vector, m_0 the free electron mass, U_e the interaction potential between electrons and $\boldsymbol{\sigma}$ the Pauli spin matrices.

The wave function $\Psi_{\mathbf{k}}$ is constructed from s and p wave functions of Tables 3.1 and 3.2. The conduction band is taken to be an s-orbital which is symbolised as $|\,S\rangle$ in Kane's (1957) notation. Fawcett *et al.* (1970), however, work with a more refined band structure for the conduction band, which we shall prefer for the pure s states when discussing scattering of electrons in the next Chapter. The wave functions of the valence electrons are constructed from the p-orbitals:

$$Y_1^{\pm 1}(\theta, \phi) = \frac{1}{2}\frac{3}{\sqrt{2\pi}}(\sin\theta\cos\phi \pm \sin\theta\sin\phi) \equiv |\,X\,\rangle \pm i|\,Y\,\rangle \tag{3.53a}$$

and

$$Y_1^0 = \frac{1}{2}\sqrt{3/\pi}\cos\theta \equiv |\,Z\,\rangle \tag{3.53b}$$

where θ and ϕ represent the polar angle and the azimuth respectively. $|\,X\,\rangle$ and $|\,Y\,\rangle$ represent thus the real and the imaginary part of $Y_1^{\pm 1}$. The eigenfunctions of the unperturbed Schrödinger equation belonging to the p states are six-fold degenerate when reckoning the two possible orientations of the spin. The conduction band state is doubly degenerate in the spin. $\Psi_{\mathbf{k}}$ has thus been spanned by the eight linearly independent base functions

$$|iS\downarrow\rangle,\ |(X - iY)\uparrow/\sqrt{2}\,\rangle,\ |Z\downarrow\rangle,\ |(X + iY)\uparrow/\sqrt{2}\,\rangle,\ |iS\uparrow\rangle,$$

$$|-(X + iY)\downarrow/\sqrt{2}\rangle,\ |Z\uparrow\rangle,\ \text{and}\ |(X - iY)\downarrow/\sqrt{2}\,\rangle \tag{3.54}$$

when **k** points in the z direction. The two spin states are denoted ↑ and ↓ respectively.

On this basis the 8 × 8 interaction matrix of Equation (3.52) becomes

$$\begin{pmatrix} H_{\mathbf{k}p} & 0 \\ 0 & H_{\mathbf{k}p} \end{pmatrix}$$

with

$$H_{\mathbf{k}\mathrm{p}} = \begin{pmatrix} E_C & 0 & kP_z & 0 \\ 0 & E_V - E'_{so}/3 & \sqrt{2}E'_{so}/3 & 0 \\ kP_z & \sqrt{2}E'_{so}/3 & E_V & 0 \\ 0 & 0 & 0 & E_V + E_{so}/3 \end{pmatrix}. \tag{3.55}$$

Here

$$P_z \equiv -i(\hbar/m_0)\langle S|p_z|Z\rangle \tag{3.56}$$

is real and

$$E'_{so} = \frac{3i\hbar}{4m_0^2c^2}\left\langle X \left| \frac{\partial U_e}{\partial x} p_y - \frac{\partial U_e}{\partial y} p_x \right| Y \right\rangle \tag{3.57}$$

represents the spin–orbit splitting of the valence band. E_V and E_C refer to the eigenvalues of the Hamiltonian, E_C corresponds to the conduction band and E_V to the valence band.

For general electronic wave vectors **k** the two spin states become $\uparrow'$ and $\downarrow'$:

$$\begin{pmatrix} \uparrow' \\ \downarrow' \end{pmatrix} = \begin{pmatrix} e^{-i\phi/2} & \cos\theta/2 \; e^{i\phi} & \sin\theta/2 \\ -e^{-i\phi/2} & \sin\theta/2 \; e^{i\phi} & \cos\theta/2 \end{pmatrix} \begin{pmatrix} \uparrow \\ \downarrow \end{pmatrix} \tag{3.58}$$

and the valence band basic functions become

$$\begin{pmatrix} |X'\rangle \\ |Y'\rangle \\ |Z'\rangle \end{pmatrix} = \begin{pmatrix} \cos\theta\cos\phi & \cos\theta\sin\phi & -\sin\theta \\ -\sin\phi & \cos\phi & 0 \\ \sin\theta\cos\phi & \sin\theta\sin\phi & \cos\phi \end{pmatrix} \begin{pmatrix} |X\rangle \\ |Y\rangle \\ |Z\rangle \end{pmatrix} \tag{3.59}$$

but the spherical conduction band function stays invariant:

$$|S'\rangle = |S\rangle. \tag{3.60}$$

From Equation (3.55) we get the band structure. For small **k** the conduction band looks like:

$$E_C = E_G + \frac{\hbar^2\mathbf{k}^2}{2m_0} + \frac{P_z^2\mathbf{k}^2}{3}\left(\frac{2}{E_G} + \frac{1}{E_G + E_{so}}\right), \tag{3.61}$$

the light hole band becomes

$$E_{lh} = \frac{\hbar^2 \mathbf{k}^2}{2m_0} - \frac{2P_z^2 \mathbf{k}^2}{3E_G}, \tag{3.62}$$

the heavy hole band becomes

$$E_{hh} = \frac{\hbar^2 \mathbf{k}^2}{2m_0} \tag{3.63}$$

and the split-off band assumes the shape

$$E_{so} = E'_{so} + \frac{\hbar^2 \mathbf{k}^2}{2m_0} - \frac{P_z^2 \mathbf{k}^2}{3(E_G + E_{so})}. \tag{3.64}$$

In these expressions the energies are reckoned positive outwards from the nucleus for the conduction band and towards the nucleus for the valence band, as customary. Here E_G represents the gap between the conduction and the valence band for $\mathbf{k} = 0$. These expressions also give the effective masses in the conduction, the light hole and the split-off bands. Only the heavy hole band mass seems to be the same as the free electron one according to these expressions. As the reader expects, these expressions are only approximate - more accurate calculations yield better results.

We can determine the values of E_G, E_{so} and P_z from spectroscopic data so that we can obtain a reasonably good band structure from Equations (3.56) and (3.57), which are based on a basis of single electronic orbitals. An improvement could be obtained by including more orbitals in the base.

Strain has not been considered here. Its inclusion means an additional term in the Schrödinger equation (3.52). Recently Bahder (1990) has re-examined Kane's theory, including the effect of stress. For stressed material the degeneracy of the light and heavy hole band is lifted at $\mathbf{k} = \mathbf{0}$.

3.5.3 Simplified analytical band structures

The ridges separating the minima of the conduction band are often taller than the gap between the conduction and the valence band. This has great practical significance: an electron entering E_G up the mountain side is very likely to cause creation of an electron–hole pair, whereby it ends down the valley again. Electrons therefore reach a summit so rarely that that part of the band structure becomes almost inaccessible to the electrons. Furthermore, when the local electric field is not too strong the electrons remain near the minima. They can only communicate with another minimum by the intercession of a phonon. When only the area around the minima is of interest a simplified analytical description of the band structure is adequate:

$$E(1 + \alpha E) = \frac{1}{2}\hbar^2(\mathbf{k} - \mathbf{k}_\alpha)\bullet\mathbf{m}^{*-1}\bullet(\mathbf{k} - \mathbf{k}_\alpha) \tag{3.65}$$

where α represents the non-parabolicity factor, describing the deviation from the

Table 3.3 Position in the Brillouin zone of the minima of the conduction band in cubic semiconductors of the zinc blende or diamond structure. ξ and η are given in units of the reciprocal wave number $2\pi/a$ where a represents the lattice constant. Values for the parameters ξ and η can be found in the literature, e.g. Landolt and Börnstein. The parameters ξ and η represent distances along the ⟨100⟩ and ⟨111⟩ directions, respectively, normalised by the factor $2\pi/a$ where a represents the lattice constant

Type of minima	Γ	X	L
Position	(0, 0, 0)	$(\xi, 0, 0)$	(η, η, η)
		$(-\xi, 0, 0)$	$(\eta, \eta, -\eta)$
		$(0, \xi, 0)$	$(\eta, -\eta, \eta)$
		$(0, -\xi, 0)$	$(\eta, -\eta, -\eta)$
		$(0, 0, \xi)$	$(-\eta, \eta, \eta)$
		$(0, 0, -\xi)$	$(-\eta, \eta, -\eta)$
			$(-\eta, -\eta, \eta)$
			$(-\eta, -\eta, -\eta)$

parabolicity, $\mathbf{m}^*$ is the effective mass tensor, $\mathbf{k}$ the wave vector and $\mathbf{k}_\alpha$ the position of a minimum. These are listed in Table 3.3. In the centre of the Brillouin zone (i.e. when $\mathbf{k}_\alpha = \mathbf{0}$) Equation (3.65) describes constant energy spheres, the effective mass tensor is diagonal with all diagonal elements equal. The other constant energy surfaces describe ellipsoids with their long axis in the direction of the straight line connecting its centre to that of the Brillouin zone and with a spherical cross-section perpendicular to it. The two minor axes of the ellipsoids are equal in length. Figure 3.8 shows the constant energy surface in the Brillouin zone around each of the minima of the conduction band for a selected energy. The use of Equation (3.65) for the band structure represents a great saving in computational effort compared to a model based on the more exact shape of the band structure. If the results of the two different models do not deviate significantly, the one with the simplest band structure is to be preferred.

The effective mass tensor in the (100) direction i.e. for the minimum with centre in $\mathbf{k}_\alpha = (2\pi/a)\ (0,0,\xi)$ is

$$\mathbf{m}^* = \begin{pmatrix} m_t^* & 0 & 0 \\ 0 & m_t^* & 0 \\ 0 & 0 & m_l^* \end{pmatrix} \tag{3.66}$$

where m_t^* and m_l^* represent the effective mass along the two minor and the one major axis of the ellipsoid respectively. They are referred to as the *transversal* and the *longitudinal effective masses* respectively. The other 13 effective mass tensors can be obtained from the operation

$$\mathbf{m}'^{*-1} = \mathbf{S}_i \cdot \mathbf{m}^{*-1} \bullet \mathbf{S}_i^{-1} \tag{3.67}$$

where $\mathbf{S}_i$ represents one of the transformation matrices (2.10) or (2.11) or a product combination of them. The transversal and longitudinal effective masses of the L-minima differ from those of the X-minima, but when the crystal has not been subjected to any stress they are equal within one type of minima. A more sophisticated presentation of effective masses and energy surfaces has been given by Rössler (1984).

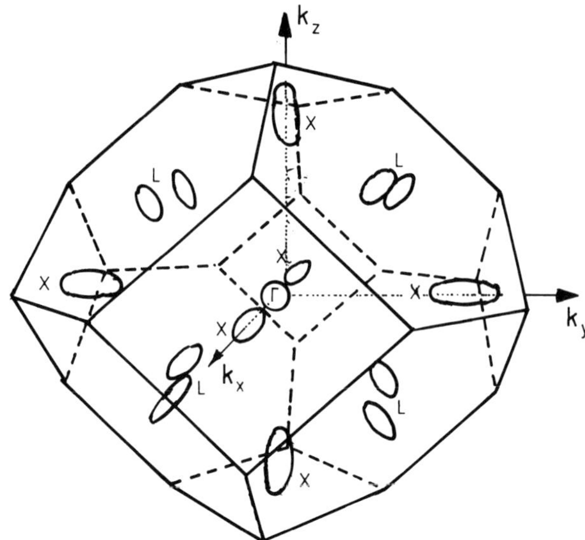

Figure 3.8 *Stereoscopic view of the Brillouin zone of a crystal of zinc blende structure. The ellipsoids and the sphere in the centre represent constant energy surfaces of the band structure. The ellipsoids have their long axis along the straight lines connecting their centres with the centre of the zone. The cross-section perpendicular to the long axis is circular*

In GaAs, when most of the carriers reside around the Γ-minimum, an isotropic description of the satellite minima may be sufficiently accurate, i.e. the three diagonal elements of the effective mass tensor can be replaced by the *effective density of states mass* $m_D^* = (m_t^{*2} m_l^*)^{1/3}$ and the off-diagonal elements by zero. The advantage of this is again saved computer time.

For holes, the band structure near the centre of the Brillouin zone is

$$E = \frac{\hbar^2 \mathbf{k}^2}{2m_0} [A' \pm \{B'^2 + C'^2(k_k^2 k_y^2 + k_y^2 k_z^2 + k_z^2 k_x^2)/k^4\}^{1/2}] \qquad (3.68)$$

where A', B' and C' are constants, the values of which can be found in the literature, e.g. Landolt and Börnstein (1982). The positive and negative signs refer to heavy and light holes respectively.

Sacrificing somewhat on accuracy, the structure of all the hole bands can be described by

$$E = \frac{\hbar^2 \mathbf{k}^2}{2m^*} \qquad (3.69)$$

with m^* representing the effective mass of the holes. Split-off holes can, in the vicinity of the centre of the Brillouin zone, almost always be described rather accurately by Equation (3.69) even when the energy of the other holes have to be calculated from Equation (3.68).

Quite generally the *effective mass* is defined by

$$\frac{1}{m_{ij}^*} = \frac{1}{\hbar^2}\frac{\partial^2 E}{\partial k_i \partial k_j} \tag{3.70}$$

where E represents the exact band structure. The effective mass already defined in Equation (3.20) is an approximation based on perturbation theory.

The *group velocity* of an electron is defined by

$$\mathbf{v} = \frac{1}{\hbar}\frac{\partial E}{\partial \mathbf{k}}. \tag{3.71}$$

$\mathbf{v} = 0$ in the minima. The group velocity should not be confused with the *phase velocity* $\hbar\mathbf{k}/m^*$, which happens to be the same for the parabolic band structure near the centre of the Brillouin zone.

3.6 FERMI–DIRAC STATISTICS

Fermions are particles or quasiparticles of half-integral spin. Electrons and holes are the most important examples met in this book. Such particles do not coexist when they possess the same set of quantum numbers. These unsocial particles therefore obey a different statistics from that of bosons. This special kind of statistics will be derived here.

Following Landau and Lifshitz (1959), consider N particles with energies ranging over a band. This band can be imagined divided into consequent narrow subranges: subrange j contains G_j quantum states, $N_j(\leqslant G_j)$ have been occupied by particles. There are

$$\Gamma_j = \frac{G_j!}{(G_j - N_j)!\,N_j!} \tag{3.72}$$

distinguishable ways of occupation within this subrange. The total number of distinguishable occupations of the entire range of energies is

$$\sigma = \prod_{j=1}^{\infty} \Gamma_j. \tag{3.73}$$

The value of j ranges to infinity when the subintervals shrink such that their total length remains unaltered. The *entropy* of the system is defined in the same way as in Section 2.8:

$$\begin{aligned} S_e &= k_B \sum_{j=1}^{\infty} \log \Gamma_j \\ &= k_B \sum_{j=1}^{\infty} \{\log G_j! - \log (G_j - N_j)! - \log N_j!\}. \end{aligned} \tag{3.74}$$

The most probable distribution of the particles over the available quantum states is the one which renders the entropy maximal subject to the conditions that the number of particles stays constant:

$$N_T = \sum_{j=1}^{\infty} N_j, \tag{3.75}$$

and the total energy is fixed:

$$E_T = \sum_{j=1}^{\infty} N_j E_j. \tag{3.76}$$

Using Stirling's formula, Equation (2.80), the entropy can be expressed as:

$$S_e = k_B \sum_{j=1}^{\infty} \{G_j \log G_j - (G_j - N_j) \log(G_j - N_j) - N_j \log N_j\}. \tag{3.77}$$

It is maximal when

$$\frac{\delta}{\delta N_j} \left[S_e + \alpha' \left(N_T - \sum_{j=1}^{\infty} N_j \right) + \beta' \left(E_T - \sum_{j=1}^{\infty} N_j E_j \right) \right] = 0 \tag{3.78}$$

where α' and β' represent Lagrange multipliers used to introduce the two constraints (3.75) and (3.76). This gives

$$f_j \equiv \frac{N_j}{G_j} = \left[\exp\left(\frac{\alpha' + \beta' E_j}{k_B} \right) + 1 \right]^{-1}, \tag{3.79}$$

the probability that sublevel N° j is occupied. The explicit form is independent of j, so that this label may be omitted.

The Lagrange multiplier α' has the significance of expressing how high up in energy the particles reach when filling all available energy states starting from the bottom one. This level is of course the *Fermi level* to which α' is related through

$$\alpha' = -E_F/T_f \tag{3.80}$$

where T_f represents the absolute temperature. We have already met the parameter β' in Section 2.8:

$$\beta' = 1/T_f. \tag{3.81}$$

Each sublevel is so narrow that the energy within it can, without making any appreciable error, be considered constant. This is exact when the subrange is of infinitesimal width. Without the label j, the *Fermi–Dirac distribution* (3.79) reads

$$f(E) = \{ \exp[(E - E_F)/(k_B T_f)] + 1\}^{-1}. \tag{3.82}$$

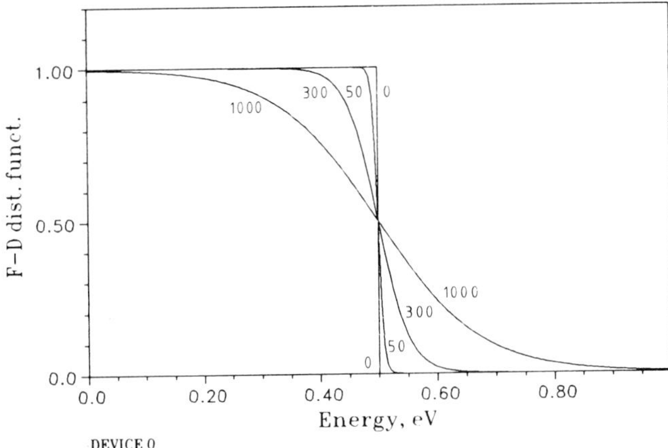

Figure 3.9 *The Fermi–Dirac distribution for temperatures of 0, 50, 300 and 1000 K*

Figure 3.9 shows the Fermi–Dirac distribution function for different temperatures. We note, regardless of temperature, that

$$f(E) \leq 1 \tag{3.83}$$

which is expected because multiple occupancy of quantum states is impossible. Furthermore

$$f(0) = 1, \tag{3.84a}$$

$$f(E_F) = 1/2 \tag{3.84b}$$

and

$$\lim_{E \to \infty} f(E) = 0. \tag{3.84c}$$

At zero temperature

$$f(E) = \begin{cases} 1 & \text{for } E < E_F \\ 0 & \text{for } E > E_f \end{cases} \tag{3.85}$$

The physical interpretation of Equation (3.85) is the already familiar one. At zero absolute temperature the particles will fill up to a certain energy, the Fermi energy: no free states will stay empty below it, all states above it are free.

This sounds well when the density of states is positive at this level. But in semiconductors this is not the case since the valence band is full, the conduction band is empty in intrinsic material and there is a gap between them. From such an argument it follows that the Fermi level must lie somewhere in the gap. The actual placing of it is ambiguous from this kind of argumentation. We shall make an exact calculation of it in Section 3.8.

The distribution, Equation (3.82), applies to electrons. The energy E is reckoned upwards, i.e. away from the nucleus or lowest state. Dealing with holes we rather look for the distribution of free states because holes represent lack of electrons. Instead of finding the probability that an energy state is occupied, we look for the chance that it is free. The alternative wording is: the chance that a hole state is occupied. This is

$$f_h(E) = 1 - f(E) = \{\exp[-(E - E_F)/(k_B T_f)] + 1\}^{-1}. \tag{3.86}$$

It is customary to reckon the energy of holes downwards from the top of the valence band towards the nucleus. E and E_F therefore change sign so that we have to rewrite Equation (3.86):

$$f_h(E) = \{\exp[(E - E_F)/(k_B T_f)] + 1\}^{-1} \tag{3.87}$$

which has the very same form as Equation (3.82). Holes and electrons obey the same statistics when we adhere to the convention regarding the orientation of energy introduced here.

3.7 DENSITY OF STATES

The *density of states* describes how many energy states there are within a small volume of the wave vector space. In Section 3.2 it was shown that the possible wave vectors form a lattice of points, Equation (3.17), each point occupying a cell of volume $8\pi^3/V_{cr}$. As the Brillouin zone for electrons and phonons are the same, the density of states $D_s(\mathbf{k})$ is defined by

$$D_S(\mathbf{k})\, d^3\mathbf{k} = \frac{1}{8\pi^3} d^3\mathbf{k}. \tag{3.88}$$

In Cartesian coordinates

$$d^3\mathbf{k} = dk_x dk_y dk_z \tag{3.89a}$$

and in polar coordinates

$$d^3\mathbf{k} = k^2 dk \sin\theta d\theta\, d\phi \tag{3.89b}$$

where θ and ϕ represent the azimuth and polar angles respectively. (Figure 2.7.) Near the centre of Brillouin zone the band structure, the isotropic version of Equation (3.65), simplifies to

$$E(1 + \alpha E) = \frac{\hbar^2 \mathbf{k}^2}{2m^*} \tag{3.90}$$

and the density of states in a sphere of radius k and thickness dk is

$$D_S(E)\, d^3\mathbf{k} = \frac{4}{4\pi^2} \left(\frac{2m^*}{\hbar^2}\right)^{3/2} (1 + 2\alpha E) \sqrt{E(1 + \alpha E)}\, dE \tag{3.91}$$

where m^* represents the effective mass of the electron and α the non-parabolicity parameter. The factor 2 comes from the summation over the two spin states. It has been assumed that no magnetic field is present. A factor 4π results from integration over the polar and azimuth angles. This expression, Equation (3.91), is the same as that given by Sze (1969), who defines the density of states as the number of states within a sphere or radius k and thickness dk. Our definition, Equation (3.88), is restricted to a volume element, and is therefore more general as it can be applied to the most general band structure determined from e.g. pseudopotential theory. Seeger (1982) prefers to define the density of states with the factor V_{cr}. Since there are some discrepancies in the details between the various authors, it is important to agree to a convention and keep to it. We shall use Equation (3.88) and refer to the result of integrating over the constant energy surface as the *density of states over a constant energy surface*.

The easiest way to calculate the density of states for the more general band structure, Equation (3.65), is to transfer to a vector space where the effective mass tensor becomes diagonal with equal diagonal elements:

$$\mathbf{k}' = \mathbf{B}_H \bullet \mathbf{S}_i \bullet (\mathbf{k} - \mathbf{k}_\alpha). \tag{3.92}$$

This transformation, the *Brooks–Herring transformation*, is most useful when discussing scattering from electrons having the band structure given by Equation (3.65). $\mathbf{S}_i$ is one of the matrices (2.10) or (2.11) or a combination of products of them, or the identity matrix

$$B_H = \begin{bmatrix} m_t^{*1/6} m_1^{*-1/6} & 0 & 0 \\ 0 & m_t^{*1/6} m_1^{*-1/6} & 0 \\ 0 & 0 & m_t^{*-1/3} m_1^{*1/3} \end{bmatrix}. \tag{3.93}$$

With this transformation the band structure, Equation (3.65), becomes

$$E(1 + \alpha E) = \frac{\hbar^2}{2m_D^*} \mathbf{k}'^2 \tag{3.94}$$

which is perfectly isotropic. The quantity

$$m_D^* = (m_t^{*2} m_1^*)^{1/3} \tag{3.95}$$

represents the *density of states effective mass*.

It would be useful to introduce lower dimensional density of states as well. At an interface between two different materials with a strong perpendicular electric field, i.e. at the silicon–silicon dioxide interface at the gate of the metal oxide semiconductor FET (MESFET) or in a modulation doped FET (MODFET) quantum well, the conduction band divides into subbands. Within each subband the wave vector perpendicular to the interface or heterojunction discretises. The electrons are free to move only parallel to the interface – i.e. in a plane, the motion becomes two-dimensional. In other words, we have a *two-dimensional electron gas*: each lattice point in the k-space now occupies a volume $4\pi^2/A$ where A represents the area of the interface or electron gas. The two-dimensional density of states is

$$D_S(k)\, d^2\mathbf{k} = \frac{1}{4\pi^2} d^2\mathbf{k}. \tag{3.96a}$$

In Cartesian coordinates

$$d^2\mathbf{k} = dk_x dk_y, \tag{3.96b}$$

and in polar coordinates

$$d^2\mathbf{k} = k dk\, d\phi \tag{3.96c}$$

where the polar angle ϕ is oriented from the k_x axis and varies anticlockwise throughout the entire circle $(-\pi \leq \phi \leq \pi)$. With the band structure (3.90) the density of states between two concentric circles of radius k and $k + dk$ is

$$D_S(E) = \frac{2}{2\pi} m^* \,(1 + 2\alpha E)/\hbar^2 \frac{dE}{d^2\mathbf{k}}. \tag{3.97}$$

A factor 2π comes from the integration over ϕ, and a factor 2 from summation over two spin states in the absence of a magnetic field. The energy levels associated with directions perpendicular to the confinement lie far apart and have to be considered individually.

With a quantising confinement in both the y and z direction only one degree of freedom in motion remains. In this case we write dk instead of $d^3\mathbf{k}$ or $d^2\mathbf{k}$ and the density of states becomes

$$D_S(k) dk = \frac{1}{2\pi} dk \tag{3.98}$$

which for the band structure given by Equation (3.90) becomes

$$D_S(E) = -\frac{m^{*\frac{1}{2}}\,(1 + 2\alpha E)}{\sqrt{2}\pi [E(1 + \alpha E)]^{\frac{1}{2}}\, \hbar} \frac{dE}{dk}. \tag{3.99}$$

In systems confined in all directions, like individual impurity atoms or quantum dots, there are no motional degrees of freedom.

3.8 IMPURITIES AND THE FERMI ENERGY

In a perfectly pure semiconductor material the Fermi level lies near the middle of the gap between the valence and the conduction bands. However the impurities are responsible for the semiconducting properties of the material and cause the Fermi level to move away from this position according to how they modify the number of carriers being available for conduction. After a short introduction to the nature of the impurities a rigorous calculation of the Fermi energy will be presented.

3.8.1 Impurities and crystal imperfections

As already mentioned the impurities are responsible for the semiconducting properties of the material. The energy of the electrons bound to the impurity atom lie in the gap between the valence and the conduction bands, though precisely where depends on the species of impurity atoms. (Figure 3.10.) The energy level of the outermost electrons from lithium in n-type silicon lie 33 meV below the conduction band edge. These atoms can easily be ionised by thermal agitation, whereby their electrons become available for conduction.

Boron in p-type germanium has three valence electrons in its neutral state. It can, because of the tetrahedral coordination of its germanium neighbours, easily

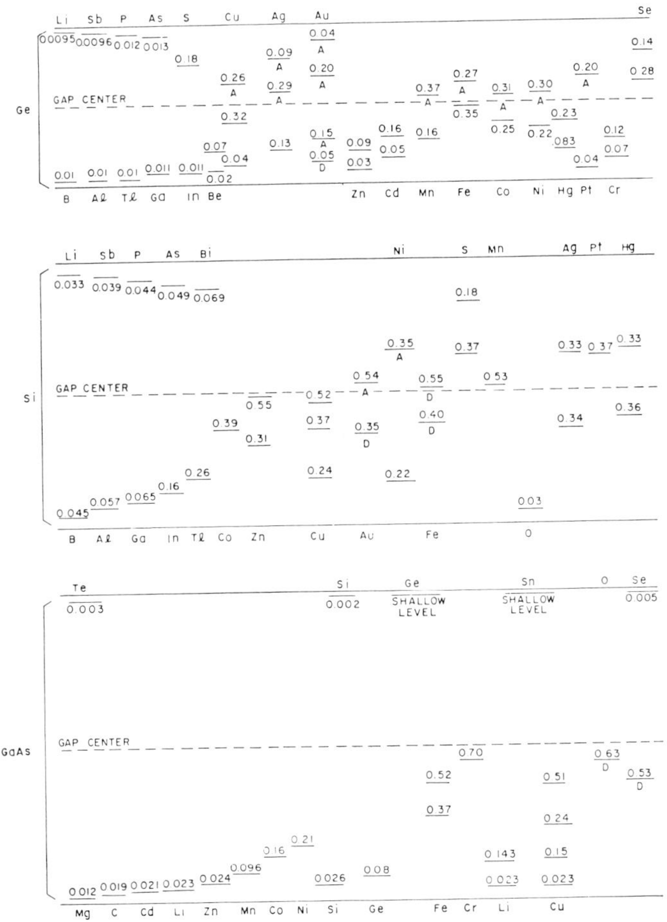

Figure 3.10 *Measured ionisation energies for various impurities in germanium, silicon and gallium arsenide. The levels below the gap centre are reckoned from the top of the valence band and are acceptor levels unless indicated by D for donor level. The levels above the gap centre are reckoned downwards from the conduction band minimum and are donor levels unless labelled A for acceptor levels. The band gaps at 300 K are 0.803, 1.12 and 1.43 eV for Ge, Si and GaAs respectively. After Sze (1969)*

take up a fourth electron to obtain a surrounding similar to that of the host atoms. It costs only 10 meV to excite an electron from the top of the valence band into the grip of the boron. The vacancy left behind, the hole, can conduct electricity.

In fact, the impurities have several energy levels in the band gap, though usually only the ground state energy is quoted in the literature. The orbits of the electrons and their energy levels resemble those of the hydrogen atom, and can be described approximately by Equations (3.24) and (3.34) respectively, with the permittivity multiplied by the dielectric constant of the host material and with Z_b equal to the number of electrons or holes the impurity atom is able to trap – usually one or two.

With a large concentration of impurities, the energy levels spread out to form bands, which may merge with the conduction or valence bands of the host. Electrons may migrate between impurity atoms by tunnelling through the potential barriers that exist between them, this is often referred to as *hopping conductivity*, and can yield a small contribution to the conductivity of the material.

Not only do foreign atoms contribute electrons to the conduction band or holes to the valence bands, most materials also contain several species of atoms in addition to those being introduced intentionally. A correct combination of donors and acceptors can cause a cancellation of the electrons and the holes released for conduction. Even if this cancellation is not quite perfect, this sort of material is referred to as *compensated. Intrinsic material*, i.e. perfectly pure material, does not exist. The material is *semi-insulating* when the contribution from the donors is balanced by that of the acceptors.

Crystal faults can also act as donors or acceptors. Often, in the vacant sites, interstitial and substituted atoms form rather complex structures. The investigation of these has received attention from many authors, both experimentally and theoretically.

3.8.2 The Fermi energy

In *thermal equilibrium* the number density of electrons is

$$n_e = \int_{E_{\min}}^{E_{\max}} D_S(E) f(E) \frac{d^3\mathbf{k}}{dE} dE \tag{3.100}$$

where $D_S(E)$ represents the density of states, Equation (3.88) and $f(E)$ the distribution probability, Equation (3.82). If n_e is given by another expression than Equation (3.100), the electron population is no longer in thermal equilibrium. We may suggest a definition of the *disequilibrium function* $\Delta(E)$ such that

$$n_e = \int_{E_{\min}}^{E_{\max}} D_S(E) f(E)\, \Delta(E) \frac{d^3\mathbf{k}}{dE} dE \tag{3.101}$$

where

$$\Delta(E) \leq 1.$$

The equality sign applies in thermal equilibrium: $\Delta(E) > 1$ is impossible. In thermal equilibrium the Fermi energy has a meaning. Otherwise the Fermi energy can only be defined meaningfully when $\Delta(E)$ is known.

In the case of a band structure given by Equation (3.90) and $D_S(E)$ by Equation (3.91) the electron density is

$$n_e = \frac{\sqrt{2}\, m^{*3/2}}{2\,\pi^2\,\hbar^3} \int_{E_{\min}}^{E_{\max}} \sqrt{E(1+\alpha E)}\,(1+2\alpha E)\, f(E)\, dE \tag{3.102}$$

where $E_{\min}$ and $E_{\max}$ represent the minimum and the maximum energy of the band. The error made replacing $E_{\max}$ by infinity is negligible because of the mathematical behaviour of $f(E)$. The lower limit is zero as we reckon the energies from the bottom of the conduction band.

Evaluating Equation (3.102) in this case with $\alpha = 0$ yields

$$n_e = \frac{2}{\sqrt{\pi}}\, n_C'\, F_{\frac{1}{2}}\left(\frac{E_F - E_C}{k_B T_{Lt}}\right) \tag{3.103}$$

where

$$n_C' \equiv 2\left(\frac{m_e^* k_B T_{Lt}}{2\,\pi\,\hbar^2}\right)^{3/2} \nu_C \tag{3.104}$$

and the other symbols have the following meaning: m_e^* effective density of states mass of the electron, assumed independent of energy; k_B Boltzmann's constant; T_{Lt} the absolute temperature of the lattice; E_F the Fermi energy level; E_C the level of the conduction band minimum; ν_C the number of equivalent minima, which for example is six for silicon, one for gallium arsenide.

$$F_j(\eta_F) \equiv \int_0^{\infty} \frac{\eta^j\, d\eta}{1 + \exp(\eta - \eta_F)} \tag{3.105}$$

represents the *Fermi integral* of order j. We should pay attention to its normalisation, which may vary between the authors. If E_F is well below E_C, then

$$n_e = n_C' \exp\left(-\frac{E_F - E_C}{k_B T_{Lt}}\right) \tag{3.106}$$

which characterises *non-degenerate semiconductors*.

The concentration of mobile holes is

$$n_h = \frac{2}{\sqrt{\pi}} \sum_{i = h, l, so} n_{Vi}'\, F_{\frac{1}{2}}\left(\frac{E_V - E_F}{k_B T_{Lt}}\right) \tag{3.107}$$

with

$$n'_{Vi} \equiv 2\left(\frac{m^*_{hi} k_B T_{Lt}}{2\pi\hbar^2}\right)^{3/2} \nu_V \tag{3.108}$$

where m^*_{hi} represents the effective density of states mass for holes, under the assumption that it is independent of energy, and ν_V the number of degenerate valence band maxima which is one.

The summation runs over heavy (h), light (l) and split-off holes (so). When the satellite minima of the conduction band also become populated, the expression for n'_C Equation (3.104) should include the summation over the various types of minima.

The crystal contains donors of type i of density n_{Di} – their donor electrons have energy E_{Di} in the neutral state. n^+_{Di} of these donors are ionised, their number given by

$$n^+_{Di} = n_{Di}\left[1 - \frac{1}{1 + \frac{1}{\nu_{Di}}\exp\left(\frac{E_{Di} - E_F}{k_B T_{Lt}}\right)}\right] = \frac{n_{Di}}{1 + \nu_{Di}\exp\left(\frac{E_F - E_{Di}}{k_B T_{Lt}}\right)} \tag{3.109}$$

Here ν_{Di} represents the ground state degeneration of the donor impurity, which is 2 for donors which can accept one electron of either spin. There are N_D different types of donors.

Similarly the crystal may contain N_A different types of acceptors. Acceptor type i is present in concentration n_{Ai} at the energy level E_{Ai}. The number of ionised acceptors is

$$n^-_{Ai} = \frac{n_{Ai}}{1 + \frac{1}{\nu_{Ai}}\exp\left(\frac{E_{Ai} - E_F}{k_B T_{Lt}}\right)}. \tag{3.110}$$

The degeneracy factor is $g_A = 4$ because in the semiconductor each acceptor can attract a light or heavy hole of either spin when the light and heavy hole bands are degenerate in the centre of the Brillouin zone.

Charge neutrality requires

$$n_e + \sum_{i=1}^{N_A} n^-_{Ai} = n_h + \sum_{i=1}^{N_D} n^+_{Di}. \tag{3.111}$$

Usually only the quantity and energy levels of the main dopants are known. This is indeed sufficient to calculate the Fermi level rather accurately. The equations for n^-_{Ai} and n^+_{Di} are strictly valid only for low impurity concentrations. With a high doping, the impurity level broadens into a band, so that the density of states for the impurity band has to be introduced and an integration over all energy levels within it has to be carried out.

For the valence and conduction bands we should, strictly speaking, use the density of states obtained from the most exact calculation of the band structure i.e. as explained in Section 3.5.1.

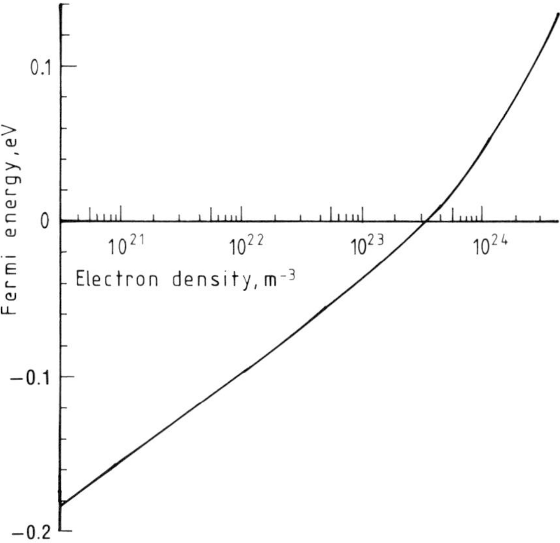

Figure 3.11 *The Fermi level against electron density for gallium arsenide*

A reasonable estimate for the Fermi energy can be obtained when assuming only one donor and acceptor level at the edge of the valence and conduction bands respectively, i.e. $E_{Di} = E_{Ai} = 0$. Figure 3.11 shows the connection between the Fermi level and donor doping for gallium arsenide calculated from Equation (3.111) with $n_{Ai} = 0$.

Finally we shall estimate the Fermi level in a pure semiconductor. Here $n_{Ai} = n_{Di} = 0$, so that the charge neutrality condition, substituting Equation (3.106) and a corresponding expression for holes in Equation (3.111), becomes

$$n_e = n_h$$

or

$$n'_C \exp\left(-\frac{E_F - E_C}{k_B T_{Lt}}\right) = \sum_{i=l,h,so} n'_{Vi} \exp\left(-\frac{E_V - E_F}{k_B T_{Lt}}\right) \tag{3.112}$$

assuming the Fermi level lies far enough from the conduction and valence bands edges that the approximation (3.106) applies. The summation over i runs over light and heavy holes. Many authors introduce a combined effective mass m_h^* for the holes. With this combined effective mass we get for the intrinsic Fermi energy:

$$E_F = \frac{3}{4} K_B T_{Lt} \log\left(\frac{m_e^*}{m_h^*}\right) + \frac{1}{2}(E_C + E_V) \tag{3.113}$$

which lies near the centre of the gap between the valence and conduction bands. This is the mathematical basis for what was stated loosely in Section 3.6.

The *quasi Fermi level* for electrons can be obtained from Equation (3.111) omitting the term n_h; that of holes by leaving out the term n_e. With only one type of carrier, the quasi Fermi level coincides with the real Fermi energy.

3.9 PLASMA OSCILLATIONS AND SCREENING

The electrons in a material tend to move in a collective manner. Their oscillation can be described by simple classical arguments: the crystal consists of positively charged ions embedded in a sea of negative charge large enough to render the total charge zero. Displacing a charge of volume density n_e a distance $\mathbf{r}$ from its equilibrium position establishes a polarisation

$$\mathbf{P} = n_e e\mathbf{r} \tag{3.114}$$

which in turn gives rise to an electric dipole field

$$\mathbf{F} = -\mathbf{P}/(\varepsilon\varepsilon_0). \tag{3.115}$$

The force attempting to restore the charge to the equilibrium position is

$$m^* \frac{d^2\mathbf{r}}{dt^2} = -\frac{n_e e^2 \mathbf{r}}{\varepsilon\, \varepsilon_0} \tag{3.116}$$

Here m_e^* represents the effective mass of the electron, ε_0 the permittivity of vacuum and ε the dielectric constant of the material. This differential equation describes a simple oscillatory motion of the charges of frequency

$$\omega_p = \left(\frac{n_e e^2}{m^* \varepsilon\, \varepsilon_0}\right)^{\frac{1}{2}} \tag{3.117}$$

known as the *plasma frequency*. The oscillations are without dispersion, which means that there is no relationship between the energy of the oscillations, $h\omega_p$, and their wave vector. The energy of the *plasmon oscillations* usually lies in the infra-red so that they can couple with optical phonons.

3.9.1 Screening

A charged impurity atom causes the surrounding electrons to rearrange themselves so that the electric field originating in it is neutralised at some distance from it. Dingle (1955) considered the electron gas in an undisturbed plasma as a charge of uniform density, in the so-called *jellium model*. At a distance r from the ion the electron concentration around it is given by

$$n_e(r) = \frac{1}{\sqrt{2}\, h^3}\left(\frac{m^* k_B T_{Lt}}{\pi}\right)^{3/2} F_{\frac{1}{2}}\left(\eta + \frac{e\, U(r)}{k_B T_{Lt}}\right) \tag{3.118}$$

where

$$\eta \equiv E_F/(k_B T_{Lt}) \tag{3.119}$$

with E_F representing the Fermi energy and $F_{\frac{1}{2}}$ the Fermi–Dirac integral of order $\frac{1}{2}$ given by Equation (3.105) with $j = \frac{1}{2}$. The potential around the impurity $\Phi_{im}(r)$ is determined from Poisson's equation:

$$\frac{\partial^2 \Phi_{im}(r)}{\mathbf{r}^2} = e[n_e(r) - n_e]/(\varepsilon\,\varepsilon_0). \tag{3.120}$$

The Expression (3.118) for $n_e(r)$ should be inserted into Equation (3.120).

When $e\Phi_{im}(r)/(k_B T_{Lt}) \ll \eta$ the function $F_{1/2}$ can be expanded in a Taylor series where only terms up to first order in $E\Phi_{im}(r)/(k_B T_{Lt})$ will be retained. Making use of the boundary conditions

$$\Phi_{im}(r \to 0) = \frac{Z_b e}{4\,\pi\,\varepsilon\varepsilon_0\, r} \tag{3.121a}$$

and

$$\Phi_{im}(r \to \infty) = 0 \tag{3.121b}$$

we obtain the screened Coulomb potential

$$\Phi_{im}(r) = \frac{Z_b\, e}{4\,\pi\,\varepsilon\varepsilon_0\, r}\,\exp(-r/\beta_s) \tag{3.122}$$

where

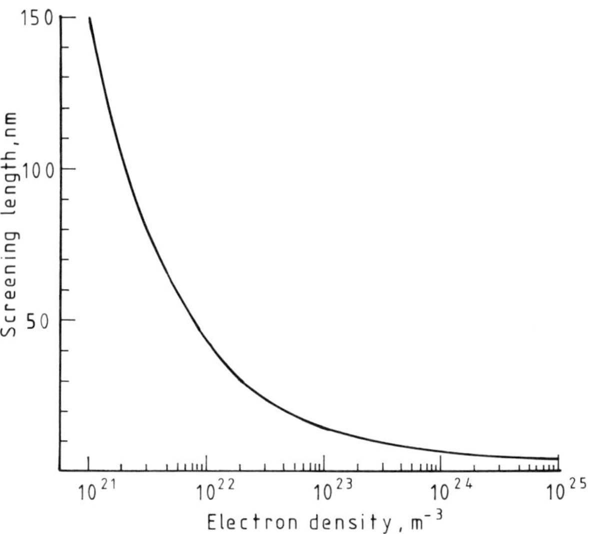

Figure 3.12 *Screening distance in GaAs against electron density*

$$\beta_s = \left\{ \frac{\varepsilon \varepsilon_0 k_B T_{Lt}}{ne^2} \frac{F_{-1/2}(\eta)}{F_{1/2}(\eta)} \right\}^{1/2} \tag{3.123}$$

represents the *screening length*. This decreases with increasing impurity concentration - in other words, a high doping density is more efficient than a low one in screening off the fields from the ionised impurities. Figure 3.12 shows the screening length versus n_e for gallium arsenide.

3.10 SUPERCONDUCTIVITY

So far no one has attempted to simulate superconductivity by means of the Monte Carlo particle model. As with all other problems we are treating, this too needs a first principles approach. According to Kittel (1971) electrons of the same quantum state but opposite spins form pairs. It requires an energy E_{cc} to break it up and this is only possible by exciting one of the carriers by an amount of energy at least equal to E_{cc}. This cannot be achieved by acoustic phonons because they are not energetic enough. Therefore the only candidates are the optical or intervalley phonons. Just as we needed a detailed knowledge of the interesting part of the band structure of the semiconductor, we also require the same for the superconductor and for the selection rules of the phonons.

Recently Takahashi (1990) presented a band structure for high temperature superconductors. (Figure 3.13.) The valence band is filled to the Fermi level, then there is a gap before the conduction band begins. The dopant impurities form another narrow band at the Fermi level which is half filled - here the density of

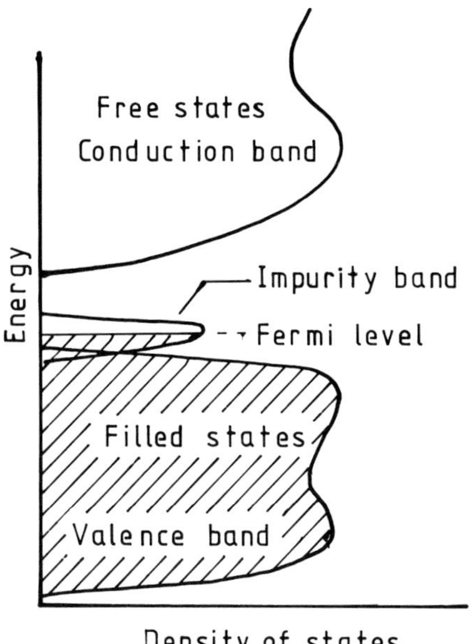

Figure 3.13 *Schematic diagram of the electronic structure of a high temperature superconductor. The shape of the band structure in the vicinity of the Fermi level has been derived from high energy spectroscopy such as photoemission. After Takahashi (1990)*

states is large. The Fermi level + E_{cc} lies in the forbidden band. While there are no phonons exciting the valence electrons into the conduction band the material remains superconducting. In the Monte Carlo simulation the state an electron is selected to scatter into is unavailable, therefore it cannot take place. The model will predict zero resistance, thus superconductivity.

3.11 DIELECTRIC BREAKDOWN

Dielectric or *Zener* (1934) breakdown can take place in solid state devices when the local electric field becomes strong enough. The electric field tilts the band structure to a slope equal to that of the field. (Figure 3.14.) If the field is sufficiently strong, the tail of the valence electron extends into the conduction band giving it a finite chance to tunnel into it creating a vacancy in the valence band - that is, the field alone is strong enough to generate an electron-hole pair.

The valence electron obeys the wave equation

$$-\frac{\hbar^2}{2m_h^*}\frac{\partial^2}{\partial x^2}\Psi_e - eFx\Psi_e = E\Psi_e \tag{3.124}$$

where Ψ_e represents the valence band electronic wave function; E the electron's energy; m_0^* its effective mass; e its charge; F the local electric field, which, since it does not vary significantly over the tunnelling length, we can consider to be uniform; and x the distance. Equation (3.124) has the *Airy function* as solution:

$$\Psi_e = Ai(\xi) \tag{3.125}$$

with

$$\xi = (2m^* eF\hbar^{-2})^{1/3}[x - E/(eF)]. \tag{3.126}$$

The general solution of Equation (3.124) is a linear combination of the $Ai(\xi)$ and the *associated Airy function* $Bi(\xi)$. Figure 3.15 shows the shape of these two

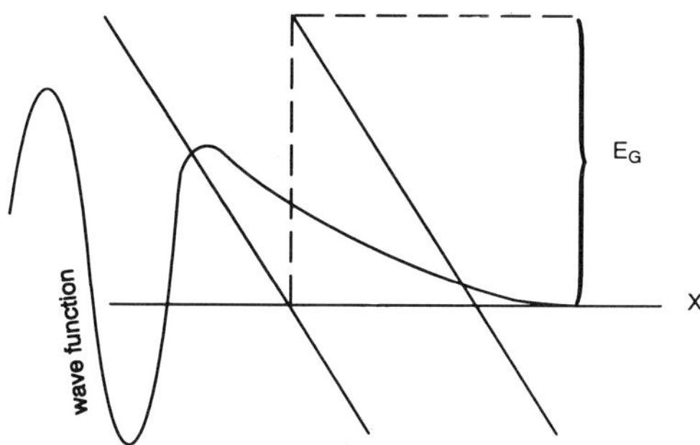

Figure 3.14 *The band structure of a semiconductor in a very high ambient electric field. The wavy line with the tail represents the wave function of a valence electron. The tail protrudes into the conduction band so that there is a finite chance that the valence electron can tunnel through the forbidden band gap into the conduction band, thus creating an electron-hole pair*

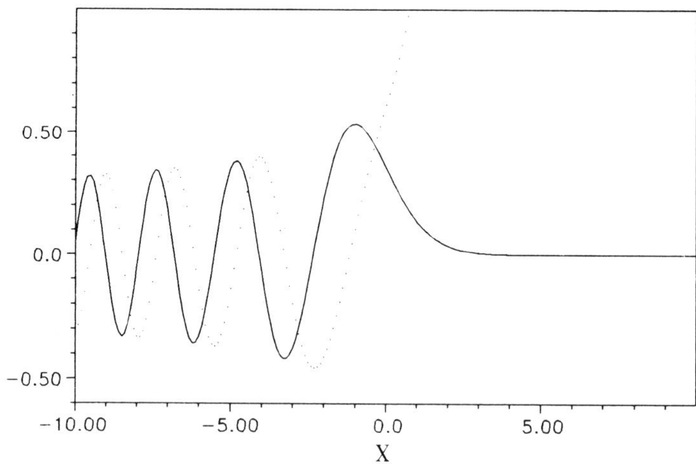

Figure 3.15 *The Airy function Ai(x) (full curve) and the associated Airy function Bi(x) (dotted) against x*

functions from which it is evident that $Bi(\xi)$ cannot be a physically meaningful solution because it diverges for growing x, i.e. into the conduction band. Therefore Equation (3.125) represents the solution of the wave equation. Airy functions have been described by Albright (1977).

4

Lattice–Electron Interaction

4.1 INTRODUCTION

We have studied the lattice and the electronic parts of the crystal Hamiltonian in the last two chapters without paying any attention to the interaction term (2.22) between them. Without explicitly expressing it we have assumed, or rather hoped, that the interaction would be small enough that its exclusion would still yield a sufficiently accurate solution for the band structure and the phonon spectrum. Our effects have indeed been crowned with success as the results compare well with experimental data.

In this chapter we shall turn our attention to the Cinderella term (2.22) which is the most interesting part of semiconductor transport theory. This term, which has been concealed in its elegant dress, contains all the different ways energy can be exchanged between the vibrating lattice and the electrons. In the next section we shall calculate this interaction by means of time-dependent perturbation theory.

The electrons interact with the lattice in several different ways. We shall distinguish between them by writing symbolically

$$\mathcal{H}_{ex} = \mathcal{H}_1 + \mathcal{H}_2 + \dots \tag{4.1}$$

where each of the Hamiltonians $\mathcal{H}_1$, $\mathcal{H}_2$, ... represent a definite type of interaction. The aim of this chapter is to discuss each of the most significant ones in detail. We are interested in the rate, or frequency, of these interactions, and in the direction of the motion of the particle afterwards. The latter, however, we shall calculate in the following chapter.

We shall first study the interaction between electrons and phonons. The phonon spectrum is complex but the laws of conservation of energy and momentum, and the crystal symmetry exclude most of the phonons from taking part. These selection rules will be given in Section 4.3. In the subsequent sections other types of interaction will be discussed individually.

An interaction does take time, for example consider an electron passing through the Coulombic field of an ionised impurity atom. The particle will spend a finite time influenced by the impurity, during which its path is deflected. It turns out, however, that this time is so short that the interaction can be considered instantaneous – this is the assumption behind Boltzmann's transport equation. In most devices the free flight, i.e. the flight free of perturbation, dominates the transport. But when we are considering very small devices, for example transistors of nanometre gate length, it may be important to consider the duration of the scattering. This will be discussed in Section 4.9.

The crystal contains imperfections in the form of foreign atoms which are easily ionised. Most of these have been introduced on purpose to make the material a semiconductor. The scattering from them is explained in Section 4.5. We shall also treat scattering of carriers from each other. The crystal also contains neutral foreign atoms and several types of stacking faults. In alloys the irregular distribution of the different atoms over the lattice sites also gives rise to scattering, which will be discussed in Section 4.6. A brief introduction to impact ionisation, and trapping and release of carriers will be given in Sections 4.7 and 4.8.

Electron scattering in solids has previously been discussed by several authors. The most comprehensive exposition with Monte Carlo simulation in mind was probably given by Fawcett *et al.* (1970) and Jacoboni and Reggiani (1979). Fawcett's work is a classic in Monte Carlo modelling: he presented all the important scattering rates in bulk gallium arsenide. Jacoboni has tabulated the various matrix elements for transition of electrons from state $\mathbf{k}$ to state $\mathbf{k}'$.

There is a wide range in the values for the different scattering rates; they range up to the order of 10^{14} s^{-1}. The higher the rate of scattering from a specified mechanism, the more likely it is to cause scattering. The importance of a scattering mechanism is measured by its rate. A particle simulator should contain all the most important ones. Usually these are scattering from optical and acoustic phonons and from ionised impurities. The inventor of a model has to judge how many different types of scattering events should be included. We shall see that it is advisable to include them in the order of their rates. In most of the modelling discussed in the following chapters it is sufficient to include only those mechanisms just mentioned. The weak mechanisms will not contribute significantly to the transport. However, the reader may decide that a combination of several weak scattering mechanisms can be significant.

Description of scattering requires knowledge both of the phonons involved and the band structure of the conduction and valence bands, and of the several constants which will be introduced in this chapter. Fortunately, this information can be found in the literature. The foremost source of such information is the tables of Landolt-Börnstein (1982). Further information, or newer values, have been published by Fischetti (1991). Fischetti and Laux (1988), Adachi (1985), Brennan and Hess (1984), Blakemore (1982), Anastassakis (1983) and Kratzer and Frey (1978) for GaAs; Brennan *et al.* (1988) and Adachi (1985) for $Al_xGa_{1-x}As$; Brennan and Park (1989) and Long *et al.* (1987) for $In_xGa_{1-x}As$; Jacoboni and Reggiani (1983) for germanium, silicon and diamond; and Brennan and Hess (1984) and Costa *et al.* (1989) for indium phosphide.

4.2 RATE OF TRANSITION BETWEEN ELECTRONIC STATES

The treatment of the interaction between the lattice and the electrons is based on time-dependent perturbation theory which is described in most textbooks on elementary quantum mechanisms e.g. Merzbacher (1961). The time-dependent Schrödinger equation reads

$$i\hbar \frac{\partial}{\partial t} \Psi^0(t) = \mathcal{H}_0 \Psi^0(t). \tag{4.2}$$

The formal solution of this time-dependent differential equation is

$$\Psi^0(t) = \sum_n \exp(-iE_n^0 t/\hbar)\,|n\rangle \tag{4.3}$$

with

$$|n\rangle = \Psi_n^0(0), \tag{4.4}$$

$|n\rangle$ representing the wave function for the quantum state n prior to the action of the perturbation. Introducing the perturbation U, we define a new Hamiltonian $\mathcal{H} = \mathcal{H}_0 + U$, and the corresponding equation of motion reads:

$$i\hbar\frac{\partial\Psi}{\partial t} = (\mathcal{H}_0 + U)\Psi. \tag{4.5}$$

Its solution can be expressed in terms of the eigenfunctions of the unperturbed Equation (4.2):

$$\Psi(t) = \sum_n c_n(t)\exp(-iE_n^0 t/\hbar)\,|n\rangle, \tag{4.6}$$

where $c_n(t)$ represents the component of the unperturbed state $|n\rangle$ found in the perturbed state at time t.

The summation runs over all quantum states of $\mathcal{H}_0|n\rangle = E^0|n\rangle$. When the quantum states form a quasi continuum the summation should be replaced by an integration. Substitution of this into Equation (4.5) and integration over the domain where the wave functions exist yields

$$i\hbar\frac{dc_k(t)}{dt} = \sum_n \langle k|U|n\rangle\, c_n \exp(i\omega_{kn}t) \tag{4.7}$$

with

$$\omega_{kn} \equiv (E_k^0 - E_n^0)/\hbar. \tag{4.8}$$

Assume that prior to the application of the perturbation, at time zero, the system was in the state $|s\rangle$. In other words $c_s(0) = 1$ and $c_k(0) = 0$ for $k \neq s$. If the perturbation is so weak or short-lived that $c_s(t) \simeq 1$ and $c_k(t) \ll c_s(t)$ for $k \neq s$ then $c_k(t)$ can be found from Equation (4.7) by integration over the time until the end of the action of the perturbation, which has been reached at time t.

$$c_k(t) = -\frac{i}{\hbar}\int_0^t \langle k|U|s\rangle \exp(i\omega_{ks}t')\,dt' \tag{4.9}$$

for $k \neq s$.

If $\langle k|U|s\rangle$ does not depend explicitly on time, this integral can be evaluated analytically:

$$c_k(t) = \frac{\langle k|U|s\rangle}{E_k^0 - E_s^0}[1 - \exp(i\omega_{ks}t)] \tag{4.10}$$

The probability that the system will be in the unperturbed eigenstate $|k>$ $(k \neq s)$ at the end of the perturbation is

$$T_{ks} = |c_k(t)|^2 = 2|\langle k|U|s\rangle|^2 \frac{1 - \cos\omega_{ks}t}{(E_k^0 - E_s^0)^2} \tag{4.11}$$

Note that this expression tells us that the probability of transfer from the unperturbed state $|s\rangle$ into $|k\rangle$ is the same as that of the transfer in the opposite direction.

In a semiconductor the states form a quasi continuum. The availability of the final state, $|k\rangle$, is $D_S^a(E_k)$. If none of the final states had been occupied prior to the perturbation, then $D_S^a(E) = D_S(E)$, the density of states. If the final state is already occupied the transfer, or scattering, cannot take place because no two electrons can occupy the same state. This possibility has to be considered, and is likely to happen, especially in degenerate semiconductors (Bosi and Jacoboni, 1976). When the scattering is of such a nature that the thermal equilibrium of the electron population to which the scattered particle belongs is conserved, then

$$D_S^a(E) = D_S(E)[1 - f(E)] \tag{4.12}$$

where $f(E)$ represents the Fermi–Dirac distribution function, Equation (3.82), and $D_S(E)$ the density of states per unit interval of energy. In areas of a rapidly changing electric field, thermal equilibrium is not present so that

$$D_S^a(E) = D_S(E)[1 - f(E)\Delta(E)] \tag{4.13}$$

where $\Delta(E)$ represents the disequilibrium function of Section 3.8.2. It may be impossible to define the Fermi energy, in which case $f(E)\Delta(E)$ has to be replaced by the actual ratio of the number of particles of energy E to the density of states. Mickevicius and Reklaitis (1987) have discussed the scattering of phonons not being in thermal equilibrium.

Considering the fact that some states may be inaccessible, the probability of scattering is obtained by summing over all possible final states:

$$\sum_{k \in f} T'_{ks} = 2\int_{k \in f} |\langle k|U|s\rangle|^2 \frac{1 - \cos\omega_{ks}t}{(E_k^0 - E_s^0)^2} D_S^a(E_k^0)\, dE_k^0. \tag{4.14}$$

The notation $k \in f$ means that the summation is only carried out over all the final states which are selected by the conservation laws for momentum and energy. The energy of the final states are considered as being so close together that the summation over them has been replaced by an integral. The temporal change of the total transition probability, or *scattering rate* is

$$\lambda_s = \frac{d}{dt}\sum_{k\in f} T'_{ks} = \frac{2}{\hbar^2}\int_{k\in f} |\langle k|U|s\rangle|^2 \frac{\sin\omega t}{\omega} D_S^a(\omega)\, d\omega. \tag{4.15}$$

The function

$$f(\omega) = \frac{\sin\omega t}{\omega} \tag{4.16}$$

has its maximum at $\omega = 0$ and decays rapidly in a diminishing oscillatory fashion for $|\omega| > 0$ (Figure 4.1). The interval between the roots nearest $\omega = 0$, i.e. between $\omega t = \pm\,\pi$ contribute by far the most to the integral. When the time $t = \pi/\omega$ is considered short in comparison with the time of the free flight, $f(\omega)$ can be replaced by Dirac's δ-function:

$$f(\omega) = \delta(\omega) = \hbar\delta(E_k^0 - E_s^0) \tag{4.16'}$$

so that

$$\lambda_s = \sum_{k\in f} \frac{2\pi}{\hbar} |\langle k|U|s\rangle|^2 D_S^a(E_k^0)\,\delta(E_f - E_s). \tag{4.17}$$

Expressing $f(\omega)$ by Equation (4.16′) means that the scattering takes place instantaneously once it has started. This is the assumption made behind the right hand side of Boltzmann's transport equation, Equation (1.16). All the final states have the same energy, E_f^0. (Note that $D_S^a(E_k^0)$ refers to all states with energy E_k^0.) This formula, Equation (4.17), expresses *Fermi's golden rule*. The function $D_S^a(E_k^0)$ expresses the density of available states with energy E_k^0, the term

$$M_{fs}^2 = |\langle f|U|s\rangle|^2 D_S^a(E_f^0) \tag{4.18}$$

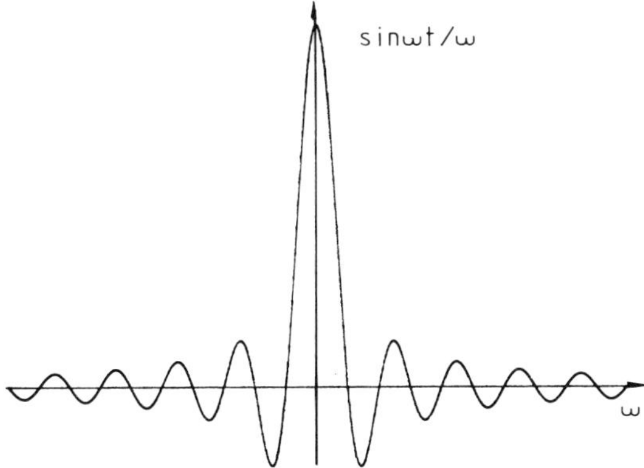

Figure 4.1 *sinωt/ω against ω. The secondary maxima lie far below the main one at ω = 0. The main contribution to the integral ∫dωsinωt/ω comes from the main peak, that from the remaining oscillations tend to cancel*

is known as the *transition matrix element* from the initial state $|s\rangle$ to the final state $|f\rangle$. The matrix element does generally depend on the wave functions – for confined unbound electrons it can be calculated from Equation (3.14). The matrix element itself does generally depend on the shape of the wave function of the initial and final particle state. For scattering between confined unbound states the matrix element becomes

$$M_{fs} = \langle f|U|s\rangle = \int \exp[i(\mathbf{k}_s - \mathbf{k}_f)\bullet\mathbf{r}]\, U u_f^*(\mathbf{r})u_s(\mathbf{r})d^3\mathbf{r}. \qquad (4.19)$$

where $\mathbf{k}_f$ and $\mathbf{k}_s$ represent the wave vector of the final and the initial state of the electron or hole respectively. When a particle is being trapped $|f\rangle$ represents the wave function of a bound electron, which resembles that of an atomic electron orbital (Table 3.1). Since tunnelling particles may also experience scattering, then either $|f\rangle$ or $|s\rangle$ represents a wave of exponential-like decaying amplitude. In both these cases the last part of Equation (4.19) assumes a different form.

In the next sections we shall meet different reasons for scattering. We shall obtain analytical or semianalytical expressions for the corresponding rates. Quite generally, these rates only depend on the quantities entering them, the most prominent parameter being the wave vector of the electron (or hole) just before scattering. Note that the electric field does not enter any of these expressions in the instantaneous scattering approximation, Equation (4.17), which means that the scattering rate is independent of it. It is important to bear this in mind because it may be tempting to believe that the field plays a role in instantaneous scattering. The fact that more scattering takes place in a high field regime than in small fields tends to support this misconception. The role of the field is only to assist the particle in obtaining a higher energy during the free flight, whereby the scattering rate in general increases.

When finite time scattering is considered, Equation (4.15), the field can play a part in that it accelerates or decelerates the particle during scattering, a phenomenon known as *field-assisted scattering*.

Expressions like Equation (4.18) enter Boltzmann's transport Equation (1.16) on the right hand side, the collision term. The expression for λ_s should be multiplied by the probability that the particle is in the initial state $|s\rangle$ in this equation.

4.3 PHONON SELECTION RULES

The exchange of energy between the vibrating lattice and the electrons consists of the electrons creating or absorbing phonons. During the exchange, both the momentum and the energy of the two systems have to be conserved. Figure 2.6 shows the phonon spectrum and the band structure along the main crystallographic directions in a typical zinc blende or diamond structure semiconductor. From this figure it should be clear from the conservation laws that only selected phonons can exchange energy with electrons. In the conduction band these phonons are: i) low energy acoustic phonons; ii) optical phonons near the centre of the Brillouin zone; iii) selected phonons causing electrons to transfer between different minima of the conduction band. These phonons are situated near the centre of the faces of the Brillouin zone.

In the valence band only the first two types of phonons can cause scattering. Acoustic phonons can also transfer holes between the degenerate light and heavy hole bands near the centre of the Brillouin zone. Low momentum optical phonons can cause holes to transfer between the split-off and the heavy or light hole band only if the difference in energy between the split-off and the other hole matches that of the optical phonon.

Phonons causing transfer between different minima of the conduction band are known as *intervalley phonons*. They must have a large wave vector because the distance between the different minima is about the same as the radius of the Brillouin zone. In analogy with the term intervalley phonons, we refer to those phonons not causing transfer between different band extremes as *intravalley phonons*. These terms are incorrect from a linguistic point of view – actually there are no phonons other than acoustic and optical ones. It spite of this we shall speak of inter- and intravalley phonons when it is convenient.

The dispersion of the relevant *acoustic phonons* can, with sufficient accuracy, be approximated by

$$E^p_{ac} = \hbar \mathbf{u}^p \bullet \mathbf{q} \tag{4.20}$$

where p labels the polarisation mode of the phonon, and $\mathbf{q}$ its wave vector. The speed of propagation of the two transversal modes of polarisation is the same unless the crystal is subjected to uniaxial strain or stress. If the scattering from acoustic phonons is of secondary importance, as it is in gallium arsenide at room temperature, the longitudinal and transversal velocities can be combined into one, so that only one amalgamated acoustic phonons branch needs to be considered.

The dispersion of the optical phonons can be neglected near the centre of the Brillouin zone. The energy spectrum of *optical phonons* of angular frequency ω^p_o and polarisation p can, sufficiently accurately within the interesting part of the spectrum, be written

$$E^p_o = \hbar\omega^p_o \tag{4.21}$$

where ω^p_o represents the angular frequency of the polar optical phonons. There are one longitudinal and two transversal modes of vibration: the latter are of equal frequency and wave vector unless stress or strain has been applied. Only the longitudinal phonons can interact with electrons (Mahan, 1972). In the optical mode of vibration neighbouring atoms move in antiphase. In the zinc blende structure semiconductors the vibrating atoms set up a dipole field by which the phonons interact with the electron through electrostatic forces. This is not the case for the diamond structure materials like silicon, carbon, germanium and alloys between them for which optical phonon scattering is not possible.

Phonons causing electrons to transfer to the vicinity of another minimum do have dispersion. (Figure 2.6.) To a good approximation this dispersion relation reads:

$$E^p_{iv} = h\omega^p_{iv,\mathbf{q}} + \hbar \mathbf{u}^p_{iv} \bullet (\mathbf{q} - \mathbf{q}^p_{iv}) \tag{4.22}$$

where iv labels the kind of intervalley transfer, e.g. from Γ to an X minimum and $\mathbf{u}^p_{iv}$ (different from $\mathbf{u}^p$) represents the velocity of that particular phonon of polarisation p and angular frequency ω_{iv}^{p}. The specific labels iv have been listed in Table 4.1. Again, there are two transversal modes which are equivalent in the absence of mechanical deformation and one longitudinal mode of polarisation. The position within the Brillouin zone of this phonon has been denoted $\mathbf{q}_{iv}^{p}$. Equation (4.22) is more general than that usually found in the literature, where the second term has been omitted, neglecting the effect of the dispersion of intervalley phonons on scattering. Although we shall not make much greater use of it here, it has been included to make the reader aware that the phonon spectrum has dispersion when the phonons do not lie just on the edge of the Brillouin zone as e.g. in silicon. The inclusion of dispersion represents possible scope for the improvement of the transport theory. In GaAs and related compounds, however, the L and X minima of the conduction band are often taken to lie at the surface of the Brillouin zone so that the dispersion of the phonon spectrum can be neglected. More accurate investigation has revealed, however, that the conduction band minima in gallium arsenide also lie a little distance from the surface (Fischetti, 1991).

Table 4.1a Complete list of possible transfers between the valleys of the conduction band. (ξ and η are given in Table 3.3)

From	To	Label 'iv'	Distance	Number of possible destinations
Γ	X	ΓX	ξ	6
Γ	L	ΓL	η	8
X	Γ	$X\Gamma$	ξ	1
X	X	f or f_X	$\sqrt{2}\xi$	4
X		g or g_X	2ξ	1
	L	XL_1	$\sqrt{(\eta-\xi)^2+2\eta^2}$	4
		XL_2	$\sqrt{(\eta+\xi)^2+2\eta^2}$	4
L	Γ	$L\Gamma$	η	1
L	X	LX_1	$\sqrt{(\eta-\xi)^2+2\eta^2}$	3
		LX_2	$\sqrt{(\eta+\xi)^2+2\eta^2}$	3
L	L	f of f_L	2η	3
		g or g_L	$2\sqrt{3}\eta$	1
		h or h_L	$2\sqrt{2}\eta$	3

Table 4.1b Simplified list for semiconductors being isotropic in all minima

From	To	Label	Number of possible destinations
Γ	X	ΓX	3
Γ	L	ΓL	4
X	Γ	$X\Gamma$	1
X	X	XX	2
X	L	XL	4
L	Γ	$L\Gamma$	1
L	X	LX	3
L	L	LL	3

The positions of the conduction band minima in the X and L directions are listed in Table 3.3. The six X and the eight L minima are *equivalent minima*. The band structure around one minimum is of the same shape as that around an equivalent one in the absence of pressure and can be derived from one by use of the symmetry relations (2.10) and (2.11). This may no longer be true when the crystal is subjected to uniaxial pressure or a strong electromagnetic field when the various minima have then to be identified by an additional label.

Table 4.1 lists all possible types of transitions between the minima of the conduction band. The transfer between opposite equivalent minima (e.g. between $(0, \xi, 0)$ and $(0, -\xi, 0)$) are referred to as *g-processes*: the other transfers between equivalent minima are known as f- or *h-processes*, depending on the distance between them. Transfer between points in the Brillouin zone farther apart than $2\pi/a$ (a denotes the lattice constant) can only be achieved by phonons leading to an electronic state outside the Brillouin zone. The electron is transferred back into it again by adding a reciprocal lattice vector $\mathbf{G}$. This is known as an *Umklapp process*.

Additional selection rules based on the crystal symmetry may further restrict the choice of scattering processes. These rules have been reviewed by Streitwolf (1969). Lax and Hopfield (1961) and Lax and Birman (1972) gave selection rules based on group theory. Specifically g_x processes are caused by longitudinal optical phonons only, while f_x processes are allowed for both longitudinal and transversal acoustic and transversal optical phonons in silicon (Streitwolf, 1969 and 1970).

In the next section we shall see that no scattering can take place if the polarisation of the phonon is perpendicular to the elastic force acting on the vibrating atoms. The polarisation is described by the polarisation vector; the propagation of the phonon can only be calculated by solving Equation (2.50).

Transfer to higher lying conduction bands can be disregarded for most semiconductors under moderate electric fields, although the student should bear this possibility in mind when intending to simulate transport of high energy carriers.

4.4 PHONON–ELECTRON INTERACTION

The most important mechanism for the scattering of electrons and holes will be discussed here. Our exposition follows mainly Madelung (1972) and Ziman (1960).

Since the mass of the vibrating nuclei is thousands of times larger than that of the electrons, we can safely assume that the electrons surrounding the nuclei follow it, although not necessarily in a rigid fashion. The potential energy of the electron at position $\mathbf{r}_i$ is

$$U_e(\mathbf{r}_i - \mathbf{r}^0_{\mathbf{l}b} - \mathbf{S}_{\mathbf{l}b}) = U_e(\mathbf{r}_i - \mathbf{r}^0_{\mathbf{l}b}) + \sum_{\mathbf{l},b} \mathbf{S}_{\mathbf{l}b} \bullet \frac{\partial U_e(\mathbf{r}_i - \mathbf{r}^0_{\mathbf{l}b})}{\partial \mathbf{S}_{\mathbf{l}b}} \tag{4.23}$$

+ higher order terms in $\mathbf{S}_{\mathbf{l}b}$. These 'higher order' terms will be neglected. The term in $\partial U_e/\partial \mathbf{S}_{\mathbf{l}b}$ represents the phonon–electron interaction. The interaction Hamiltonian reads, to the first order in the phonon amplitude:

$$\mathcal{H}_{el-ph} \equiv \sum_{b,\mathbf{l}} \mathbf{S}_{\mathbf{l}b} \bullet \frac{\partial U_e(\mathbf{r}_i - \mathbf{r}^0_{\mathbf{l}b})}{\partial \mathbf{S}_{\mathbf{l}b}}. \tag{4.24}$$

Substituting for the phonon amplitude, $\mathbf{S}_{\mathbf{l}b}$ Equation (2.61a), this reads:

$$\mathcal{H}_{el-ph} = -i \sum_{\substack{b,\mathbf{l} \\ \mathbf{q},p}} \left(\frac{\hbar}{2N_{cr}M_b\omega^p_{\mathbf{q}}}\right)^{\frac{1}{2}} \xi^p_{\mathbf{q}b} \exp(-i\mathbf{q}\bullet\mathbf{r}_{\mathbf{l}}) \, (\mathbf{a}^{+p}_{\mathbf{q}} - \mathbf{a}^p_{-\mathbf{q}}) \bullet \frac{\partial U_e}{\partial \mathbf{S}_{\mathbf{l}b}} \tag{4.25}$$

where:

p: (as superscript) represents the polarisation of the vibrational mode
N_{cr}: the number of Bravais cell of the crystal
$\xi^p_{\mathbf{q}b}$: the unit vector in the direction of polarisation of the phonon
$\mathbf{q}$: the phonon wave vector
$\omega^p_{\mathbf{q}}$: the corresponding phonon angular frequency
M_b: the mass of ion at site b in the elementary unit cell
$\mathbf{r}^0_{\mathbf{l}b}$: the equilibrium position of ion b in Bravais cell $\mathbf{l}$
$\mathbf{S}_{\mathbf{l}b}$: the displacement of ion b of Bravais cell $\mathbf{l}$
$\mathbf{r}_i$: the position of electron i.

We choose a particular phonon of wave vector $\mathbf{q}$ and polarisation p. The matrix element describing the transfer between the electronic states $\mathbf{k}$ and $\mathbf{k}'$ is

$$M_{\mathbf{kk}'} = \Big\langle n'^p_{\mathbf{q}}\mathbf{k}' \Big| -i \sum_{\substack{b,\mathbf{l} \\ \mathbf{q},p}} \left(\frac{\hbar}{2N_{cr}M_b\omega^p_{\mathbf{q}}}\right)^{\frac{1}{2}} \xi^p_{\mathbf{q}b} \bullet \exp(-i\mathbf{q}\bullet\mathbf{r}_{\mathbf{l}}) \frac{\partial U_e}{\partial \mathbf{S}_{\mathbf{l}b}}$$

$$\times \, (a^{+p}_{\mathbf{q}} - a^p_{-\mathbf{q}}) \, c^+_{\mathbf{k}'} c_{\mathbf{k}} \Big| n^p_{\mathbf{q}'} \mathbf{k} \Big\rangle \tag{4.26}$$

where $|n^p_{\mathbf{q}}, \mathbf{k}\rangle = |n^p_{\mathbf{q}}\rangle \, |\mathbf{k}\rangle$ represents the wave function consisting of the product of that of an electron of wave vector $\mathbf{k}$, $|\mathbf{k}\rangle$, and that of $n^p_{\mathbf{q}}$ phonons each of wave vector $\mathbf{q}$ and polarisation p, $|n^p_{\mathbf{q}}\rangle$. The electron will be replaced either by one of wave vector $\mathbf{k} + \mathbf{q}$ by absorption of a phonon or by one of wave vector $\mathbf{k} - \mathbf{q}$ by phonon emission. The number of phonons of wave vector $\mathbf{q}$ reduces or increases correspondingly by one. Below, we shall calculate the matrix element for phonon emission or creation: the other case is analogous. For simplicity we have assumed $\Delta(E) = 1$, though in a more rigorous Monte Carlo model the possibility that $\Delta(E) < 1$ should be considered. $\Delta(E)$ represents the disequilibrium function defined in Section 3.8.2.

The electronic wave function is given by Bloch's theorem, Equation (3.14):

$$|\mathbf{k}\rangle = \exp(-i\mathbf{k}\bullet\mathbf{r}) \, u_{\mathbf{k}}(\mathbf{r}) \tag{4.27}$$

so that we get for the matrix element of phonon emission or creation:

$$M^E_{\mathbf{kk}'} = -i \sum_{\substack{b \\ \mathbf{q},p}} \left(\frac{\hbar(n^p_{\mathbf{q}} + 1)}{2N_{cr}M_b\omega^p_{\mathbf{q}}}\right)^{\frac{1}{2}} \sum_{\mathbf{l}} \exp[i(-\mathbf{q}\bullet\mathbf{r}_{\mathbf{l}} - \mathbf{k}\bullet\mathbf{r}_{\mathbf{l}} + \mathbf{k}'\bullet\mathbf{r}_{\mathbf{l}})]$$

$$\times \int u^*_{\mathbf{k}+\mathbf{q}}(\mathbf{r})\xi^p_{\mathbf{q}b} \bullet \frac{\partial U_e}{\partial \mathbf{S}_{1b}} \mathbf{u}_k(r)\mathbf{d}^3 r \tag{4.28}$$

(Note that the matrix element $M^E_{kk'}$ has been given an additional superscript, E, in Equation (4.28) to indicate that it refers to phonon creation.)

The cosmological principle says that the integral has to be the same whatever cell the crystal is viewed from. The integral over $\mathbf{r}$ can be taken outside the summation sign, only the exponential terms remain inside it. When the crystal contains more than a few hundred atoms, the δ-function approximation to the sum is sufficiently accurate, so that the sum becomes $N_{cr}\delta(\mathbf{k}' - \mathbf{k} - \mathbf{q} + \mathbf{G})$ where $\mathbf{G}$ represents a reciprocal lattice vector (which may be zero). The factor N_{cr} will be absorbed in the wave function which is normalised accordingly:

$$M^E_{\mathbf{kk}'} = -i \sum_{\substack{b \\ \mathbf{q},p}} \left(\frac{\hbar(n^p_{\mathbf{q}} + 1)}{2N_{cr}M_b\omega^p_{\mathbf{q}}}\right)^{\frac{1}{2}} \int u^*_{\mathbf{k}+\mathbf{q}}(\mathbf{r})\xi^p_{\mathbf{q}b} \bullet \frac{\partial U_e}{\partial \mathbf{S}_{1b}} \mathbf{u}_k(r)\mathbf{d}^3 r \tag{4.29a}$$

The matrix element for phonon absorption or annihilation is likewise

$$M^A_{\mathbf{kk}'} = -i \sum_{\substack{b \\ \mathbf{q},p}} \left(\frac{\hbar n^p_{\mathbf{q}}}{2N_{cr}M_b\omega^p_{\mathbf{q}}}\right)^{\frac{1}{2}} \int u^*_{\mathbf{k}-\mathbf{q}}(\mathbf{r})\xi^p_{\mathbf{q}b} \bullet \frac{\partial U_e}{\partial \mathbf{S}_{1b}} \mathbf{u}_k(r)\mathbf{d}^3 r \tag{4.29b}$$

From the Golden rule, Equation (4.17), the rate of transmission from the electronic state of wave vector $\mathbf{k}$ is

$$\lambda_s = \frac{\pi}{\omega^p_{\mathbf{q}}} \sum \frac{1}{M_b} \int \left| u^*_{\mathbf{k}\pm\mathbf{q}}(\mathbf{r})\ \xi^p_{\mathbf{q}b} \bullet \frac{\partial U_e}{\partial \mathbf{S}_{1b}} \mathbf{u}_k(r)\mathbf{d}^3 r \right|^2$$
$$\times\ (n^p_{\mathbf{q}} + \tfrac{1}{2} \pm \tfrac{1}{2})\delta(E_{\mathbf{k}'} - E_{\mathbf{k}} \pm \hbar\omega^p_{\mathbf{q}})D^a_S(E_{\mathbf{k}'}). \tag{4.30}$$

The upper and lower sign apply to phonon creation and annihilation respectively. The integral vanishes when $\partial U_e/\partial \mathbf{S}_{1b}$ and $\xi^p_{\mathbf{q}b}$ are perpendicular to each other. When $\xi^p_{\mathbf{q}b}\bullet\partial U_e/\partial \mathbf{S}_{1b}$ does not depend on the electronic coordinates, it can be taken outside the integral so that the overlap integral between the periodic parts of the Bloch function becomes

$$G(\mathbf{k}',\mathbf{k}) = \int |u^*_{\mathbf{k}'}(\mathbf{r})u_{\mathbf{k}}(\mathbf{r})d^3\mathbf{r}|^2 \tag{4.31}$$

For free electrons or for pure s-states $G(\mathbf{k}',\mathbf{k}) = 1$. However, at larger $\mathbf{k}$ there is a growing admixture of p-type wave functions into the conduction band states so that G will differ from unity. To a good approximation Fawcett *et al.* (1970) and Kane (1957) have calculated $G(\mathbf{k}',\mathbf{k})$:

$$G(\mathbf{k}',\mathbf{k}) = (a_{\mathbf{k}}a_{\mathbf{k}'} + c_{\mathbf{k}}c_{\mathbf{k}'}\cos\theta)^2 \tag{4.32}$$

where

$$a_{\mathbf{k}} = \sqrt{\frac{1 + \alpha E(\mathbf{k})}{1 + 2\alpha E(\mathbf{k})}} \tag{4.33a}$$

and

$$c_{\mathbf{k}} = \sqrt{\frac{\alpha E(\mathbf{k})}{1 + 2\alpha E(\mathbf{k})}} \tag{4.33b}$$

where α represents the non-parabolicity factor and θ the angle between the wave vector of the incident, $\mathbf{k}$, and the scattered electron, $\mathbf{k}'$; θ is the same polar angle of scattering as shown in Fig. 2.7.

The valence electrons are mainly p electrons for small $\mathbf{k}$. This implies that for transitions within the heavy and light hole bands

$$G(\mathbf{k}', \mathbf{k}) = \tfrac{1}{4}(1 + 3\cos^2\theta) \tag{4.34}$$

and for transitions between the light and the heavy hole bands

$$G(\mathbf{k}', \mathbf{k}) = \frac{3}{4}\sin^2\theta = \frac{3}{4}(1 - \cos^2\theta) \tag{4.35}$$

Below we shall work with a combination of all these cases, generally writing:

$$G(\mathbf{k}', \mathbf{k}) = a_g + b_g \cos\theta + c_g \cos^2\theta \tag{4.36}$$

where the values of the three dimensionless parameters a_g, b_g and c_g have been listed in Table 4.2. In this table we have allowed for the possibility that the non-parabolicity factor α differs for the wave vectors $\mathbf{k}$ and $\mathbf{k}'$.

4.4.1 Acoustic phonon scattering

The dispersion relation for acoustic phonons is given by Equation (4.20). The electrons rigidly follow the slowly vibrating nuclei. The lattice vibrations represent a propagating pressure wave paired with a local relative change in the volume by

Table 4.2 Values of a_g, b_g, and c_g for the overlap integral $G(k', k) = a_g + b_g\cos\theta + c_g\cos^2\theta$

Parameter electrons	Within light or heavy hole valence band	Between light and heavy hole valence band
$a_g \dfrac{(1 + \alpha E)(1 + \alpha' E')}{(1 + 2\alpha E)(1 + 2\alpha' E')}$	$\frac{1}{4}$	$\frac{3}{4}$
$b_g \dfrac{2\sqrt{\alpha\alpha' EE'(1 + \alpha E)(1 + \alpha' E')}}{(1 + 2\alpha E)(1 + 2\alpha' E')}$	0	0
$c_g \dfrac{\alpha\alpha' EE'}{(1 + 2\alpha E)(1 + 2\alpha' E')}$	$\frac{3}{4}$	$-\frac{3}{4}$

$\Delta V_{cr}/V_{cr}$. The phonon represents a uniaxial dilation of the lattice in or perpendicular to the direction of propagation:

$$\frac{\Delta V_{cr}}{V_{cr}} = \frac{\partial \mathbf{S}_{\mathrm{l}b}}{\partial \mathbf{r}_{\mathrm{l}}} \tag{4.37}$$

where $\mathbf{r}_{\mathrm{l}}$ represents the displacement from equilibrium. This pressure wave also produces a change in the size of the unit crystallographic cell, which in turn causes the conduction and the valence band to shift correspondingly by ΔE_{α}. To the first order this can be written in the form:

$$\Delta E_{\alpha} = \frac{\partial E_M}{\partial V_{cr}} \Delta V_{cr} = V_{cr} \frac{\partial E_M}{\partial V_{cr}} \frac{\Delta V_{cr}}{V_{cr}} = V_{cr} \frac{\partial E_M}{\partial V_{cr}} \frac{\partial \mathbf{S}_{\mathrm{l}b}}{\partial \mathbf{r}_{\mathrm{l}}} \tag{4.38}$$

where the label α represents the valence or conduction band. The last step has been derived by means of Equation (4.37). The acoustic phonon *deformation potential* is defined by

$$\Xi_{ac} = V_{cr} \frac{\partial E_M}{\partial V_{cr}}. \tag{4.39}$$

This deformation potential is, as we have just seen, related to the shift in the band gap which can be measured by luminescence or Raman spectroscopy. The value of the deformation potential can also be obtained by observing the shift in the conduction or valence band minimum with pressure (Thomas, 1961).

The deviation from the equilibrium position oscillates as

$$\mathbf{S}_{\mathrm{l}b} = \exp[i(\mathbf{q} \bullet \mathbf{r}_{\mathrm{l}} - \omega_{\mathbf{q}}^{p} t)], \tag{4.40}$$

where $\omega_{\mathbf{q}}^{p}$ represents the angular frequency of the acoustic phonon, $\mathbf{q}$ its wave vector and t the time. The interaction Hamiltonian, Equation (4.24), becomes

$$\begin{aligned} \mathcal{H}_{el-ph} &= \mathcal{H}_{ac} = \mathbf{S}_{\mathrm{l}b} \bullet \frac{\partial U_e}{\partial \mathbf{S}_{\mathrm{l}b}} = i\Xi_{ac}\mathbf{q} \bullet \mathbf{S}_{\mathrm{l}b} \\ &= \Xi_{ac} \sum_{\mathbf{q},p} \left(\frac{\hbar}{2N_{cr} M_b \omega_{\mathbf{q}}^{p}} \right)^{\frac{1}{2}} \exp(-i\mathbf{q} \bullet \mathbf{r}_{\mathrm{l}}) \xi_{\mathbf{q}b}^{p} (a_{\mathbf{q}}^{+p} - a_{-\mathbf{q}}^{p}) \bullet \mathbf{q}, \end{aligned} \tag{4.41}$$

substituting the expression for $\mathbf{S}_{\mathrm{l}b}$ given by Equation (2.61a). The relationship between the deformation potential and the elastic properties of the material are implicit in Equation (4.37). We are not going to discuss this relationship further, instead we refer the reader to the work of Wiley (1970), Christensen (1984) and Herring and Vogt (1956).

The matrix element for the transition from the state $\mathbf{k}$, where the electron's wave vector is $\mathbf{k}$, to the state with wave vector $\mathbf{k}' = \mathbf{k} \pm \mathbf{q}$ by an acoustic phonon of polarisation p is

$$M_{\mathbf{kk}',ac} = \left\langle n_{\mathbf{q}}^{p} \middle| e^{-i\mathbf{k}'\bullet\mathbf{r}}\, \Xi_{ac} \sum_{\mathbf{q},p} \left(\frac{\hbar}{2N_{cr} M_b \omega_{\mathbf{q}}^{p}} \right)^{\frac{1}{2}} \xi_{\mathbf{q}b}^{p} \bullet \mathbf{q} G(\mathbf{k}, \mathbf{k}') \right.$$

$$\left. \times\ (a_{\mathbf{q}}^{+p} - a_{-\mathbf{q}}^{p}) \exp(-i\mathbf{q}\bullet\mathbf{r})\, e^{i\mathbf{k}\bullet\mathbf{r}} \middle| n_{\mathbf{q}}^{p} \right\rangle \tag{4.42}$$

where $|n_{\mathbf{q}}^{p}\rangle$ represents the wave function of n phonons each of wave vector $\mathbf{q}$ and polarisation p, discussed in Section 4.2, and the explicit form of the electronic wave function, Equation (3.14), which has already been introduced.

The scattering rate from acoustic phonons can now be evaluated:

$$\begin{aligned} \lambda_{ac} &= \frac{2\pi}{\hbar} \sum_{\mathbf{k}'} |M_{\mathbf{kk}',ac}|^2 D_S^a(E')\delta(E - E') \\ &= \frac{2\pi}{\hbar} \sum \frac{\hbar \Xi_{ac}^2}{2\rho V_{cr} \omega_{\mathbf{q}}^{p}} q^2 G(\mathbf{k}, \mathbf{k}')\, (n_{\mathbf{q}}^{p} + \tfrac{1}{2} \pm \tfrac{1}{2}) D_S^a(E_{\mathbf{k}'})\delta(E - E') \\ &\quad \times [1 - f(E')] \end{aligned} \tag{4.43}$$

where $n_{\mathbf{q}}^{p}$ represents the phonon occupation number which in thermal equilibrium is given by the Bose-Einstein distribution function:

$$n_{\mathbf{q}}^{p} = \{\exp[\hbar\omega_{\mathbf{q}}^{p}/(k_B T_{Lt})] - 1\}^{-1} \tag{4.44}$$

where $\pm$ applies to phonon creation and annihilation respectively, and $f(E)$ stands for the Fermi–Dirac distribution function. When the electron gas is not in thermal equilibrium, Expression (4.44) has to be modified, introducing the disequilibrium function. The temperature T_{Lt} is a macroscopic parameter. When there is a strong local heating of the lattice or large thermal gradients, we cannot consider the phonon system to be in thermal equilibrium which means that this equation is no longer correct. In other words temperature no longer has a useful meaning. In such a case we have to keep account of the phonons as well to obtain $n_{\mathbf{q}}^{p}$.

When the dispersion relation of acoustic phonons is given by Equation (4.20), $\omega_{\mathbf{q}}^{p}$ can be replaced by $\mathbf{u}^{p}\bullet\mathbf{q}$. Furthermore when the phonon energy is small compared to $k_B T_{Lt}$, which is the case at room temperature,

$$n_{\mathbf{q}}^{p} = \frac{k_B T_{Lt}}{\hbar \mathbf{u}^{p} \mathbf{q}} \tag{4.44$'$}$$

so that

$$\lambda_{ac} = \frac{\pi \Xi_{ac}^2 k_B T_{Lt}}{\rho V_{cr} \mathbf{u}^{p2h}} \sum_{\mathbf{k}'} G(\mathbf{k}, \mathbf{k}') D_S^a(E) \tag{4.45}$$

Any Umklapp process has been neglected under the assumption that it is unlikely that $\mathbf{k}' = \mathbf{q} + \mathbf{k}$ will fall outside the Brillouin zone because $\mathbf{q} \simeq 0$.

The final states are so close together that the summation can be replaced by an integral, which in three-dimensional $\mathbf{k}$-space becomes

$$\sum \to \int D_S(\mathbf{k}')d^3\mathbf{k}' = \frac{V_{cr}}{8\pi^3}\int_0^\infty k'^2 dk' \int_0^\pi \sin\theta\, d\theta \int_0^{2\pi} d\phi \tag{4.46}$$

which contains the density-of-states function (3.88). For scattering in confined spaces of lower dimensions, the corresponding density of states functions (3.96) or (3.98) have to be used. With the band structure (3.65)

$$k'^2 dk' = \sqrt{2}(m_t^{*2} m_l^*)^{\frac{1}{2}} \sqrt{E'(1+\alpha E')}\ (1+2\alpha E')\, dE/\hbar^3 \tag{4.47}$$

so that, when the state $\mathbf{k}'$ is free,

$$\begin{aligned}\lambda_{ac} = {} & \frac{\sqrt{2}\Xi_{ac}^2 (m_t^{*2} m_l^*)^{\frac{1}{2}} k_B T_{Lt}}{2\pi\rho u^{p2}\hbar^4}(a_g + c_g/3)\sqrt{E(1+\alpha E)} \\ & \times \left\{1 + 2\alpha E \pm E_{ac}^p \left[2\alpha + \frac{(1+2\alpha E)^2}{2E(1+\alpha E)}\right]\right\} \\ & \times [1 - f(E \mp E_{ac}^p)]\end{aligned} \tag{4.48}$$

where E_{ac}^p represents the energy of the acoustic phonon, Equation (4.20). The upper and lower sign apply to phonon annihilation and creation respectively. The density of available states, $D_S^a(E)$, still applies because the state $|\mathbf{k}'\rangle$ may be occupied. This formula is also valid for isotropic holes, Equation (3.69), when $\alpha = 0$ and $b_g = 0$, so that

$$\lambda_{ac} = \frac{\sqrt{2}\Xi_{ac}^2 m^{*\frac{3}{2}} k_B T_{Lt}}{4\pi\rho u^{p2}\hbar^4}\sqrt{E}\left(1 \pm \frac{E_{ac}^p}{2E}\right)[1 - f(E \mp E_{ac}^p)]. \tag{4.49}$$

When the valence band maxima are degenerate the hole and electron excitations have a deformation coupling to the transverse phonons (Mahan, 1965) for both interband and intraband scattering. The scattering rate for anisotropic holes becomes

$$\begin{aligned}\lambda_{ac} = {} & \frac{\sqrt{2}\Xi_{ac}^2 k_B T_{Lt} m^{*\frac{3}{2}}}{8\pi^2\rho u^{p2}\hbar^4}\sqrt{E}\left(1 \pm \frac{E_{ac}^p}{2E}\right)\int_0^\pi d\theta \sin\theta \\ & \times \int_0^{2\pi} \frac{2\phi}{\{A' \pm \sqrt{B'^2 + C'^2(\sin^2\theta\cos^2\theta + \sin^4\theta\cos^2\phi\sin^2\phi)}\}^{\frac{3}{2}}} \\ & \times [1 - f(E \pm E_{ac}^p)].\end{aligned} \tag{4.50}$$

For the general band structure the expression for the scattering rate has to be computed numerically. The phonon energy, E_{ac}^p, is small compared to the energy of the electron. Phonon scattering therefore does not significantly change the energy of the electron, but the change of its direction of motion is important.

Such scattering is often referred to as *quasielastic*. Hesto *et al.* (1984) have studied quasielastic acoustic scattering. Neglecting E_{ac}^p, the term in it vanishes which implies that the exchange of energy with the lattice is being omitted. Adding

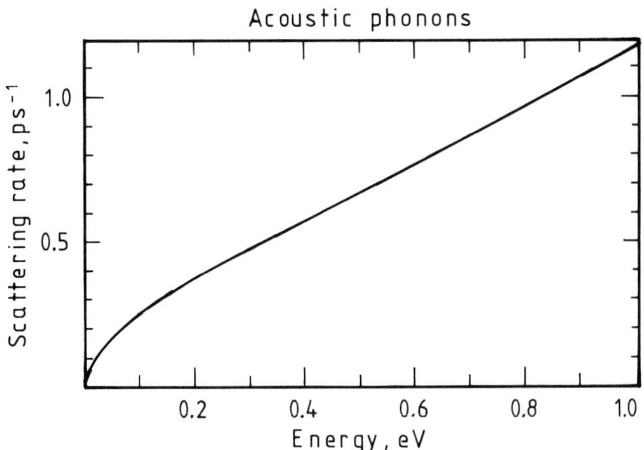

Figure 4.2 *Bulk acoustic phonon scattering rate for electrons in the Γ minimum of the conduction band in GaAs*

the rates for phonon absorption and emission yields the total acoustic quasielastic scattering rate:

$$\lambda_{ac} = \frac{\sqrt{2}\Xi_{ac}^2 (m_t^{*2} m_1^*)^{\frac{1}{2}} k_B T_{Lt}}{\pi \rho u^{p2} \hbar^4} \frac{\sqrt{E(1 + \alpha E)}}{1 + 2\alpha E}$$

$$\times \{(1 + \alpha E)^2 + \frac{1}{3}(\alpha E)^2\} [1 - f(E)] \tag{4.51}$$

which is the same expression as given by Fawcett *et al.* (1970). Figure 4.2 shows the scattering rate for acoustic phonons within the Γ minimum of the conduction band in GaAs calculated from this expression.

For two- and one-dimensional scattering, Expressions (3.96) and (3.98) respectively have to be used for the density of states in Relation (4.46). For non-parabolic bands the calculation of λ_{ac} is relatively simple. Lower dimensional phonon scattering has been discussed by Ridley (1982), Mon *et al.* (1981) and Kelly and Hanke (1981). Recently Bockelmann and Bastard (1990) have discussed scattering in zero, one and two dimensions extensively.

The acoustic phonon scattering rate in a quantum well which confines the electrons to a sheet of thickness w is

$$\lambda_{ac} = \frac{\Xi_{ac}^2 k_B T_{Lt} m^*}{4\rho u^{p2} \hbar^3 w} (2a_g - c_g) [1 + 2\alpha(E \pm E_{ac}^p)] [1 - f(E \pm E_{ac}^p)] \tag{4.52}$$

where the ± sign has the same meaning as above. The angle, θ, in Equation (4.36) varies from $-\pi$ to π.

In one dimension, i.e. scattering in quantum wires, the carrier can either continue in the original direction or start motion in the opposite direction after scattering. This means either $\theta = 0$ or π in Equation (4.36). The one-dimensional acoustic scattering rate is

$$\lambda_{ac} = \frac{\Xi_{ac}^2 k_B T_{Lt}}{\sqrt{2}\rho u^{p2}\hbar^2 A} \frac{\sqrt{m^*}(a_g + c_g)}{[E(1+\alpha E)]^{\frac{3}{2}}} [1 - f(E \pm E_{ac}^p)]$$
$$\times [(1 + 2\alpha E)E(1 + \alpha E) \pm E_{ac}^p(4\alpha^2 E^2 + \alpha E + 1)] \qquad (4.53)$$

where A represents the cross-sectional area of the wire. Analogous expressions can be calculated for more complex band structures.

4.4.2 Intervalley scattering

Acoustic and optical phonons of short wavelength or large wave vectors cause the electron to transfer between different minima. This type of scattering is also described in the deformation potential approximation. The interaction Hamiltonian reads

$$\mathcal{H}_{iv} = \Xi_{iv} \sum_{b,\mathbf{q}} \left(\frac{\hbar}{2M_b \omega_{iv,\mathbf{q}}^p} \right)^{\frac{1}{2}} (\mathbf{a}_{\mathbf{q}}^{+p} \mathbf{e}^{iq\cdot r} + \mathbf{a}_{-\mathbf{q}}^p \mathbf{e}^{-iq\cdot p}) \qquad (4.54)$$

where Ξ_{iv} denotes the deformation potential and the subscript iv labels the type of intervalley transition. Ξ_{iv} may also depend on the polarisation.

There is a number of different phonons causing transfer between band extrema: for example in unstrained silicon there are six different possibilities. The corresponding interaction Hamiltonians differ in the choice of value for the deformation potential Ξ_{iv}. Both these deformation potentials and the quasielastic acoustic phonon scattering deformation potential discussed in the last subsection have been treated as phenomenological constants. However recent theory (Fischetti and Higman, 1991) indicates that these deformation potentials can be calculated from pseudopotential theory. They thus have a sound physical foundation on first principles. Gram and Jørgensen (1973) have demonstrated the relationship between the deformation potentials and elastic constants of the material – the latter can be obtained from quantum theory (McLellan, 1991). Mon *et al.* (1981) have evaluated deformation potentials for several compound semiconductors. Hackenberg and Fasol (1989), Mirlin *et al.* (1988) and Wang *et al.* (1989), just to mention a few experimental investigators, have measured the scattering rates for intervalley transfers by means of time-resolved spectroscopy. Such measurements are valuable in helping to verify the correctness of the calculated rates, and many more papers like these are expected to appear in the near future. Recently Zollner *et al.* (1989) calculated the phonon momentum dependency of the intervalley deformation potential for certain types of phonon scattering.

The square of the matrix element of the transition is

$$M_{kk'iv}^2 = Z_{iv} \frac{\Xi_{iv}^2 \hbar}{2\rho V_{cr} \omega_{iv,\mathbf{q}}^p} (n_{iv,\mathbf{q}}^p + \tfrac{1}{2} \pm \tfrac{1}{2}) G(\mathbf{k}, \mathbf{k}') \qquad (4.55)$$

where the + and − of the ± sign apply to phonon creation and annihilation respectively, and $n_{iv,\mathbf{q}}^p$ represents the phonon occupation number, which for thermal equilibrium is

$$n^p_{iv,\mathbf{q}} = \{\exp[\hbar\omega^p_{iv,\mathbf{q}}/(k_B T_{Lt})] - 1\}^{-1}. \tag{4.56}$$

For transport where no thermal equilibrium exists $n^p_{iv,\mathbf{q}}$ should be replaced by the actual phonon number by keeping account of the created and annihilated phonons. This involves more housekeeping; it may even be necessary to include phonon transport in the model. Collins and Yu (1984) have performed an experimental study of non-equilibrium phonons in GaAs. Z_{iv} represents the number of equivalent minima the electron can be scattered into, which is one less than the actual number of equivalent minima when the electron transfers into an equivalent minima.

The rate of intervalley scattering, assuming the overlap integral $G(\mathbf{k}, \mathbf{k}')$ is given by Equation (4.36), is

$$\begin{aligned}\lambda_{iv} &= \frac{\sqrt{2}Z_{iv}\Xi^2_{iv}(m_t^{*2}m_1^*)^{\frac{1}{2}}}{2\pi\rho\omega^p_{iv,\mathbf{q}}\hbar^3}\frac{\sqrt{E'(1+\alpha E')}}{1+2\alpha E'}(n^p_{iv,\mathbf{q}} + \tfrac{1}{2} \pm \tfrac{1}{2})\\ &\times\left\{1 + \alpha E + \alpha' E' + \frac{2}{3}\alpha\alpha' EE' \pm E^p_{iv}\left[\frac{1+2\alpha' E'}{2[E'(1+\alpha' E')]}\right.\right.\\ &\times\left.\left.(1 + \alpha E + 3\alpha' E' + \frac{2}{3}\alpha\alpha' EE' + 2\alpha'^2 E'^2 + \frac{3}{4}\alpha\alpha'^2 EE'^2\right]\right\}\\ &\times[1 - f(E')].\end{aligned} \tag{4.57}$$

This cumbersome expression for λ_{iv} is probably not to be found in the literature. When $E^p_{iv} = 0$ Equation (4.57) reduces to the intervalley scattering rate quoted by Fawcett *et al.* (1970). $E^p_{iv} = 0$ implies zero phonon dispersion or a momentum transfer being exactly that connecting the band extremum, $\mathbf{k}_\alpha$, the particle resided in prior to scattering with that it enters into, $\mathbf{k}'_{\alpha'}$. Reckoning the energies from the extrema,

$$E' = E + E(\mathbf{k}'_0) - E(\mathbf{k}_0) \pm \hbar\omega^p_{iv,\mathbf{q}}. \tag{4.58}$$

The + and − of the ± sign apply to phonon absorption and emission respectively. The correction term in Equation (4.57), E^p_{iv}, allows for phonon dispersion, analogous to that of the acoustic phonons, Equation (4.48).

We have to pay attention to the meaning of Z_{iv}. Fawcett *et al.* (1970) state that for the X valleys $Z_{iv} = 3$, which is half of the actual number of X minima in the conduction band. This convention, which is the usual one, has been followed by later authors. Values for Z_{iv} according to this convention have been given in Table 4.1b. If in Fawcett's expression we had chosen Z_{iv} as the actual number of equivalent minima then the value of the deformation potential would have to be divided by $\sqrt{2}$ to yield the correct scattering rate. The reader should ascertain that they use compatible values for Z_{iv} and Ξ_{iv}.

Figure 4.3 shows the rate of scattering from the Γ minimum to the L minima of GaAs as calculated from Equation (4.57) with non-parabolicity, $E^p_{iv} = 0$ and $f(E') = 0$. The threshold for this scattering is equal to the difference in energy between the L and Γ minima. Phonon absorption can start at this energy but phonon emission requires that the electron has the additional energy equal to that of the intervalley phonon.

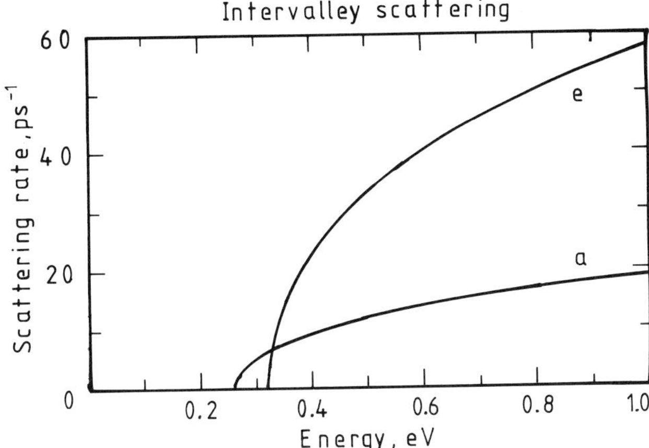

Figure 4.3 *Rate of bulk intervalley scattering of* Γ *minimum electrons into the L minima of the conduction band in GaAs – e: phonon emission; a: phonon absorption*

Two-dimensional scattering has received attention by several authors with application to the inversion layer near the oxide in the silicon MOSFET, in superlattices and in heterostructures. Substituting the two-dimensional scattering rates, neglecting phonon dispersion:

$$\lambda_{iv} = \frac{\Xi_{iv}^2 (m_1 m_t)^{\frac{1}{2}} (2a_g + c_g)}{4w\rho\omega_{iv,\mathbf{q}}^p \hbar^2} (n_{iv,\mathbf{q}}^p + \tfrac{1}{2} \pm \tfrac{1}{2}) I_{ov} [1 - f(E')] \tag{4.59}$$

In Equation (4.59) we have to introduce the factor I_{ov}, the *overlap integral* defined by

$$I_{ov} = \int \Psi_{t\mathbf{k}'}^*(z) \Psi_{t\mathbf{k}}(z)\, dz \tag{4.60}$$

where $\Psi_{t\mathbf{k}}(z)$ and $\Psi_{t\mathbf{k}'}(z)$ represent the envelope of the wave function of the initial, **k**, and final, **k**′, states respectively, of the scattered particle. The label t is there to indicate that the wave function represents the envelope function perpendicular to the quantum well, i.e. in the z direction – it has been obtained by solving the one-dimensional Schrödinger equation perpendicular to the quantum well. Transitions between states where $I_{ov} = 0$ are impossible. This point has been discussed extensively by Ando *et al.* (1982) for silicon.

Note that the two-dimensional scattering rate depends on the width of the quantum well through the quantity w. At the silicon–silicon dioxide interface in MOSFETs or at single heterojunctions where the potential perpendicular to the interface is quasitriangular, w depends on the energy level in the well and increases for higher energy levels leading to a corresponding reduction in the scattering rate.

4.4.3 Polar optical phonon (Fröhlich) scattering

The optical mode of vibration of III–V or II–VI semiconductors gives rise to local electrical polarisation; only the longitudinal mode phonons can cause electrons to scatter (Mahan, 1972) because only these set up strong electric fields.

The corresponding interaction Hamiltonian for the electron and the polarisation is (Ziman, 1960):

$$\mathcal{H}_{po} = -i\left(\frac{e^2\hbar\omega_o^p}{2\varepsilon_0 V_{cr}}\right)^{\frac{1}{2}}\left(\frac{1}{\varepsilon_\infty}-\frac{1}{\varepsilon}\right)^{\frac{1}{2}}\sum_q \frac{1}{q}\left(\mathbf{a}_\mathbf{q}^{+p}\mathbf{e}^{-i\mathbf{q}\cdot\mathbf{r}} - \mathbf{a}_{-\mathbf{q}}^{p}\mathbf{e}^{i\mathbf{q}\cdot\mathbf{r}}\right) \tag{4.61}$$

where ε represents the static dielectric constant or permittivity and ε_∞ the dielectric function at the optical phonon frequency. The latter is often referred to as the *high-frequency dielectric constant* and is found quoted for several different semiconductors, among others by Landolt-Börnstein (1982). Its value can be measured in the infra-red from the refractive index.

The scattering rate of electrons from polar optical phonons can be calculated from Equation (4.17) with $U = \mathcal{H}_{po}$:

$$\lambda_{po} = \frac{2\pi}{\hbar}\sum_\mathbf{q}\frac{e^2\hbar\omega_o^p}{2\varepsilon_0 V_{cr}}\left(\frac{1}{\varepsilon_\infty}-\frac{1}{\varepsilon}\right)\frac{1}{q^2}\left(n_o^p+\tfrac{1}{2}\pm\tfrac{1}{2}\right)\delta(E-E')$$
$$\times D_S^a(E')\left[1-f(E')\right] \tag{4.62}$$

with

$$n_o^p = \{\exp[\hbar\omega_o^p/(k_B T_{Lt})] - 1\}^{-1} \tag{4.63}$$

The scattering causes the wave vector of the electron or hole to change from $\mathbf{k}$ to $\mathbf{k}'$ such that

$$\mathbf{k}' = \mathbf{k} + \mathbf{q} \tag{4.64}$$

giving

$$q^2 = k^2 + k'^2 - 2kk'\cos\theta \tag{4.65}$$

where θ represents the angle between $\mathbf{k}$ and $\mathbf{k}'$. The energy of the scattered particle changes from E to E':

$$E' = E \mp \hbar\omega_0^p \tag{4.66}$$

The $-$ and $+$ of the $\mp$ sign here and the $+$ and $-$ of the $\pm$ in Equation (4.63) refer to phonon creation and annihilation respectively.

As with acoustic phonons, the summation over $\mathbf{q}$ can be replaced by an integral over $\mathbf{k}'$, Expression (4.46):

$$\lambda_{po} = \frac{\sqrt{2}e^2\omega_o^p m^{*\frac{1}{2}}}{8\pi\varepsilon_0\hbar}\left(\frac{1}{\varepsilon_\infty}-\frac{1}{\varepsilon}\right)\frac{1+2\alpha E'}{\sqrt{E(1+\alpha E)}}$$

$$\times \left\{ 2[4a_g + 2\gamma_o b_g + \gamma_o^2 c_g] \log \left| \frac{\sqrt{E(1+\alpha E)} + \sqrt{E'(1+\alpha' E')}}{\sqrt{E(1+\alpha E)} - \sqrt{E'(1+\alpha' E')}} \right| \right.$$

$$\left. - (2b_g - c\gamma_o) \right\} (n_o^p + \tfrac{1}{2} + \tfrac{1}{2}) [1 - f(E')] \tag{4.67}$$

with

$$\gamma_o = \frac{E(1+\alpha E) + E'(1+\alpha' E')}{\sqrt{EE'(1+\alpha E)(1+\alpha' E')}} \tag{4.68}$$

This has been derived from Equation (4.36). The result applies to electrons as well as to holes subject to the relevant choice of parameters a_g, b_g and c_g, Table 4.2.

Especially for electrons with the non-parabolic band structure, Equation (3.65),

$$\lambda_{po} = \frac{\sqrt{2}e^2\omega_o^p(m_t^{*2}m_1^*)^{\frac{1}{6}}}{8\pi\varepsilon_0\hbar}\left(\frac{1}{\varepsilon_\infty} - \frac{1}{\varepsilon}\right)\frac{1+2\alpha' E'}{\sqrt{E(1+\alpha E)}}(n_{po}^p + \tfrac{1}{2} \pm \tfrac{1}{2})$$

$$\times \left[\{2(1+\alpha E)(1+\alpha' E') + \alpha[E(1+\alpha E) + E'(1+\alpha' E')]\}^2 \right.$$

$$- 2\,2\alpha\sqrt{E(1+\alpha E)E'(1+\alpha' E')}\,[4(1+\alpha E)(1+\alpha' E')$$

$$\left. + \alpha[E(1+\alpha E) + E'(1+\alpha' E')] \right]$$

$$\times \{(1+\alpha E)(1+\alpha' E')(1+2\alpha E)(1+2\alpha' E')\}^{-1}$$

$$\times (n_o^p + \tfrac{1}{2} \pm \tfrac{1}{2})[1 - f(E')] \tag{4.69}$$

which differs from the one given by Fawcett *et al.* (1970) by a factor $1/(4\pi\varepsilon_o)$ because they work in the cgs system of units, while we use the SI system. Considering this, there is after all agreement between us.

Figure 4.4 shows that polar optical phonon absorption can take place at any energy; creation, however, can obviously not start until the electron has an energy at least equal to that of the optical phonon. At the threshold energy the scattering rate rises sharply to a maximum and abates slowly for higher energy.

Usually, only intraband polar optical scattering from holes can take place: the relevant scattering rate is

$$\lambda_{po} = \frac{\sqrt{2}e^2\omega_o^p m^{*\frac{1}{2}}}{64\pi\varepsilon_0\hbar}\left(\frac{1}{\varepsilon_\infty} - \frac{1}{\varepsilon}\right)(n_o^p + \tfrac{1}{2} \pm \tfrac{1}{2})[1 - f(E')]$$

$$\times \left\{ 2[10EE' + 3(E^2 + E'^2)] \log \left| \frac{\sqrt{E} + \sqrt{E'}}{\sqrt{E} - \sqrt{E'}} \right| \right.$$

$$\left. + 3(E + E')\sqrt{EE'} \right\} / (EE'^{\frac{3}{2}}) \tag{4.70}$$

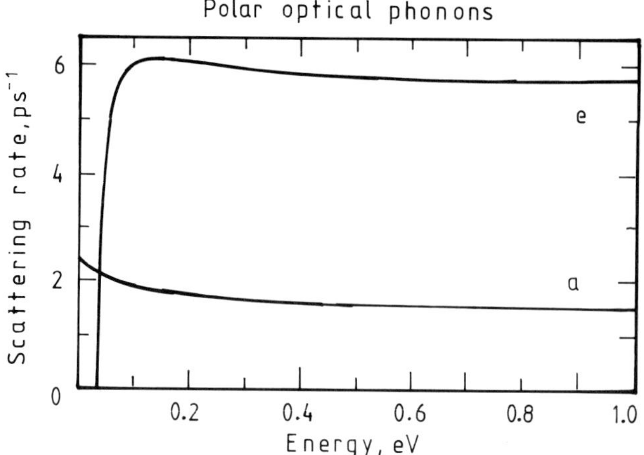

Figure 4.4 *Bulk optical phonon scattering rate for electrons in the Γ minimum of the conduction band in GaAs – e: phonon emission; a: phonon absorption*

Expressions for more complex band structures or lower dimensional scattering can be calculated along the same lines as discussed for acoustic phonon scattering.

Ferry (1978) has discussed polar optical scattering of electrons in two dimensions, quoting expressions for a parabolic band. Hess (1979) has found that polar optical scattering is enhanced in quantum wells.

4.4.4 Acoustic piezoelectric scattering

Many semiconductors are piezoelectric. Piezoelectricity originates from the local lattice deformation caused by acoustic phonons. Group IV semiconductors (silicon, diamond, germanium etc.) are not piezoelectric because they lack an inversion centre. The III–V semiconductors, such as GaAs, are weakly piezoelectric, so that piezoelectric scattering is unimportant. However the II–VI semiconductors like CdS and ZnO are extremely piezoelectric – this type of scattering should therefore be included in the transport equation for these materials. Piezoelectric scattering has received attention from several authors, including Zook (1964), Fedders (1983) and Dasgupta and Sengupta (1985).

The relationship between the stress S_{ij} and the electric field, **F**, it generates in the piezoelectric crystal has Cartesian components

$$F_k = \sum_{i,j} M_{ijk} S_{ij} \tag{4.71}$$

where M represents the *piezoelectric tensor*. Following Mahan (1972), the periodic electric field is derived from the periodic potential $\Phi^p_{\mathbf{q}}$:

$$F_k = -\frac{\partial}{\partial x_k}\Phi^p_{\mathbf{q}} = -\frac{1}{\sqrt{V_{cr}}}\sum_{\mathbf{q}} i\mathbf{q}\Phi^p_{\mathbf{q}} e^{i\mathbf{q}\cdot\mathbf{r}} \tag{4.72}$$

and the interaction Hamiltonian becomes

$$\mathcal{H}_{pe} = e\Phi_{\mathbf{q}}(\mathbf{r}) = ie\sum_{\mathbf{q},p}\left(\frac{\hbar}{2\rho V_{cr}\omega_{\mathbf{q}}^{p}}\right)^{\frac{1}{2}} M_p(a_{\mathbf{q}}^{+p}e^{i\mathbf{q}\cdot\mathbf{r}} + a_{-\mathbf{q}}^{p}e^{-i\mathbf{q}\cdot\mathbf{r}}) \qquad (4.73)$$

The piezoelectric interaction is very anisotropic, but the matrix element M_p does not depend on the magnitude of $\mathbf{q}$, only on its direction and polarisation.

Introducing the acoustic phonon dispersion relation, Equation (4.20), and the acoustic phonon occupation, Equation (4.44′), the scattering rate for a phonon of wave vector $\mathbf{q}$ and polarisation p becomes

$$\lambda_{pe} = \frac{\sqrt{2}e^2M_p^2k_BT_{Lt}(m_t^{*2}m_l^*)^{1/6}(1+\alpha'E')}{16\pi\hbar^2\rho\mathbf{u}^{p2}\sqrt{E(1+\alpha E)}}$$
$$\times\left\{(4a_g + 2\gamma_o b_g + \gamma_o^2 c_g)\log\left|\frac{\sqrt{E(1+\alpha E)}+\sqrt{E'(1+\alpha'E')}}{\sqrt{E(1+\alpha E)}-\sqrt{E'(1+\alpha'E')}}\right|\right.$$
$$\left. - (2b_g - \gamma_0 c_g)\right\}[1-f(E)] \qquad (4.74)$$

with M_p representing the value of M averaged over the direction of polarisation of the phonon and γ_o given by Equation (4.68). Note the resemblance with the scattering rate for polar optical phonons, Equation (4.67).

4.4.5 Higher order scattering

In most semiconductors used today the most important scattering mechanisms are the interactions between the phonons and carriers. In silicon, which is a non-polar material, optical phonon scattering is prohibited by symmetry (Streitwolf, 1969, 1970). However, this is strictly only valid for phonons in the centre of the Brillouin zone. Ferry (1976a) has pointed out that second order scattering from optical phonons is possible. Terashima *et al.* (1985) studied first order inter-valley scattering. The square of the matrix element for first order phonon scattering is

$$M^2_{\mathbf{kk},2o} = \frac{\Xi^2_{2o}\hbar\mathbf{q}^2}{2V_{cr}\rho\omega_o^p}(n_o^p + \tfrac{1}{2} \pm \tfrac{1}{2})G(\mathbf{k},\mathbf{k}') \qquad (4.75)$$

where the second order deformation potential, Ξ_{2o}, has dimension energy. This matrix element differs from the first order intervalley one, Equation (4.55), by the extra factor $\mathbf{q}^2$. The second order scattering rate is

$$\lambda_{2o} = \frac{\sqrt{2}\Xi_{2o}(m_t^{*2}m_l^*)^{\frac{5}{6}}}{\pi\rho\omega_o\hbar^5}\sqrt{E'(1+\alpha'E')}\,(1+2\alpha E')\,(n_o^p + \tfrac{1}{2} \pm \tfrac{1}{2})$$
$$\times\left\{E(1+\alpha E) + E'(1+\alpha'E')]\,(a_g + c_g/3)\right.$$
$$\left. -\frac{2}{3}b\sqrt{E(1+\alpha E)E'(1+\alpha'E')}\right\}[1-f(E')] \qquad (4.76)$$

which is a more general expression than that given by Ferry (1976a) as the non-parabolicity and anisotropy of the conduction band have been accounted for. Ferry's expression is recovered by letting $a_g = 1, b_g = c_g = 0$ and $m_t^* = m_l^*$.

4.4.6 *Refinement of the scattering rate formulae*

Although the formulae presented here for phonon scattering have proven sufficiently accurate for most modelling cases encountered in the literature there is always scope for improvement. The process of refinement ceases when no further improvement can be achieved within the experimental uncertainty of available measured data used for comparison, or when nothing new can be learnt from it. Continuing beyond this point is meaningless. Possible ways of improvement have been discussed by several authors; a few cases will be mentioned here.

Screening can be incorporated in the rate for acoustic scattering, Equation (4.43), by replacing the magnitude of the phonon wave vector, q, by $(q^2 + \beta_{ph}^{-2})/q$ where β_{ph} represents the phonon screening length which is caused by the phonon rearranging the electrons as it propagates. Giner and Comas (1988) have derived expressions for scattering rates using the Huang approach, obtaining an interaction Hamiltonian which accounts for the electron screening. The Fröhlich Hamiltonian, Equation (4.61), represents the limiting case of zero screening.

The possibility of multiphonon scattering has not been considered by us, although this too has been the subject of investigation, e.g. Trakhtenberg and Flerow (1983). In principle, such scattering can be included in our Monte Carlo model, but it will prove to have a low rate compared to those we have discussed above. However, when we include weak scattering mechanisms we have to make sure that no stronger or more significant ones have been omitted from our model.

Usually, the X minima of the conduction band are degenerate, they have the same energy. This also applies to the L-minima. This degeneracy lifts when the crystal has been strained or exposed to a magnetic field or when a quantum well has been formed. The formula of the scattering rates should then be amended accordingly. Stress is also expected to change the anisotropy of the band structure. Strained crystals occur at junctions where the lattices do not match as in e.g. InGaAs/GaAs heterojunctions.

Scattering in lower-dimensional systems has received attention by many writers. The quantum well formed at the inversion layer under the gate of the MOSFET and in modulation doped heterostructures represent examples of two-dimensional systems because the transport is mainly restricted to the plane perpendicular to the well. The description of the transport is complex through the formation of electronic sub-bands. Transfer between them is only possible if the energy exchange matches the difference in the sub-band levels and if the wave function of the two levels overlap. Krowne (1983) has studied transport in such inversion layers. Furthermore, phonons may split in three at a junction, one passing it, one reflected from it and one following it. This makes the phonon structure rather complex as Toennies (1990) has pointed out in a humorous fashion. Ezawa *et al.* (1974) have discussed scattering from phonons moving parallel to interfaces, *surfons*, in great detail. Ridley (1982) and Riddoch and Ridley (1983) have performed an extensive study of two-dimensional scattering of electrons. In any structure where the carriers are restricted to a small volume, e.g. quantum wells or thin films of a few atomic layers, two-dimensional scattering has to be considered. Arora and Awad (1981) said that quantum effects are significant when the width of the confinement is small compared to the deBroglie wavelength of the thermal electron. Leburton (1984) considered both one- and two-dimensional scattering.

4.5 SCATTERING FROM ELECTRIC CHARGES

Foreign atoms represent a break in periodicity of the atomic arrangement of the crystal. The foreign atoms, the impurities, may replace host atoms at their lattice sites, or take interstitial positions. This causes perturbation of the electronic orbitals, giving rise to scattering without loss or gain of energy of the electron. Stacking faults and vacancies also cause scattering. The rate of scattering from ionised atoms is comparable to that of phonons. Inelastic scattering can only take place when it is coupled with a phonon.

4.5.1 Ionised impurity scattering

Scattering from ionised impurity atoms is far the strongest or most frequent scattering due to lattice imperfections. If the ionised atoms are not too close together, the electric field generated by the ion is

$$\Phi_{ii} = \frac{Ze}{4\pi\varepsilon\varepsilon_0 r}\exp(-r\beta_s) \tag{4.77}$$

where β_s represents the screening length, given by Equation (3.123), r the distance from the ion and Z the number of elementary electronic charges the ion carries. The other symbols have the same significance as in Section 3.3. The screening is established by a local rearrangement of the electrons reacting to the field from the ion. By close together we mean that, when the foreign atom in its neutral state can be considered as a hydrogen-like atom with one electron, this electron will assume an orbit of a radius given by Equation (3.33) with ε_0 replaced by $\varepsilon\varepsilon_0$ and m_0 by the effective mass of the bulk electrons of the host. For silicon and gallium arsenide this means a radius of about 100 Å. If the impurities lie more than 200 Å apart, which corresponds to an impurity density of about 10^{23} m^{-3}, the orbitals will not overlap, and our approach should hold. In this case the matrix element describing the interaction with an ionised impurity is, according to Brooks–Herring theory,

$$M_{kk',ii} = \frac{N_{ii}^{+}e^2}{4\pi\varepsilon\varepsilon_0 V_{cr}}\int_{V_{cr}} \frac{1}{r}\exp(-i\mathbf{k}'\bullet\mathbf{r})\,u_{\mathbf{k}'}^{*}(\mathbf{r})$$

$$\times \exp(-\beta_s r)u_{\mathbf{k}}(\mathbf{r})\exp(ik\bullet\mathbf{r})\,\mathbf{d}^3\mathbf{r} \tag{4.78}$$

where r represents the magnitude of the position vector $\mathbf{r}$, and N_{ii}^{+} the number of ionised impurities in the crystal. The integral over $\mathbf{r}$ extends over the entire crystal. The scattering rate becomes

$$\lambda_{ii} = \frac{2\pi}{\hbar}\sum_{\mathbf{k}'}\frac{N_{ii}^{+2}e^4G(\mathbf{k},\mathbf{k}')}{(\varepsilon\varepsilon_0)^2V_{cr}^2(q^2+\beta_s^{-2})^2}\delta(E'-E)\,D_S^q(E'). \tag{4.79}$$

The δ-function in the energy takes care of energy conservation. The scattered particle keeps the original energy, but changes its direction of motion on scattering. Chattopadhyay and Queisser (1981) have calculated the screening length, β_s, to be given by

$$\beta_s^{-2} = \frac{n_{ii}^+ e^2}{2\varepsilon\varepsilon_0 k_B T_{Lt}} \frac{F_{-\frac{1}{2}}(\eta)}{F_{\frac{1}{2}}(\eta)} \tag{4.80}$$

where n_{ii}^+ represents the number density of the ionised impurities ($n_{ii}^+ = N_{ii}^+/V_{cr}$) and $F_j(\eta)$ the Fermi integral given by Equation (3.105). The screening is caused by the carriers rearranging near the ion. Resta (1977) has made a study of the variation of the dielectric constant in the vicinity of one ionised impurity; Dingle (1955) has discussed the screening in the presence of both electrons and holes. The scattering rate for an anisotropic, non-parabolic band structure with $G(\mathbf{k}, \mathbf{k}')$ given by Equation (4.36) is

$$\lambda_{ii} = \frac{\sqrt{2}e^4 n_{ii}^+ (1 + 2\alpha E)}{16\pi(\varepsilon\varepsilon_0)^2 (m_t^{*2} m_1^*)^{\frac{1}{6}} [E(1 + \alpha E)]^{\frac{3}{2}}}$$

$$\times \left\{ 2 \frac{a_g \gamma_r^2 + b_g \gamma_r + c_g}{1 - \gamma_r^2} (b_g + 2c_g/\gamma_r) \log \left| \frac{4k^2 + \beta_s^{-2}}{\beta_s^{-2}} \right|^2 + 2c_g \right\} \tag{4.81}$$

with

$$\gamma_r = \frac{2k^2}{2k^2 + \beta_s^{-2}} \tag{4.82a}$$

and

$$k = \sqrt{2m^* E(1 + \alpha E)}/\hbar. \tag{4.82b}$$

The values for a_g, b_g and c_g are listed in Table 4.2. For the light and heavy hole band structure given by Equation (3.68) the expression for the scattering rate becomes more complicated.

Note that Equation (4.81) applies to both positively and negatively charged ionised impurities. The parameter n_{ii}^+ entering Equation (4.81) represents the sum of all types of ionised atoms. There is also ionised impurity scattering in compensated semiconductors because there are just as many positively as negatively charged impurities.

Chattopadhyay and Queisser (1981) have given an extensive review of ionised impurity scattering, with refinements of the expression for the scattering rate. These authors have not, however, considered non-parabolicity and crystal anisotropy, but they discuss the validity of the Born approximation used to derive the scattering rates and compare the relative importance of the various refinements of the Brooks–Herring theory they present. The extension of the calculation of the scattering rate in heavily doped semiconductors is also considered. This has also been discussed by Kay and Tang (1991a) for scattering from clusters of ionised impurities using phase-shift analysis. The results of the Brooks–Herring

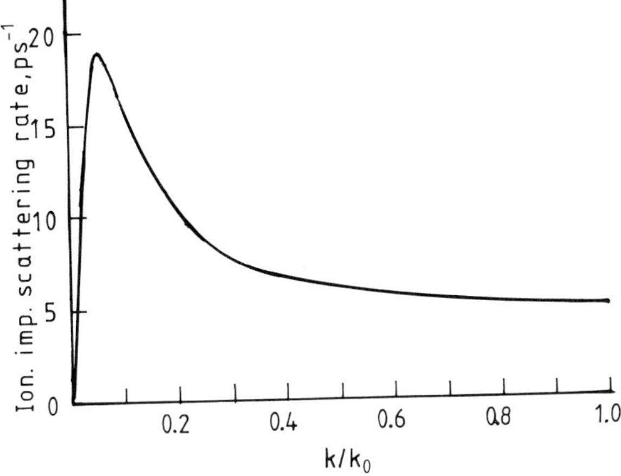

Figure 4.5 *Bulk ionised impurity scattering rate of* Γ *minimum electrons in GaAs with an ionised impurity concentration of* 10^{23} m^{-3}

theory presented here are recovered for small impurity concentrations in the sense explained above. The work of Kay and Tang, however, is restricted to isotropic material. Using the improved ionised impurity scattering model Kay and Tang (1991b) have calculated the low-field mobility in heavily doped silicon. Kayanuma and Fukuchi (1982) have considered scattering from a pair of ionised ions, Gubernatis (1987) approached impurity scattering from many-body theory and Theodorou and Queisser (1979) considered the variation in screening length on the scattering from the dielectric function.

Figure 4.5 shows that the scattering rate reaches a maximum around 35 meV, then decays with increasing energy. This rate has been calculated for non-parabolic electrons in the Γ minimum of gallium arsenide with an ionised impurity concentration of $10^{23}\,m^{-3}$. Ionised impurity scattering is therefore of relatively minor importance for very hot electrons or holes. Thus ionised impurities reduce the drift velocity of the carriers, but the saturation drift velocity reached at high electric field is hardly affected by it. Walukiewicz *et al.* (1979) have studied the mobility against the density of ionised dopants in GaAs. The distribution function is also affected by ionised impurity scattering (Das Sarma and Vinter, 1981). Van de Roer and Widdershoven (1986) have studied the effect of ionised impurity scattering on the distribution function using the Brooks–Herring model of scattering.

The scattering rate for two-dimensional systems can be calculated from Equation (4.78) using the two-dimensional density of states, Equation (3.97). Hess (1979) reports that ionised impurity scattering can be enhanced in quantum wells.

4.5.2 Remote polar optical and ionised impurity scattering

Modulation doped heterostructures, as shown in Fig. 10.14, are designed such that the majority of the current flows through an undoped channel. The idea is to reduce the resistance to the current by eliminating ionised impurity scattering. If the material on either side of the channel immediately adjacent to the heterojunction interface is doped the ionised atoms nearest to the channel cause remote ionised impurity scattering of the particles in the channel. This scattering can be eliminated by leaving the slice of the confining material on both sides of the channel undoped. The ionised atoms are the source of a potential given by Equation (4.77) with a position-dependent screening length so that the matrix element, $M_{\mathbf{kk}',ii}$, Equation (4.78), will yield a modified expression for the scattering rate. The two-dimensional nature of the channel should of course be considered in the calculations. Lassnig (1988) discusses remote impurity scattering but does not produce a practical expression for the scattering rate. The scattering from remote polar phonons is more important because it cannot be eliminated by introducing an undoped layer adjacent to the channel. According to Hess and Vogl (1979) the relevant squared matrix element reads

$$M^2_{kk',rp} = \frac{e^2\hbar\omega_o{}^p}{2\varepsilon_0 A}\left(\frac{1}{\varepsilon_\infty + \varepsilon'_\infty} - \frac{1}{\varepsilon' + \varepsilon'_\infty}\right)\frac{b^6_{rp}}{(b_{rp} + q)^6_q} \tag{4.83}$$

with ε'_∞ and ε' representing the optical phonon frequency and the static dielectric constant respectively of the material adjacent to the stratum carrying the current and ε_∞ the optical phonon frequency dielectric constant for the material of the channel. Furthermore, A denotes the area of the interface; $b_{rp} = 3/d_{rp}$ where d_{rp} represents the average distance of the electrons from the interface, which depends on the sub-band it occupies; and $\omega_o{}^p$ the phonon angular frequency. The vibrational mode of the phonon is transverse.

Moore and Ferry (1980a) find that the mobility in the inversion layer forming at the gate in MOSFET is not significantly affected by remote impurity scattering, but it does have an important effect on the energy relaxation in the electron gas. The same authors (1980b) also state that scattering from remote phonons in the oxide should be considered.

4.5.3 Carrier-carrier scattering

Electrons repel each other because of the electrostatic charge they carry. An electron at position $\mathbf{r}$ represents a source of electrostatic potential, which at position $\mathbf{r}'$ is

$$\Phi_{rr'} = -\frac{e^2}{4\pi\varepsilon\varepsilon_0|\mathbf{r} - \mathbf{r}'|}\exp(-\beta_s|\mathbf{r} - \mathbf{r}'|). \tag{4.84}$$

An electron there feels the presence of the other electron so that the course of both alters; they are scattered. The matrix element describing their interaction can, strictly speaking, not be obtained from the Born approximation like the other scattering mechanisms discussed so far. The condition for this approach is that the scattering potential falls off rapidly at large distances and the carriers have large energy. When this is not the case the phase shift approach has to be used to calculate the matrix element. The result for the Coulomb interaction, however, turns out to be the same as that obtained from Born's approximation (Merzbacher, 1961). Carrier scattering has been reviewed extensively by Kaveh and Wiser (1984). These authors touch many aspects of interelectron scattering in systems of different dimensions. Lugli and Ferry (1985) give scattering formulae for scattering rates in a form which is better suited for Monte Carlo modelling.

For wave vectors larger than $k \simeq 2\pi/\lambda_D$, where λ_D represents the Debye length, Equation (8.8), the particles interact and scatter more or less like individual particles. For smaller wave vectors the system behaves collectively, the particles are scattered from plasmons, the quantised collective motion of the particles through their electrostatic interaction.

The scattering rate for particles subjected to Coulombic screening is (Osman and Ferry, 1987; Inoue and Frey, 1980; Takenaka *et al.*, 1979):

$$\lambda_{ee} = \frac{m_r^* e^4 n_e}{4\pi(\varepsilon\varepsilon_0)^2\hbar^3 N_e \beta_s^{-2}} \sum_{k_2} \frac{|\mathbf{k} - \mathbf{k}'|}{|\mathbf{k} - \mathbf{k}'|^2 + \beta_s^{-2}} f(\mathbf{k}') \tag{4.85}$$

where the carriers have been vector $\mathbf{k}$ and $\mathbf{k}'$ prior to scattering. The carriers cannot enter another band extremum on scattering so they have to swap momentum and energy. Because of this the scattering will always take place regardless of the availability of states. m_r^* represents the *reduced mass*

$$m_r^* = \frac{m_1^* m_2^*}{m_1^* + m_2^*}. \tag{4.86}$$

Here m_1^* and m_2^* represent the effective mass of the two particles and n_e the carrier density. As usual ε and ε_0 denote the dielectric constant and permittivity of vacuum respectively, and β_s the screening length, Equation (4.80). N_e represents the number of particles the carrier can scatter from and $f(\mathbf{k}')$ their distribution.

Inoue and Frey (1980), have evaluated the sum, assuming the carrier gas is in thermal equilibrium. The function $f(\mathbf{k}')$ in their paper can be interpreted as a Maxwellian distribution. We shall see later that this is often not the case in semiconductor devices so we have to use the local distribution which is calculated dynamically by the Monte Carlo particle simulator. The best estimate is obtained by summing over the wave vectors of all the particles of effective mass m_2^* which are in the vicinity of the place where the scattering takes place. As will be explained in Chapter 8, the simulated particles represent only a sample of all the particles in the semiconductor; there will therefore be relatively few (often less than ten) simulated particles in the vicinity of the scatterer. Even then a fairly good estimate of the scattering rate can be obtained.

The collision partner, the particle with which the actual exchange of energy and momentum takes place, is chosen within an enclosure surrounding the particle. In real life the colliding particle selects its partner within a screening length's distance.

Any particle within reach can be chosen as a collision partner. The choice should be made between all of them by means of a random number without any discriminative weighting.

The Monte Carlo particle simulation of devices is in practice carried out by following each particle in turn for a short span of time. Each particle moves independently of the other in the field created by the presence of the other particles. By this scheme the effect of the scattering on the partner particle will not be calculated. In order to conserve the momentum and energy of the carrier gas, it is necessary to compensate for the change in momentum and energy of the scattered particle by updating the motion of the partner accordingly. But this implies that the partner suffers one additional scattering – the amount of inter-carrier scattering doubles. To compensate for this the scattering rate, Equation (4.85), has to be multiplied by $\frac{1}{2}$. This factor has, however, not been incorporated into Expression (4.85).

A Monte Carlo particle simulation involves a large number of particles that have to be simulated. In Chapter 8 we shall see that this is done by following each particle for a set time, the field adjusting time step. After determining the local field distribution, the particles are simulated for another period of time; this process continues throughout the entire simulation. The particles which are involved in electron–electron scattering are listed. During the next time step this list is referred to for every particle to find the collision partner with which to exchange energy and momentum. A new list is set up for each time step and the old one is discarded when we no longer need it. This scheme is faster than attempting to update it. An alternative way to find the partner is to make use of the P^3M algorithm we shall explain in Chapter 11.

The interparticle scattering rate (Equation 4.85) is general because it allows scattering from a carrier of a different effective mass. This expression thus applies to electron–electron, electron–hole or hole–hole scattering. It also describes scattering between electrons from different minima in the conduction band or different valence bands, e.g. between Γ and L electrons, or between light and heavy holes.

In many cases, intercarrier scattering is relatively unimportant. However, many authors have seen minor changes in the carrier distribution function due to such scattering, including Inoue and Frey (1980) and Osman and Ferry (1987). Brey and Tejedor (1985) have calculated the effect of such scattering on the band structure by means of the many-body theory. Choi *et al.* (1986) also used many-body theory to investigate this effect. Röpke and Höhne (1981) studied electron–electron interaction from a correlation function approach, Katsnel'son and Sadovskii (1983) from a Green's function approach. Collet and Amand (1984) discussed the influence of screening on intercarrier scattering.

We now return to the case of scattering from the collective action of the carriers, the *plasmons*. Guha and Ghosh (1979) found that the longitudinal optical phonons interact with plasmons. The angular frequency of the plasmons is given by Equation (3.117):

$$\omega_p = \sqrt{\frac{n_e e^2}{m_\alpha^* \varepsilon_\infty \varepsilon_0}} \tag{4.87}$$

In a homogeneous plasma there is one frequency for each effective mass, that means that in e.g. gallium arsenide when the Γ, X, and the L minima are occupied the electron gas oscillates at three different frequencies – m_α^* represents the effective mass of the electron residing in an $\alpha(=\Gamma, X, L)$ minimum of the conduction band or of a hole in any of the three valence bands. The density of carriers in the s-band is n_α – ε and ε_0 represent the permittivity of vacuum and the dielectric constant of the material respectively.

The electron–plasma scattering rate has been calculated by Lugli and Ferry (1985):

$$\lambda_{ep} = \frac{m_\alpha^* e^2 \omega_p}{4\pi\varepsilon_0 \hbar^2 k} \log\left(\frac{q_c/k}{\sqrt{[1 \mp h\omega_p/E']} - 1}\right)(n_{\mathbf{q}}^p + \tfrac{1}{2} \pm \tfrac{1}{2}). \qquad (4.88)$$

Here $n_{\mathbf{q}}^p$ represents the Bose–Einstein occupation number of the plasmons:

$$n_{\mathbf{q}}^p = \{\exp[\hbar\omega_p/(\mathbf{k}_B \mathbf{T}_{Lt})] - 1\}^{-1} \qquad (4.89)$$

where $\mathbf{k}$ represents the wave vector of the scattering electron and $E' = E \pm h\omega_p$. The + and − of the ± sign stand for plasmon absorption and creation respectively. The scattering takes place for wave vectors below $\mathbf{q}_c$ which is of the order $2\pi/\lambda_D$.

Here we have considered plasmas consisting of only one species of carriers, electrons. In photodetectors, lasers and bipolar transistors the plasma consists of electrons and holes. Vasconcellos and Luzzi (1980) show that longitudinal optical phonons couple with such a plasma. Kotel'nikov (1983) studied interaction between optical phonons and excitons.

4.6 SCATTERING FROM NEUTRAL IMPERFECTIONS

In a perfect crystal each atom of the same type has a lattice coordination which is the same everywhere in the crystal. However, the laws of thermodynamics favour flaws in the form of lacking occupation of lattice sites, atoms placed between lattice sites and other stacking faults or dislocations. Furthermore, the crystal contains foreign atoms which have been introduced deliberately or not. When two or more different species of atoms are randomly distributed over the lattice sites, these give rise to alloy scattering. This happens in any non-stoichiometric compound like Si_xGe_{1-x}, $Al_xGa_{1-x}As$, $In_xGa_{1-x}P$ etc., where x represents mole fractions. In this section we shall present various types of scattering from electrically neutral irregularities in the lattice.

4.6.1 Alloy scattering

Suppose the anion sites in the zinc blende structure is occupied by one species of atoms and the cation sites by two species, denoted A and B, which are distributed randomly throughout the material. There are N_A and N_B atoms of type A and B respectively, such that

$$N_A + N_B = N_C \qquad (4.90)$$

where N_C is the number of cation sites in the crystal. Harrison and Hauser (1976) have derived the squared matrix element

$$M_{\mathbf{kk}',al}^2 = \frac{C_A(1 - C_A)}{V_{cr}} \left| \int_{V_{cr}} \Psi_{\mathbf{k}'}^*(\mathbf{r}) U_{al} \Psi_{\mathbf{k}}(\mathbf{r})\, d^3\mathbf{r} \right|^2 \qquad (4.91)$$

for alloy scattering using the hard sphere approximation to the scattering potential U_{al}

$$U_{al} = \begin{cases} E_{al} & \text{for } r \leqslant r_{al} \\ 0 & \text{for } r > r_{al} \end{cases}. \tag{4.92}$$

Here r_{al} represents the radius of a sphere occupying the same volume as the Wigner–Seitz cell, V_{cr}, and Ψ_k the electronic wave function. This gives

$$\mathrm{M}^2_{\mathbf{kk}'al} = C_A(1 - C_A)G(\mathbf{k}, \mathbf{k}')U^2_{al} \tag{4.93}$$

where $G(\mathbf{k}, \mathbf{k}')$ is given by Equation (4.36) and $C_A = N_A/N_c$, the concentration of type A atoms. There is some uncertainty about the meaning of E_{al}. Harrison and Hauser (1976) suggest it represents the difference in the work function between the two materials A and B. Values for E_{al} for different ternary and quaternary alloys have been published by Littlejohn *et al.* (1978), who have also derived an expression for the alloy scattering in quaternary compounds.

The corresponding scattering rate is

$$\lambda_{al} = \frac{3\sqrt{2}\pi C_A(1 - C_A)E^2_{al}(m^*_t m^*_1)^{\frac{1}{2}}}{16 n_c \hbar_4} \sqrt{E(1 + 2\alpha E)}\,(1 + 2\alpha E)\,(a_g + c_g/3). \tag{4.94}$$

This formula applies both to isotropic holes and to electrons. For anisotropic holes a more complicated expression has to be derived. m^*_t and m^*_1 denote the transverse and longitudinal effective mass respectively, and n_c represents the number density of cation sites.

4.6.2 Neutral impurity scattering

The rate for neutral impurity scattering is of an order of magnitude below that of the rate for ionised impurity scattering. Neutral impurity scattering was studied by Erginsoy (1950), but has not attracted any interest since. The matrix element reads (Jacoboni and Reggiani, 1979):

$$M_{ni} = \frac{20\pi\, n_{ni}\hbar^4 a_B}{V_{cr} m^{*2} k} \tag{4.95}$$

where n_{ni} represents the concentration of neutral impurities, and a_B the Bohr radius, Equation (3.33), with the permittivity of vacuum multiplied by the dielectric constant and m_0 replaced by the effective mass of the host material. The scattering rate is

$$\lambda_{ni} = \frac{5 n_{ni}\hbar a_B}{\pi(m^{*2}_t m^*_1)^{\frac{1}{3}}}\,(a_g + c_g/3)\,(1 + 2\alpha E). \tag{4.96}$$

This type of scattering may be significant at very low temperatures and should therefore be considered in this case.

4.6.3 Dislocation scattering

The two main types of dislocations are edge and screw dislocations, both representing stacking faults in the otherwise periodic arrangement of the atoms constituting the lattice. The *edge dislocation* can be considered as being formed by inserting an additional half plane with a straight edge into the crystal. The plane is placed so that the lattice matches perfectly everywhere except at the edge. The *screw dislocation* can be imagined formed by cutting the crystal halfway along a crystallographic plane and then put together again by displacing the two faces of the incision one unit cell in the direction of the edge of the cut so that the lattice matches again.

These types of crystal faults tend to follow certain crystallographic directions e.g. Onga *et al.* (1976). In a cubic crystal they are parallel or form right angles with one another. They are usually of the order of tens or hundreds of atoms long. However crystals with a very low dislocation density have been grown.

The dislocations are surrounded by a large strain or stress which causes electrons to scatter. Brown (1977) formulated the *golden rule* for such scattering from state **k** to **k**′. The potential surrounding the dislocation is

$$U_d(k, k') = \frac{2\pi V_{cr}}{8\pi^3\hbar}\left| U_d(k_z; k', k)\right|^2 \frac{k'd\phi}{v_x(k')} \tag{4.97}$$

and $k' = k_z + k_x$

$$v_x(\mathbf{k}') = \frac{1}{\hbar}\left\{\frac{\partial E(\mathbf{k}')}{\partial \mathbf{k}'} - \mathbf{k}_z \bullet \frac{\partial E(\mathbf{k})}{\partial \mathbf{k}} \bullet \mathbf{k}_1\right\} \tag{4.98}$$

with k_z and k_x representing unit vectors along and perpendicular to the dislocation respectively, and U_α oriented perpendicular to it.

We shall not discuss this type of scattering further as it is usually a weak one, and the interpretation of Equation (4.98) is questionable. The density of the dislocation is difficult to ascertain and depends very much on the way the crystal has been grown and treated afterwards. Modern crystal growers endeavour to get as few dislocations as possible because these degrade the performance of the semiconductor components made from that ingot. This of course reduces the importance of this type of scattering accordingly.

Yamaguchi (1984) discussed defect scattering in two-dimensional systems.

4.7 IMPACT IONISATION

A carrier with sufficient energy is able to liberate an electron from the valence band and bring it into the conduction band. Hereby a vacancy will be left behind in the valence band so that the process actually creates an additional hole and electron. The particle initialising this process must have an energy at least equal to that of the band gap. The conservation of momentum and energy requires

$$\mathbf{k} = \mathbf{k}_e + \mathbf{k}_h \tag{4.99a}$$

and

$$E(\mathbf{k}) = E_e(\mathbf{k}_e) + E_h(\mathbf{k}_h) + E_G \tag{4.99b}$$

where $E_e(\mathbf{k})$ and $E_h(\mathbf{k})$ represent the band structures of the conduction and valence bands respectively, and E_g the band gap. In direct semiconductors such as GaAs, the impact ionisation starts just above the band gap energy; in indirect semiconductors the process needs a much higher energy or assistance from a phonon. The rate of impact ionisation of particles trapped by the dopants is much lower than the direct one and can be included in the model if needed. We may also consider phonon-assisted impact where the phonon carries away the excess momentum to facilitate the ionisation. This is the only possible mechanism in silicon.

Many authors say that the impact ionisation rate is proportional to $\exp(-E/F)$ where F represents the electric field. This phenomenological model cannot be used in Monte Carlo modelling. Indeed the impact rate has nothing to do with the electric field – only the energy of the particle counts. The only effect of the field is to help the particle gain sufficient energy.

The Keldysh expression represents a simple approach to the impact ionisation rate:

$$\lambda_{\text{imp}} = P_{\text{imp}}[(E - E_{\text{imp}})/E_{\text{imp}}]^2, \tag{4.100}$$

where E_{imp} represents the minimum energy required for impact ionisation. Sano *et al.* (1990) have determined values for the parameter P_{imp} empirically.

In many semiconductors the energy required for impact ionisation lies above the range where the non-parabolic approximation to the band structure, Equation (3.65), starts to deviate significantly from the correct one. It is therefore necessary to introduce the more exact numerical expressions for the band structure. The most thorough analysis of impact ionisation has been presented by Shichijo and Hess (1981) who found that the impact ionisation in GaAs can only take place in certain crystallographic directions.

Even with the simplified band structure, Equation (3.65), and use of the formula for λ_{imp}, Equation (4.100), good qualitative results for avalanche breakdown of field effect transistors can be obtained.

4.8 TRAPPING AND RELEASE OF CARRIERS

Trapping and release of carriers has usually a very low rate, usually of the order up to 10^9 s^{-1}, which corresponds to a trapping or release time of more than a nanosecond, which in most practical cases is considered too long to simulate. Such events are rare and, when they occur, the effect on the carrier distribution is usually negligible. Martin *et al.* (1977) published diagrams showing the trapping rates for various impurities in gallium arsenide. The most important effect of trapping in devices is to modify the local electric field due to the trapped charges. A possible guide when calculating the effect of the trapped charges is to simulate with an exaggerated trapping rate for a short time and then revert to the physically correct one. This possibly has to be repeated a few times to get the correct distribution of the trapped charges.

It is now possible to grow semiconductor devices by vapour deposition at rather low temperatures. The material will then contain a large number of lattice vacancies which give rise to trapping at a frequency of the order of picoseconds. In this case the dynamics of release and carrier trapping becomes significant, and should be considered in the particle model.

The trapping and release of carriers can be treated as an additional scattering mechanism. Reggiani *et al.* (1990) have suggested a phenomenological trapping rate of

$$\lambda_{tr} = c_{tr}(n_{\mathbf{q}} + \tfrac{1}{2} \pm \tfrac{1}{2}) \tag{4.101}$$

where $n_{\mathbf{q}}$ represents the phonon occupation number - the + and − of the ± sign apply to electrons and holes respectively. Trapping electrons involves the creation of a phonon, of holes phonon absorption.

The rate of release is correspondingly

$$\lambda_{rl} = c_{tr}(n_{\mathbf{q}} + \tfrac{1}{2} \pm \tfrac{1}{2}) \tag{4.102}$$

where now the + and − of the ± sign apply to holes and electrons respectively.

The physics of trapping is complex. *Trapping* means that an electron enters an energy level between the valence and the conduction band. The only way of getting there for the electron or the holes is respectively to emit or absorb a phonon. The rate of this is given by

$$\lambda_{tr} = \frac{2\pi}{\hbar} \left| \int \Psi_{nlms} H_{tr} e^{i\mathbf{k}\cdot\mathbf{r}} u_{\mathbf{k}}(\mathbf{r}) d^3\mathbf{r} \right|^2 \delta(E - E_{tr})\,(n_{\mathbf{q}} + 1) \tag{4.103}$$

where Ψ_{nlms} is an orbital for the trapped particle, which, to a good approximation, is one of the functions given in Table 3.1. Usually $\Psi_{nlms} \simeq \Psi_{100s}$ or another orbital of low principal quantum number. n, l, m and s represent the principal, angular, magnetic and spin quantum number respectively and $u_{\mathbf{k}}(\mathbf{r})$ the Bloch function lattice modulating the wave function of the untrapped electron. $\mathcal{H}_{tr}$ represents the perturbation Hamiltonian causing the trapping or release of the carrier. $\mathcal{H}_{tr}$ is the acoustic, Equation (4.54), or polar optical phonon Hamiltonian, Equation (4.61). Only electrons of energy less than $\hbar\omega_o^p - E_{tr}$, where E_{tr} represents the trapping level, can be trapped. A similar argument applies to holes. If $\Psi_{nlms} = \Psi_{100s}$ the trapped particle is in its ground state, otherwise it can decay to a lower state by the action of the electrostatic field of the trapping centre, in a manner analogous to energy transition in a free atom, or by the exchange of phonons.

Release of trapped particles is the opposite process. The trapped electron has to absorb a phonon to get free. The release rates are computed from the same formula, Equation (4.101), but they turn out to be slightly different due to the factor $n_{\mathbf{q}}$ occurring in the expressions for the relevant Hamiltonians (Equations (4.54) and (4.61)).

4.9 TIME-DEPENDENT SCATTERING

So far we have only been interested in instantaneous scattering, which is a basic assumption behind Boltzmann's transport equation. However, the Monte Carlo method can also be extended to interactions of finite duration. In Section 4.2 we have derived an expression, Equation (4.15), which contains a term $\sin(\omega_{ks}t)/\omega_{ks}$ which we later replaced by a delta function under the assumption that its width in the time domain was negligible. A justification for this is that the duration of scattering is usually at least of an order of magnitude shorter than the time of free flight. Barker (1973) has estimated that an ionised impurity scattering event lasts for about $2m^*\beta_s^2/\hbar$ where m^* represents the effective mass of the scattered carrier and β_s the screening length. Quasielastic acoustic deformation scattering lasts about $\hbar/(k_B T_{Lt})$ where k_B represents Boltzmann's constant and T_{Lt} the lattice temperature; the duration of a polar optical interaction is $1/\omega_{po}$, ω_{po} being the optical phonon frequency. The corresponding numerical values are, respectively, of the order of 2, 1000 and 10 fs. The quasielastic scattering thus takes the longest time. This we could also expect from Equation (4.15) because in this case we have a small ω_{ks}; it takes longer to make $\omega_{ks}t = \pi$, which is the position of the first zero outside the origin of the sine. For very low temperatures the acoustic scattering will almost be a continuously ongoing process.

Finite duration of scattering means that the electric field accelerates the particle during scattering. The particle may either lose energy to the field, in which case the scattering rate reduces compared to the instantaneous one, or gain from the field so that the scattering rate increases. In the latter case the threshold for phonon emission lowers (Till and Herbert, 1983). The scattering starts even when the particle has insufficient energy to emit a phonon, provided it acquires the needed amount of energy during scattering. This is known as *field-assisted scattering*.

We may use Equation (4.15) to calculate the scattering rate, remembering that ω_{ks} depends on time. We are not going to derive these scattering rates here. The problem has been addressed by various authors using different approaches. Barker and Ferry (1979) have done it by means of many-body theory; Bertoncini *et al.* (1989) using spectral theory. Another practicable approach may be to describe scattering by means of advanced and retarded Green's functions. Evaluating scattering rates we may even have to consider the possibility that the next process starts before the previous one has finished. This means that multiphonon processes and combinations of e.g. ionised impurity and phonon scattering should be considered. This makes room for many challenging problems which have to be solved when studying really small electronic components.

Lugli *et al.* (1986) found, using Monte Carlo simulation, that the finite duration of scattering, collision broadening, augments the tail of the distribution function of the carriers. Therefore effects due to hot electrons will also become more prominent. Pollak and Miller (1984) interpreted the finite collision time as the real part of the complex flux correlation function and the time it takes to tunnel through a barrier as the imaginary part of it. Bertoncini *et al.* (1989) studied scattering by means of the spectral density function. Ferry and Barker (1979) approached the finite duration of interaction from the balance equations derived from Boltzmann's transport equation and using the form of scattering rates given by Equation (4.15). In the same year Barker and Ferry (1979) discussed scattering based on evaluation of the retarded path integral. Nougier *et al.* (1981) showed that in some circumstances the duration of scattering is comparable to the time of free flight. Finally Lowe (1985) pointed out the importance of defining the precise meaning of scattering time.

5

The Monte Carlo Method

5.1 INTRODUCTION

Knowing the band structure, the phonon spectra and the interaction between them it is possible to make a realistic simulation of the transport in semiconductors. It turns out that the duration of the free flights, the lattice–electron interaction and the direction of the subsequent free flight after scattering are distributed stochastically. Selecting random numbers of the same distributions, possible transport histories of individual particles can be calculated. This is the essence of the Monte Carlo particle model.

We shall open this chapter by discussing ways of obtaining random numbers. Then we move on to the calculation of the time of free flight, the selection of the scattering event terminating it and finally the determination of the initial direction of the next free flight. Readers who want to write their own simulation code will experience that the machine will spend the lion's share of the time doing these things. It is therefore important that this part of the code is written with processing efficiency in mind. A few hints towards this will be given.

The chapter terminates by calculating the position of the particles at the end of the free flight.

5.2 GENERATION OF RANDOM NUMBERS

The most popular and simplest method of generating a sequence of *random numbers* with a uniform distribution is to use the recursive formula (Hammersley and Handscomb, 1964)

$$x'_{i+1} = a_r x'_i + c_r \text{ (modulo } n\text{)}. \tag{5.1}$$

The next number makes use of the previous one. The computed sequence has to start from a freely chosen number x'_0 known as the *seed*. Here n is a large integer selected considering the design of the computer - usually it is a large power of 2 or 10. a_r and c_r are integers between 0 and $n-1$. Equation (5.1) yields a number between 0 and $n-1$, the condition modulo n' means that if $x'_{i+1} \geqslant n$ then x'_{i+1} should be replaced by $x'_{n+1} - n$. Furthermore:

$$\begin{aligned}
&\text{i) } c_r \text{ and } n \text{ should have no common divisor}\\
&\text{ii) } a_r \equiv 1 \text{ (modulo } p\text{) for every prime factor } p \text{ of } n\\
&\text{iii) } a_r \equiv 1 \text{ (modulo 4) if } n \text{ is a multiple of 4.}
\end{aligned} \tag{5.2}$$

The further calculation

$$r_{i+1} = x'_{i+1}/n \tag{5.3}$$

normalises the random number to a value in the range zero to one. Such a number is known as a *flat random number*. Such numbers have, as the name indicates, a uniform distribution, which means that any number between 0 and 1 is equally probable. In other words, having calculated a large number, N_r, of random numbers, we should find that if the interval between 0 and 1 is divided into k equally long subintervals each subinterval contains the same number of values when N_r approaches infinity. This, however, is neither a practical nor a valid test of randomness. The sequence is random when there is no pattern in the numbers nor any autocorrelation between them. The interested reader can find the proof of randomness in the literature, e.g. Hull and Dobell (1962).

In fact, the numbers generated by Equations (5.1) and (5.3) are strictly not random in the statistical sense although they are nearer to it than any numbers generated by other simple algorithms. In a sequence of random numbers it should not be possible to predict the next one, but our algorithm does per definition make it possible to predict it when x'_i is known. The generated sequence will have a period of n – the numbers are therefore referred to as *pseudorandom*. If n is large, of the order of 2^b where b is the number of bits in the computer representation of the mantissa, this sequence will be too long for most simulation runs to be a problem. If, however, the simulation should need more than n random numbers, it is highly likely that the computer will pass through the same code with the same data when the sequence of random numbers is passed through it again so that we get a repetition of previous results. One way of breaking the periodicity of the sequence, however, is to stop the simulation before n numbers have been used, and resume with a different seed (value for x_0).

Any number can be used as a seed, e.g your tax number, social security code, telephone number or the date and time of the day. The latter ensures a different number sequence for every computer run. However, when testing the code, it is advisable to use the same seed every time.

Usually the computer manufacturer provides a random number generator which should be reliable. The programmer should check this with the installation manager.

5.3 NON-UNIFORM RANDOM NUMBERS

Simulation of transport histories requires a non-uniform distribution of random numbers. This can be achieved by a transformation of the distribution generated from the recursion formula (5.1). Let $f(t)$ be a single-valued function of the variable t defined for $t_s \leqslant t \leqslant t_f$ – $f(t)$ should have the properties that

$$f(t_1) \leqslant f(t_2) \text{ for } t_1 < t_2 \tag{5.4}$$

and

$$\int_{t_s}^{t_f} f(t)dt = 1. \tag{5.5}$$

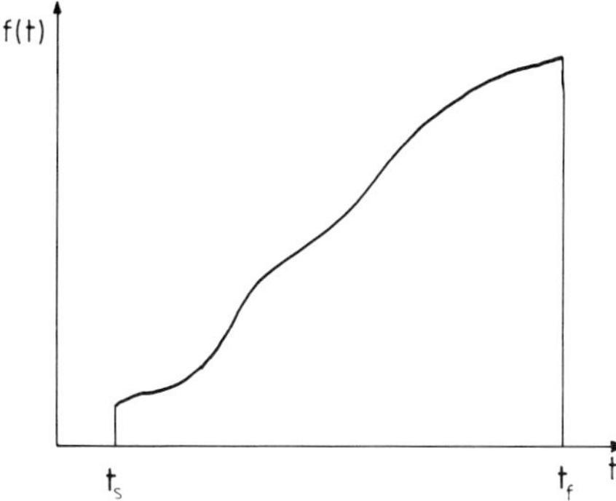

Figure 5.1 *Possible shape of a monotonously increasing function of the independent variable t defined in the interval* $t_s \leq t \leq t_f$

Condition (5.4) says that $f(t)$ is a monotonously growing function of the independent variable t and Condition (5.5) says that the function can be normalised. Figure 5.1 shows a possible form of $f(t)$. Note that $f(t)$ may be horizontal for some range of t, but never falling – it does not need to be continuous.

If r_i is a flat random number of uniform distribution then r_i' is a number of stochastic density $f(t)$:

$$\int_{t_s}^{r_i'} f(t)dt = r_i. \tag{5.6}$$

This can be generalised to a single-valued function $f(t_1,t_2,\ldots,t_n)$ of more variables defined for $t_{ks} \leqslant t_k \leqslant t_{kf}$ for $k = 1,2,\ldots,n$:

$$f(t_1,t_2,\ldots,t_{kl},\ldots,t_{kn}) \leqslant f(t_1,t_2,\ldots,t_{k2},\ldots,t_{kn}) \tag{5.4'}$$

for $t_{k1} < t_{k2}$ and

$$\int_{t_{1s}}^{t_{1f}} dt_1 \int_{t_{2s}}^{t_{2f}} dt_2 \ldots \int_{t_{ns}}^{t_{nf}} dt_n f(t_1,t_2,\ldots,t_n) = 1. \tag{5.5'}$$

If r_k is a flat random number, another random number r_k' is obtained from

$$\int_{t_{s1}}^{t_{f1}} dt_1 \ldots \int_{t_{sk}}^{r_k} dt_k \ldots \int_{t_{sn}}^{t_{fn}} dt_n f(t_1,t_2,\ldots,t_k,\ldots,t_n) = r_k. \tag{5.6'}$$

A step function, i.e. a function which is constant in intervals on the domain in which it has been defined, is of special interest. Figure 5.2 shows a possible example of such a step function and is defined by splitting the interval $t_s \leqslant t \leqslant t_f$ into smaller intervals $\{t_0, t_1, t_2, \ldots, t_m)$ such that

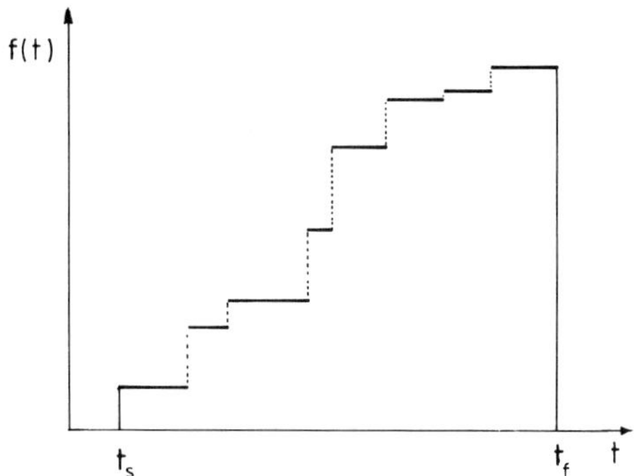

Figure 5.2 *Possible shape of a piecewise constant function of the independent variable t defined in the interval* $t_s \leq t \leq t_f$

$$f(t) = c_k \text{ for } t_{k-1} < t \leqslant t_k \tag{5.7}$$

where c_k is a constant, $t_0 \equiv t_s$ and $t_m \equiv t_f$.

A die represents the simplest possible example of such a function: it takes the value 1/6 in the entire interval on which it has been defined, i.e. $f(t)$ is defined by

$$f(t) = \frac{1}{6}\,. \tag{5.8}$$

Let t^{sup} represent the smallest integer greater than or equal to t (e.g. $1.56^{\text{sup}} = 2$; $0.03^{\text{sup}} = 1$; $5^{\text{sup}} = 5$). $f(t)$ has the properties described by Equations (5.4) and (5.5). A random number r_i' in the range 1–6 is obtained from

$$\int_0^{r_i'} f(t)dt = \int_0^{r_i'} \frac{1}{6}\,dt = r_i \tag{5.9}$$

where $0 \leqslant r_i \leqslant 1$. The value r'^{sup} chosen in this way represents the face of the die.

The reader may find this exposition a cumbersome way of selecting an integral random number rather than actually tossing a die. This method, however, is the simplest way of simulating the outcome of casting the die by means of the Monte Carlo method.

Monte Carlo methods have now been used for more than a century. The earliest account of the use of random numbers was to calculate π, published by Hall (1873). A famous application was made in connection with the Manhattan Project to calculate the distribution of neutrons in a nuclear fission reactor. Leslie and Chitty (1951) studied the biometrics of capture and recapture by means of the Monte Carlo method. They used a tin of lotto tokens to generate random numbers. By means of Equations (5.7) and (5.9) it is possible to computer simulate games with the roulette wheel. This method will also be used to select particle scattering.

5.4 THE TIME OF FREE FLIGHT

The duration of the free flights of particles is stochastically distributed. Their distribution is analogous to that of more mundane situations like a switchboard receiving telephone calls. Of course we assume that the traffic through the switchboard is not saturated. There is always a chance that someone will call. The probability per unit time that this happens is known as the *traffic density*. The situation is analogous for particles in free flight in a semiconductor. There is always a chance that the flight terminates, in other words that the particle gets scattered. The probability per unit time that it happens is the scattering rate. The scattering rate corresponds to the traffic density.

Assume that we keep track of the time when someone calls or when a particle is being scattered. We have a stop-watch which we reset every time this happens. To calculate the probability that it takes place when our stop-watch reads t, the time span since the last event is divided into small intervals which are numbered sequentially in chronological order. Interval N° k has length Δt_k. Obviously, when the next event happens during this interval it should not have happened during any previous interval. Thus the probability that it happens during the kth interval of time is

$$P_k \Delta t_k = \Gamma(t_k) \Delta t_k \tag{5.10}$$

where the time t_k lies in this interval. $\Gamma(t)$ represents the traffic density or the scattering rate. The chance that the event does not happen in a previous interval j is one minus the chance that it does happen:

$$P_j \Delta t_j = 1 - \Gamma(t_j) \Delta t_j \tag{5.11}$$

(where the time t_j lies in interval N° j, and $j = 1, 2, \ldots, k - 1$).

An event in one interval is stochastically independent from what takes place in any other one. The probability that it happens just in interval N° k is the product of the Expression (5.10) and the $k - 1$ expressions of the form (5.11):

$$P \Delta t_k \equiv \prod_{j=1}^{k-1} P_j \Delta t_j P_k \Delta t_k = \prod_{j=1}^{k-1} (1 - \Gamma(t_j) \Delta t_j) \Gamma(t_k) \Delta t_k . \tag{5.12}$$

To see whether the event occurs in a given interval is known as a *Bernoulli trial.* In elementary textbooks on statistics Γ is constant. In our case this is not true: chance of scattering depends on the energy of the particle. The particle loses or gains kinetic energy to the local electromagnetic field during the flight so that the scattering probability changes as it moves. Therefore our Bernoulli trials are trials with variable probability, also known as *Poisson trials* (Feller, 1961).

If the intervals shrink, then, of course, k increases but in such a way that $\sum_{j=1}^{k} S \Delta t_j = t$ remains unaltered. To evaluate Equation (5.12) take the natural logarithm:

$$\ln(P\Delta t_k) = \sum_{j=1}^{k-1} \ln[1 - \Gamma(t_j)\Delta t_j] + \ln[\Gamma(t_k)\Delta t_k]. \tag{5.13}$$

When Δt_j is sufficiently small it is possible to expand the logarithm in a Taylor series:

$$\ln[1 - \Gamma(t_j)\Delta t_j] \simeq -\Gamma(t_j)\Delta t_j + \text{higher orders in } \Delta t_j \tag{5.14}$$

so that

$$\ln(P\Delta t_k) = -\sum_{j=1}^{k-1} \Gamma(t_j)\Delta t_j + \log[\Gamma(t_k)\Delta t_k] \tag{5.15}$$

Letting each interval shrink to infinitesimal length this expression becomes

$$\ln[P(t)dt] = -\int_0^t \Gamma(t')dt' + \ln[\Gamma(t)dt] \tag{5.16}$$

and

$$P(t)dt = \Gamma(t)\exp[-\int_0^t \Gamma(t')dt']dt \tag{5.17}$$

represents the probability density that the next event occurs at time t after the previous one. $P(t)$ has the property that

$$\int_0^\infty P(t)dt = 1 \tag{5.18}$$

and is monotonously increasing with t. $P(t)$ therefore satisfies Equations (5.4) and (5.5).

The time of free flight, τ, is calculated using

$$\int_0^\tau \Gamma(t)\exp[-\int_0^t \Gamma(t')dt']dt = r_t \tag{5.19}$$

where r_t represents a flat random number.

Unfortunately the process of selection turns out to be rather time consuming because of the complicated relationship between $\Gamma(t)$ and t. Figure 5.3 shows a typical Γ against energy or wave vector. The energy is related to the time by the dynamics of the particles, Section 5.8, and Γ represents the sum of all scattering rates because any event can terminate the free flight, just as you may receive any type of call, business or private, local or long distance. From the traffic point of view we are only interested in the fact that someone calls.

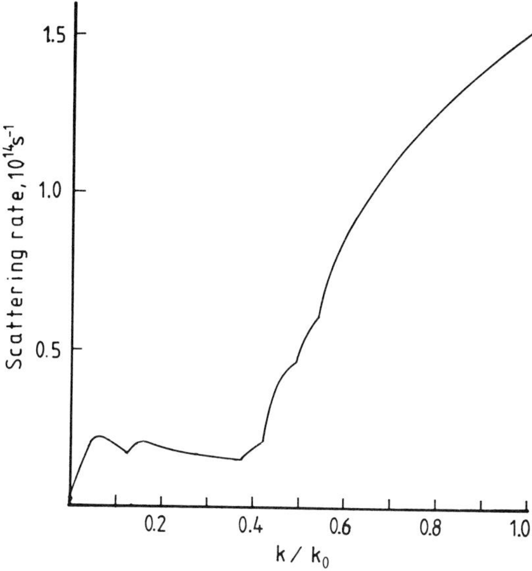

Figure 5.3 *Total scattering rate against energy for electrons in the Γ minimum of the conduction band of GaAs. The abscissa represents the ratio k/k_0 where k_0 denotes the wave vector corresponding to the energy of 1 eV*

The scattering rate shown in the figure is typical. The rate reaches a maximum, decreases due to the decrease in ionised impurity scattering with increasing energy and finally increases again due to phonons.

5.5 SELECTION OF SCATTERING EVENTS

The total scattering rate Γ is a sum of all possible scattering rates. For computational technical reasons each scattering mechanism has been numbered sequentially from one to N_Γ, the total number of such mechanisms considered. Once the numbering has been decided it should not change during simulation. Phonon absorption and emission are, for this purpose, considered as different scattering mechanisms because the rates differ by factors $n_{\mathbf{q}}^p$ and $n_{\mathbf{q}+1}^p$ where $n_{\mathbf{q}}^p$ represents the phonon occupation given by Equation (4.56). It is advisable, as far as practically possible, to number the mechanisms in order of decreasing strength. However, a definite mechanism may be strong for one range of energies and weak for another, yet the same mechanism must, for the purposes of easy housekeeping, have the same number for all energies or wave vectors. For example, optical phonon emission starts at an energy of $h\omega_0^p$ where ω_0^p represents the frequency of the optical phonon. Below this energy the rate is zero, but should still be considered as a proper rate. If the rate of this scattering is large, above $h\omega_0^p$ it should be given a low ordinal number.

The total scattering rate is

$$\Gamma[\mathbf{k}(t)] = \sum_{i=1}^{N_\Gamma} \lambda_i[\mathbf{k}(t)] \tag{5.20}$$

where t represents time, which is related to the energy or wave vector of the scattering particle through Equations (5.42), and λ_i the rate of scattering mechanism i. Figure 5.4 shows a family of curves numbered sequentially, starting with the bottom one. The difference between curve $i-1$ and i represents the scattering rate for scattering mechanism i. Thus curve i represents the sum of the i first scattering mechanisms: $\sum_{j=1}^{i} \lambda_j[k(t)]$. The topmost curve represents the total scattering rate Γ given by Equation (5.20).

The scattering mechanism terminating the free flight is selected by means of a random number

$$r_\gamma = \Gamma r_s \tag{5.21}$$

where r_s represents a flat random number. That scattering mechanism j is selected satisfying the inequality

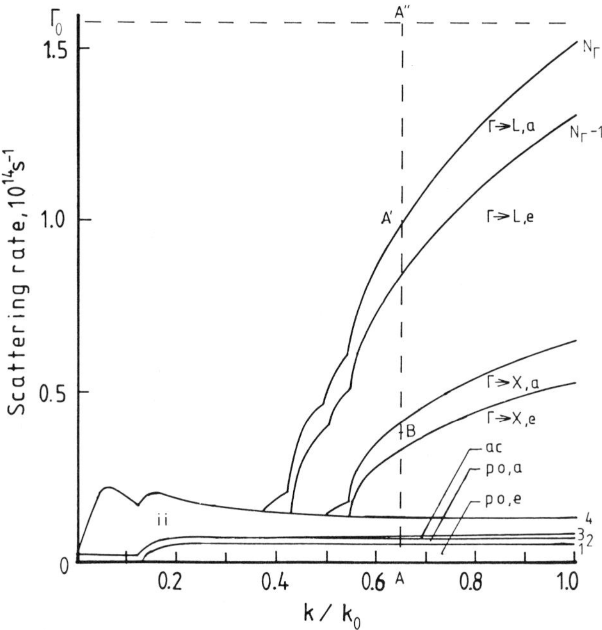

Figure 5.4 *Repetition of Fig. 5.3, but now with curves for $\Sigma_{i=1}^{j}\lambda_i$, $j = 1, 2, \ldots, N_\Gamma$ having been added. The curves have been numbered starting from the bottom one. The difference between two adjacent curves represents a scattering rate of the indicated mechanism – e.g. 'Γ → X, e' indicates transfer from Γ into an X minimum by emission of a phonon (a, phonon absorption; e, phonon emission; po, polar optical phonon; ac, acoustic phonon; ii, ionised impurity). The scattering rates have been calculated assuming an isotropic spherical band structure and an ionised donor density of 8×10^{23} m^{-3}. A–A' represents a possible wave vector at the start of scattering of an electron of wave vector* **k**, **k**$_0$ *represents the wave vector corresponding to an electron of energy 1 eV. B represents a possible choice of scattering mechanism according to the scheme (5.22). A'–A" is an extension of A–A' to Γ_0, the sum of the real and the Rees self-scattering rates*

$$\sum_{i=1}^{j-1} \lambda_i[\mathbf{k}(t)] \leqslant r_\gamma < \sum_{i=1}^{j} \lambda_i[\mathbf{k}(t)] \tag{5.22}$$

Here we should interpret

$$\sum_{i=1}^{0} \lambda_i[\mathbf{k}(t)] \equiv 0. \tag{5.23}$$

Thus, the various scattering mechanisms are passed through in the sequence they have been ordered until $\Sigma\ \lambda_i$ exceeds r_γ. The search terminates sooner the smaller the value of r_γ, which is why we have given the above-mentioned advice on numbering the various scattering mechanisms. This scheme makes a weighted selection such that, over a long period of time, the relative amount of scattering from a particular type equals its fractional part of the total scattering rate or cross-section.

In Fig. 5.4 the number Γ is represented by the height of the bar A–A' where A' lies on the topmost curve. B represents a possible selected value for r_γ. Particularly in Fig. 5.4 the point B lies between the curves representing the rate of scattering from the Γ into an X minimum, absorbing (curve above B) and emitting (curve below B) a phonon respectively. In this example the selected scattering mechanism is the transfer of the electron from the Γ to an X minimum of the conduction band by absorption of a phonon.

5.6 ALGORITHMS TO SHORTEN THE SEARCH FOR TIME OF FREE FLIGHT

The numerical integration to find the time of free flight τ from Equation (5.19) is unnecessarily lengthy even if the values for Γ have been tabulated in advance. To simplify the determination of τ Rees (1986) introduced an additional fictive scattering, the *Rees*, *null* or *self-scattering* where neither the momentum nor the energy of the particle changes. In fact, Rees scattering does nothing to the particle, but the rate of this additional scattering mechanism, λ_0, is chosen to depend on the energy in such a way that the sum of the real and the fictive scattering becomes a constant in energy or carrier momentum:

$$\Gamma + \lambda_0 = \Gamma_0 \tag{5.24}$$

Γ_0 should be chosen equal to or larger than the maximum likely Γ to avoid a negative self-scattering rate which is unphysical and may even lead to wrong results. Introducing this new rate into Equation (5.17) we can evaluate the integral analytically:

$$P(t)dt = \Gamma_0 \exp(-\Gamma_0 t)dt. \tag{5.25}$$

The duration of the free flight can now be chosen from

$$\int_0^{\tau} \Gamma_0 exp(-\Gamma_0 t)dt = r_\tau \tag{5.26}$$

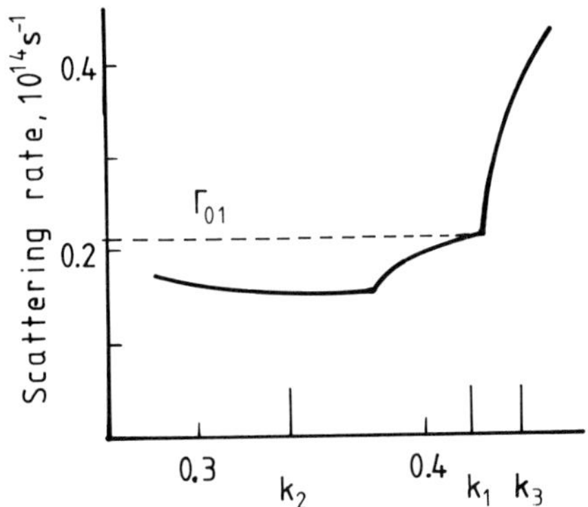

Figure 5.5 *Detail of Fig. 5.3 illustrating the method of the iterative Gamma*

which we integrate to yield

$$\tau = -\log(r_\tau - 1)/\Gamma_0 \tag{5.27}$$

In practical applications we prefer to choose r_τ rather than $r_\tau - 1$ as a random number to obtain τ.

The price to pay for this is the risk of selecting a null event as scattering. This does not affect the result of our simulation, it only implies a needless calculation of a mid-flight position of the particle and that the whole process of reselection has to be repeated from this position.

The inclusion of the Rees scattering means that the line in A–A' in Figure 5.4 has been extended to A''. It is obvious from the figure that the chance of selecting null scattering is overwhelmingly large in a significant part of the energy range. However, there is a further refinement to our selection procedure, known as the method of the *iterative Gamma*: if we knew that the wave vector for a particular carrier would remain between $\mathbf{k}_1$ and $\mathbf{k}_2$ (see Fig. 5.5) it would be sufficient to select $\Gamma_0 = \Gamma_{01}$, which significantly reduces the chance of selecting self-scattering and therefore speeds up the simulation. If the particle has wave vector $\mathbf{k}_1$ at the start of the flight, and $\mathbf{k}_2$ at the end of it (it has lost energy during the flight according to the figure) this yields a correct choice of τ. But if the particle has wave vector $\mathbf{k}_3 > \mathbf{k}_1$ at the end of the flight, scattering mechanisms are excluded from the choice yielding a negative scattering rate which we have to avoid. A new selection has therefore to be made with

$$\Gamma_{02} = \Gamma_{01} + \Delta\Gamma. \tag{5.28}$$

If $\Gamma_{02} < \Gamma(\mathbf{k}_3)$ the process has to be repeated by adding another increment $\Delta\Gamma$ to Γ_{01}. This process usually converges after a few steps. The selection of $\Delta\Gamma$ is a matter of experience – usually $\Delta\Gamma = 0.1\Gamma_0$ is a reasonable choice. The best choice represents a compromise between the computational speed of selecting a real scattering event and loss of time from calculating a mid-flight position unnecessarily. The author's experience is that the iterative Γ represents a saving of a factor between two and three in computing speed compared to the scheme of Equation (5.27) with a fixed Γ_0.

Of course the time of free flight can be determined directly from Equation (5.19) by numerical integration. Even with tabulated values for the scattering rates this is slower than the iterative Gamma scheme. Rockett (1987) has suggested a scheme based on this which is claimed to be rather efficient.

There also exists the possibility of fitting a polynomial or any other manageable analytical form to $\Gamma(k)$, and using this as a basis for determining τ. A combination of any of the three schemes can also be applied. The optimal choice depends on the material and on the architecture of the computer. The author's personal experience is that the iterative Gamma is the fastest to compute.

If looking up tables is slower than recalculating scattering rates, recalculating these rates each time they are needed may be carried out instead of collecting them from the computer memory.

Once a scattering mechanism has been selected it may be necessary to re-examine the decision when the free flight extends beyond the end of the field-adjusting time step which we shall define in Chapter 7.

The expression for the scattering rates contains a factor expressing the probability that the state to be scattered into is free. This may be position dependent through local variations in the Fermi level. When the scattering rates are calculated and tabulated in preparation for the simulation it is advisable to calculate these rates without the factor $1 - f(E)$. This simplifies the tables as they will be applicable everywhere in the device and means that the tabulated scattering rates may be higher than the actual ones. An additional random number representing the chance that the state after scattering is available, $1 - f(E)$, has to be selected. If this number is greater than the probability that the scattering can take place, the selected scattering mechanism should be considered a null event and the duration of the next free flight calculated as described above.

A similar argument also applies to ionised impurity and intercarrier scattering. The above-mentioned tables of scattering rates have been calculated with the maximum anticipated concentration of scattering centres (impurities or carriers). According to Equation (4.81) the rate is almost proportional to the ionised impurity concentration. When the ionised impurity scattering has been selected, its tabulated rate has to be scaled by the ratio between the local ionised impurity concentration and that assumed to calculate the rate. Along with the possibility that the goal of scattering is already occupied, an additional random number representing this ratio has to be generated to determine whether the scattering actually should take place. These two random numbers can be combined into one representing their product.

This approach to ionised impurity scattering may lead to a large number of null events especially where there are few, if any, dopants. To avoid this, the actual scattering rate can instead be calculated each time according to need. This may save computer time, though the amount actually saved depends on the architecture and organisation of the computer and the software. In such a scheme, Γ, Equation (5.20), or Γ_0, Equation (5.24), has been calculated *without* the ionised impurity rate, λ_{ii}, or the intercarrier scattering rates, λ_{ee}. The times of free flight and scattering have then to be selected as explained above with

$$\Gamma' = \Gamma + \lambda_{ii} + \lambda_{ee} \tag{5.29}$$

or with

$$\Gamma_0' = \Gamma' + \lambda_0' \tag{5.30}$$

with Γ' given by Equation (5.29) and the Rees' scattering rate λ_0' defined such that Γ_0' is constant, instead of Γ or Γ^0.

5.7 CHOICE OF SCATTERING ANGLES

Once the scattering mechanism has been selected, the scattering angles have to be calculated in order to determine the initial direction of the next flight. In three-dimensional scattering this involves two additional flat random numbers, namely r_θ for choice of the polar angle and r_ϕ for the azimuth.

The rate is calculated from the matrix element for the transition between the initial, $|s\rangle$, and the final state, $|f\rangle$, Equation (4.17), which is obtained by summing over all possible final states. The summation is generally substituted by an integral, Transition (4.46), over the polar angle, θ, and the azimuth, ϕ, i.e. we may formally write:

$$s(\theta,\phi) \equiv |\langle f|U|s\rangle|^2 \tag{5.31}$$

for the matrix element entering the expression for the scattering rate, Equation (4.18). Calculating λ_s thus involves evaluating the integral

$$S_\lambda = \int_0^\pi d\theta \sin\theta \int_0^{2\pi} d\phi s(\theta,\phi). \tag{5.32}$$

The scattering rate is $\lambda_s = c_\lambda S_\lambda$ where c_λ represents that part of the expression for the scattering rate which does not involve the angles. This integral has been discussed and evaluated for different scattering mechanisms in Chapter 4, e.g. Equations (4.48), (4.50) and (4.51) for acoustic phonon scattering, Equation (4.67) for Fröhlich scattering, Equation (4.57) for intervalley scattering and Equation (4.81) for ionised impurity scattering. In all cases the integration has been carried out over the entire range of the angles.

The polar angle is selected as that angle θ satisfying

$$\frac{1}{S_\lambda}\int_0^\theta d\beta \sin\beta \int_0^{2\pi} d\phi s(\beta,\phi) = r_\theta \tag{5.33}$$

where r_θ denotes a flat random number. The azimuth, ϕ, is selected likewise from

$$\frac{1}{S_\lambda}\int_0^\pi d\theta \sin\theta \int_0^{\phi} d\beta s(\theta,\beta) = r_\phi \tag{5.34}$$

where r_ϕ also represents a flat random number. These two expressions satisfy Conditions (5.4′) and (5.5′). Often we use expressions for s which are explicitly independent of ϕ, in which case Equation (5.33) yields

$$\frac{2\pi}{S_\lambda}\int_0^\pi d\theta \sin\theta s(\cos\theta) = r_\phi \tag{5.35}$$

which yields the following choice of ϕ:

$$\phi = 2\pi r_\phi . \tag{5.36}$$

This applies to electron scattering by phonons and ionised impurities, and most of the scattering when the band structure is given by Equation (3.65).

The evaluation of θ, however, is almost always more complex, since we rarely encounter a simple analytical formula from which θ can be calculated. In the case of electrons with a non-parabolic band structure, the quasielastic acoustic phonon scattering rate involves the calculation of the integral

$$S_\lambda = \int_0^\theta d\phi \sin\phi \,(a_g + b_g \cos\phi + c_g \cos^2\phi), \tag{5.37}$$

an expression we have already met computing Equation (4.48) with π as the upper limit. Evaluating it again with an upper limit different from π yields

$$\frac{a_g(1-\cos\theta) + b_g(1-\cos^2\theta)/2 + c_g(1-\cos^3\theta)/3}{2(a_g + c_g/3)} = r_\theta \tag{5.38}$$

which is an equation of the third degree in $\cos\theta$. In the case $b_g = c_g = 0$ this simplifies to

$$1 - \cos\theta = 2r_\theta \tag{5.38'}$$

giving

$$\cos\theta = 1 - 2r_\theta \tag{5.39}$$

or

$$\theta = \arccos(1 - 2r_\theta). \tag{5.39'}$$

The determination of θ by analytical or fast numerical means may be impossible or too slow. Krowne (1987) has developed a simplified algorithm for determining the scattering angles analytically, though his expressions are not exact. In general, the angles can be found using a scheme deriving from von Neumann (1951) when the determination of θ makes use of the function $s(\cos\theta)$ of Equation (5.35). When $s(\cos\theta)$ is a monotonously increasing function of $\cos\theta$ in the range -1 to 1 with a maximum value of $s(1)$ for $\cos\theta = 1$, we can generate a pair of flat random numbers r_1 and r_2, the first one representing $\cos\theta$ in the range -1 to 1, the second $s(\theta)$ between $s(-1)$ and $s(1)$. If $r_2 < s(r_1)$ then r_1 is taken to be the selected value of $\cos\theta$, otherwise a second pair has to be generated. The accepted values of r_1 follow the required probability distribution $s(r_1)$. The scheme also works for a monotonously decreasing function $s(\cos\theta)$ when $1 - r_\theta$ represents $\cos\theta$.

The scheme is fast and usually only a handful of pairs have to be generated before a scattering angle has been selected. However, if s is small everywhere except at the end of the range, the process may take a long time before it gives any result. If s is not increasing monotonously, the interval of $\cos\theta$ has to be divided into subintervals such that s is monotonous within each subinterval. If one of the intervals has to be selected by means of a random number, then θ has to be determined from this interval as described above.

The process can be extended to a function of more than one variable. Evaluating the scattering rates for holes with the band structure (3.68) involves an $s(\theta,\phi)$ which is not separable. Assume that $s(\theta,\phi)$ is increasing monotonously both in $\cos\theta$ and in ϕ for $-1 \leqslant \cos\theta \leqslant 1$ and $0 \leqslant \phi \leqslant 2\pi$. The selection of scattering angles is carried out by generating a triplet of random numbers. r_1 represents $\cos\theta$ in the range -1 to 1; r_2 represents ϕ in the range 0 to 2π; r_3 represents s in the range between the minimum value $s(-1,0)$ and the maximum value $s(1,2\pi)$. If $r_3 < s(r_1, r_2)$ then r_1 and r_2 represent the selected angles $\cos\theta$ and ϕ respectively, otherwise this procedure has to be repeated for a new triplet.

This scheme can be extended to functions of many variables in an obvious way.

5.8 MOTION OF PARTICLES IN THE LOCAL ELECTROMAGNETIC FIELD

During the free flight the particle is subject to the *Lorentz force*

$$\hbar d\mathbf{k}/dt = e[\mathbf{F} + \mathbf{v}(\mathbf{k}) \times \mathbf{B}] \tag{5.40}$$

where, as before, **F** represents the electric field, **B** the magnetic field, **v** the velocity of the particle, defined by Equation (3.71). **m***, the effective mass, defined by Equation (3.70). Assuming the effective mass **m***, which is a tensor defined by Equation (3.70), does not change during the free flight, the force can be formulated as:

$$\mathbf{F}_L = \hbar d\mathbf{k}/dt = e(\mathbf{F} + \hbar \mathbf{m}^{*-1} \bullet \mathbf{k} \times \mathbf{B}) \tag{5.41}$$

($\mathbf{m} \bullet \mathbf{v} = \hbar\mathbf{k}$). Selecting a Cartesian coordinate system where $\mathbf{m}^{*-1}$ is diagonal and **B** is directed along the z axis, the set of the three differential Equations (5.40) becomes rather elementary. The solution is

$$\begin{aligned} k_x &= \frac{m_x^*}{\hbar B_z}\left\{F_x\left(1 - \cos\frac{eB_z t}{m_x^*}\right) + F_y \sin\frac{eB_z t}{m_x^*}\right\} + k_{0x} \\ k_y &= \frac{m_y^*}{\hbar B_z}\left\{F_y\left(1 - \cos\frac{eB_z t}{m_y^*}\right) + F_x \sin\frac{eB_z t}{m_y^*}\right\} + k_{0y} \\ k_z &= \frac{eF_z t}{\hbar} + k_{0z} \end{aligned} \tag{5.42}$$

where k_0 with the Cartesian components k_{0x}, k_{0y} and k_{0z}, represents the wave vector of the particle at the beginning of the flight, F_x, F_y and F_z the Cartesian components of the electric field which has been assumed not to change during the flight and B_z the magnetic flux. The subscript z is there to indicate it is oriented in the z direction. The justification for assuming a constant field will be discussed in Section 8.2.

The position of the particle at the beginning of the flight is represented by $\mathbf{r}_0$ with components r_{0x}, r_{0y} and r_{0z}. At the next scattering event it is $\mathbf{r}$ with components

$$r_x = \frac{F_y t}{B_z} + \frac{m_t^*}{eB_z}\left(v_x - \frac{F_y}{B_z}\right)\sin\frac{eB_z t}{m_x^*} + \left(v_y + \frac{F_x}{B_z}\right)\left(1 - \cos\frac{eB_z t}{m_x^*}\right) + r_{0x}$$

$$r_y = -\frac{F_y t}{B_z} + \frac{m_y^*}{eB_z}\left(v_x - \frac{F_y}{B_z}\right)\left(\cos\frac{eB_z t}{m_y^*} - 1\right) + \left(v_y + \frac{F_x}{B_z}\right)\sin\frac{eB_z t}{m_y^*} + r_{0y}$$

$$r_z = \frac{eF_z t^2}{2m_z^*} + \frac{\hbar k_{0z}}{m_z^* t} + r_{0z}. \tag{5.43}$$

The reader may correctly argue against the motion being treated by classical Newtonian mechanics – a proper quantum mechanical approach, treating the particles as wave packets, would be more appropriate. We have not considered transport in magnetic fields in any of the examples of modelling we are going to present later. This has to be included when studying Hall mobility. Ganguly *et al.* (1982) have discussed free flights in electrical and magnetic fields.

Because of the wave nature of the particle the position of it cannot be established with a greater accuracy than $\Delta\mathbf{r}$ with a Cartesian component Δr_j. The wave packet nature does not allow a definition of the momentum with a greater accuracy than $\Delta\mathbf{p}$ with Cartesian component Δp_j. The momentum and position are conjugate coordinates in quantum theory (e.g. Merzbacher, 1961), the products of two Cartesian components of the uncertainties in them follow the law:

$$\Delta p_j \Delta x_j \geqslant \hbar \tag{5.44}$$

or approximately

$$\Delta k_j \Delta x_j \geqslant 1 \tag{5.45}$$

with j standing for x, y or z. When the position of the particle can be defined to within 10 nm, the uncertainty in the wave vector, Δk_j would correspondingly amount to $10^8\,\mathrm{m}^{-1}$, which represents a fairly small energy. In semiconductors of practical dimensions which are of the order of micrometres, this does not play any significant role. This approach therefore seems adequate, except for particles of very low kinetic energy; though since these are relatively few in number because of the density of states the error made by classical consideration is negligible. However, proper quantum mechanical transport has to be considered when the particle is confined to spaces measuring a few angströms. We shall discuss this further in Section 10.5. When we are going to simulate future ultrasmall semiconductor components, the particles have to be treated like waves rather than as classical points.

The classical approach has proven accurate enough and is very fast in computer time.

6

Simulation of Bulk Properties of Solids

6.1 INTRODUCTION

When the band structure, the phonon spectrum and the interaction between them are known, the Monte Carlo method can be used to calculate the bulk transport properties of a semiconductor. We need to know the shape of the energy bands, the values of the effective masses, the phonon energies, phonon deformation potentials and the dielectric constant. These values have been collected in the semiconductor tables edited by Landolt-Börnstein (1982) or from several individual publications mentioned in Chapter 4. Several authors have calculated the drift velocity and mobility of the carriers for bulk material, the population ratio of the various minima of the conduction band, the energy and momentum relaxation times and the mean free paths by means of Monte Carlo simulation. Comparing the physical constants they use, we note some differences in the choice of values between the authors. Some of these values, e.g. some deformation potentials, may be difficult to ascertain because the transport is not very sensitive to their value. However, when we have decided on a set yielding macroscopic bulk quantities in agreement with measured ones, our chosen value can be considered representative. The author's set for silicon and gallium arsenide have been quoted in Tables 6.1 and 6.2. Once a set has been established, it should not be changed during the simulation of devices. The selection of basic constants is referred to as *priming the model.*

Values of deformation potentials, effective masses etc. are considered microscopic – they do not fluctuate. Quantities such as mobility, relaxation times, diffusion constants etc. are macroscopic because they represent an average over a large number of particles. These macroscopic parameters must never be used as input to a Monte Carlo simulation. Often macroscopic quantities obtained from Monte Carlo calculations are used as input for hydrodynamic modelling.

A framework consisting of the basic semiconductor physics has been constructed in the first five chapters. Here, and in subsequent chapters, this frame will be filled in. In this chapter we shall be concerned with the simulation technique as well as with the physical interpretation of our calculations. We shall study electrons and holes in gallium arsenide and electrons in silicon as illustrative examples. The examples will also include transport at the interface between silicon and its dioxide under a strong transversal electric field. The practical application of this is to describe the current flowing under the gate of a metal oxide semiconductor field-effect transistor (MOSFET).

There are technically important distinctions to be made between electrons and holes in GaAs and electrons in bulk Si. In GaAs it is customary to consider all the minima of the conduction band as spherical. The reason for this is that the

Table 6.1 Material parameters for GaAs

Velocity of sound	u	5220 m s^{-1}
Density	ρ_e	5370 kg m^{-3}
Lattice temperature	T_{Lt}	300 K
Dielectric constant	ε	13.18
Optical frequency dielectric constant	ε_∞	10.89
Optical phonon energy	E_0	36.3 eV

Extremum	Γ	X	L	light hole	heavy hole	split-off
Energy(eV)	1.42	1.91	1.73	0	0	0.33
Effective mass(m_0)	0.067	0.85	0.56	0.067	0.62	0.15
Non-parabolicity (eV^{-1})	0.576	0	0	0	0	0
Acoustic deformation potential (eV)	7.0	7.0	7.0	3.5	3.5	3.5
Intervalley of inter-band deformation potential (10^8 eV m^{-1})	1.6	1.6	1.6	1.6	1.6	1.6
Intravalley deformation potential (10^8 eV m^{-1})	1.6	1.6	1.6			
Intervalley phonon (eV)	29.9	29.9	29.9	29.9	29.9	29.9
Intravalley phonon (eV)	29.9	29.9	29.9			

Table 6.2 Material parameters for silicon*

Non-parabolicity	α	0.5 eV^{-1}
Band gap energy	E_g	1.113 eV
Longitudinal effective mass	m^*_l	0.9160 m_0
Transversal effective mass	m^*_t	0.1906 m_0
Velocity of sound	u	9037 m s^{-1}
Density	ρ_e	2329 kg m^{-3}
Lattice temperature	T_{Lt}	300 K
Dielectric constant	ε	11.7
Opt. freq. dielectric function	ε_∞	10.82
Quasielastic acoustic Deformation potential	Ξ_a	9.00 eV
Intervalley, acoustic deformation potential	Ξ_{iv}	
Type of phonon deformation	energy	
LA f_1 phonon	47 meV	4.80×10^{-9} J m^{-1}
TO f_2 phonon	59 meV	6.40×10^{-9} J m^{-1}
LO g-phonon	62 meV	6.28×10^{-8} J m^{-1}

Energy† of Γ minimum 2.20 eV
X minima 3.52 eV
L minima 1.15 eV

Position parameter of X minima (see Table 3.3) $\xi = 0.85$

* Canali *et al.* (1975). † Fischetti (1991)

population of the secondary minima is so sparse that it has not been possible, until recently, to measure the transversal and longitudinal masses in the L and the X minima. The band structure for holes can be considered isotropic but the more refined form, Equation (3.68), will be used as a basis for our calculations. In silicon the electrons reside in the X minima, for which the values for longitudinal and transversal effective masses are well-established. We shall use the anisotropic model for this material. Such a model should also be used for GaAs but we shall follow common practice by using an isotropic description. The reader may of course use an anisotropic model for GaAs as well. The effective masses needed for the anisotropic model have only recently been determined (Fischetti, 1991).

The drift velocity of the carriers against the homogeneous electric field will be calculated in Section 6.2 and compared with published experimental data. We shall also calculate the occupancy of the various extrema of the bands and examine the meaning of the observed drift velocity in the light of this. This brings us to the problem of overshoot velocity. It is a trivial exercise to derive the mobility from the calculated drift velocity.

The distribution function, i.e. the number density of particles against the energy for various fields, will be introduced in Section 6.3. In silicon this distribution function turns out to be the familiar Maxwellian one, but for gallium arsenide, even at the lowest fields, such an approximation is rather poor due to scattering from polar optical phonons.

Section 6.4 discusses energy and momentum relaxation, which prove to be field-dependent, while in Section 6.5 we study the path of free flight. The chapter ends with an exposition of transport near the silicon–silicon dioxide interface in MOSFETs in a condition of strong inversion, which is a true quantum mechanical problem.

6.2 THE DRIFT VELOCITY

Lacking any better knowledge of the electron distribution, the simulation starts by assuming a Maxwellian velocity distribution. The choice of the initial distribution is immaterial, the correct one will always evolve. Here we shall calculate the drift velocity for gallium arsenide, including acoustic and optical phonon scattering and intervalley scattering in the model. Ionised impurity scattering and intracarrier scattering will be neglected because we shall restrict ourselves to intrinsic material. A constant uniform electric field is assumed. As we only work in momentum space, the actual geometrical position of the particles need not be calculated. Figure 6.1 shows the calculated drift velocity of electrons in gallium arsenide. The Monte Carlo model used to obtain these results is essentially the same as that of Jacoboni and Reggiani (1983), Fawcett *et al.* (1970).

The drift velocity shows a maximum of about 2.5×10^5 m s^{-1} at 0.3 MV m^{-1}; for higher fields it decreases to saturate near 1.0×10^5 m s^{-1}. Before reaching the maximum, the relationship between the drift velocity and the electric field, $\mathbf{F}$, is linear:

$$\mathbf{v}_{dr} = \mu_e \mathbf{F} \tag{6.1}$$

where μ_e represents the *mobility*, which is 0.7 m^2 s^{-1} V^{-1} for intrinsic gallium

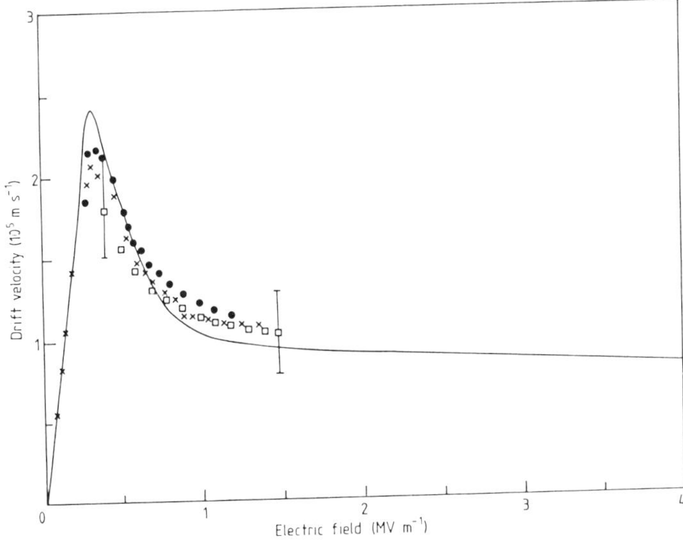

Figure 6.1 *The steady state drift velocity of electrons in bulk GaAs against the electric field, as calculated by means of the Monte Carlo technique. The result are in full agreement with experimental data of Houston and Evans (1977) and Ruch and Kino (1968) taken along the [111] (x) and the [100] axis (•) respectively*

arsenide. Multiplying Equation (6.1) by the electronic charge, e, and the density of particles, n_e, gives Ohms law:

$$n_e e\mathbf{v}_{dr} = \sigma_e \mathbf{F} \tag{6.2}$$

with

$$\sigma_e = e n_e \mu_e \tag{6.3}$$

representing the conductivity. For fields above 0.3 MV m^{-1}, the conductivity is no longer constant, the material becoming non-Ohmic for larger fields. Here μ_e and σ_e have been treated like scalars, though in more refined theory they should be tensors.

Figure 6.1 also shows the agreement between the calculated and the measured drift velocity. Within the uncertainty of the measurements there is agreement with Houston and Evans (1977) and Ruch and Kino (1968). The latter authors found a small velocity dependency on the crystallographic direction, though this lay well within the uncertainty of the data. Any further refinement of the model is without purpose because of the uncertainty of the experimental data.

The explanation for the reduction in the drift velocity lies in the steady state population of the various minima of the conduction band. (Figure 6.2.) In the absence of an electric field all carriers reside in the centre of the Brillouin zone. Applying an increasingly strong field, the L minima before populated. The L population culminates around 3 MV m^{-1}, then the higher lying X minima are

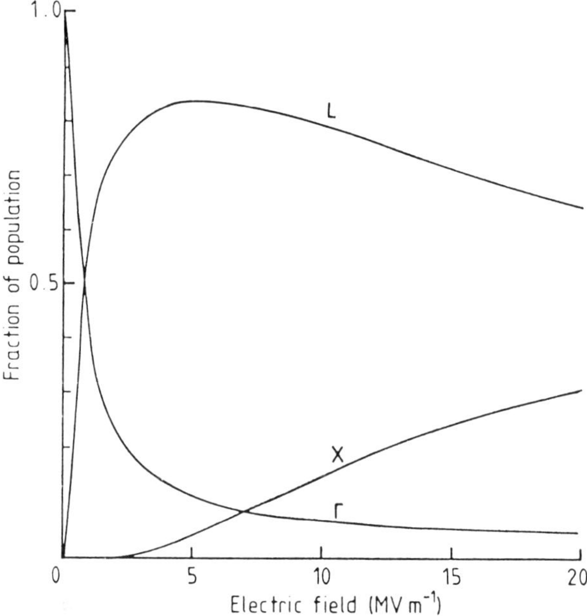

Figure 6.2 *Calculated steady state population ratio for the three sets of equivalent band minima in bulk GaAs against the electric field*

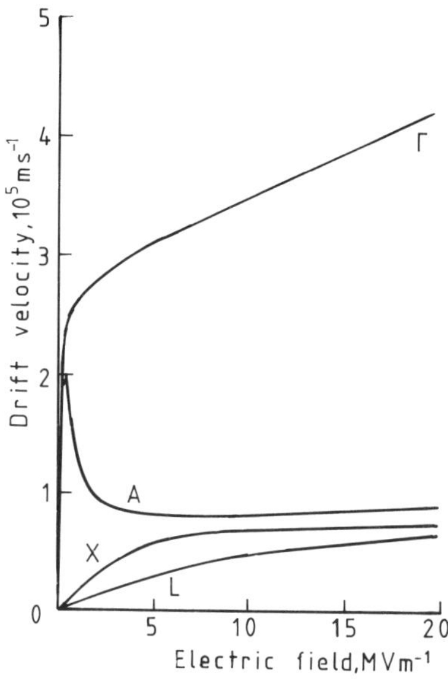

Figure 6.3 *Calculated average steady state drift velocity at room temperature against the electric field in GaAs for each set of minima (Γ, X and L) and for the entire population (A). The latter curve is a reproduction of that in Fig. 6.1*

filled largely because of the L minima. Figure 6.3 shows that the drift velocity increases steadily with the field within each conduction band minimum. This increase abates with a strengthening field and becomes linear above a threshold field which is different for the different types of minima. The slope of the velocity curve is determined by the amount of scattering. For the L minima there is more scattering from optical phonons and phonons causing transfer between equivalent minima than for any other valley. (Figure 6.4.)

The scattering depends only on the wave vector of the carrier, *not* on the field. The only effect of the field is to accelerate the particle, in other words, to furnish it with energy. In Chapter 4 we saw that most of the scattering rates increase with energy and only the strength of the scattering defines the population of the various extrema of the bands. (Private communication, H. Gray, Naval Research Laboratory, Washington DC.) This idea should be kept in mind in order to understand why devices where the electric field is, in places, strong enough to expect breakdown still survive. If the carriers do not spend sufficient time in these areas to attain sufficient energy from the field they cannot cause impact ionisation so the device will therefore not break down.

The cause of the fall in the average drift velocity lies in the greater effective mass of the carriers in the L and the X minima. (Table 6.1.) The net effect of increasing the fraction of the electrons in these minima is a reduction in the average drift velocity. As the reader understands from Figs. 6.3 and 6.2, hardly any electron moves at the average drift velocity for electric fields above $0.3\,\text{MV m}^{-1}$. The concept of the average drift velocity is almost analogous to the average score of casting a die, which is $3\frac{1}{2}$, a value which cannot be realised in any individual toss. The heavy, viz. the X and the L, electrons move slower, the Γ electrons considerably faster; even at $10\,\text{MV m}^{-1}$ a significant number of electrons reside in the Γ minimum during steady state transport. These electrons are the cause of the celebrated overshoot effect.

Conventional transport physics makes a distinction between thermal and hot carriers. The thermal ones are supposed to have a thermal viz. (displaced) Maxwellian distribution and drift with the local electric field at a speed given by Equation (6.1); the hot ones move much faster and therefore require special attention in the models based on average drift velocity of the carriers, e.g. drift diffusion and hydrodynamical approaches. This is the reason for making the distinction between hot and thermal carriers, though it is not necessary in Monte Carlo particle modelling which automatically takes care of all the particles. However when there are very few hot carriers, their effect on the transport and carrier distribution may be overlooked in the particle model, especially when we use large superparticles. Phillips and Price (1977) have carried out Monte Carlo simulation of the high energy tail of the carrier distribution. Hot electron effects can be significant in heterostructures (Hirakawa and Sakaki, 1988) and in silicon–silicon dioxide interfaces in MOSFETs (Ferry, 1976b, c). Poli *et al.* (1989) and Rota *et al.* (1989) have investigated weighted schemes to enhance the occurrence of hot electron effects.

The interaction with low-momentum acoustic and optical phonons represents the most important scattering rates for holes. Acoustic and non-polar optical phonons are responsible for transfer between light and heavy hole bands (Wiley and di Domenico, 1970). The interaction with the split-off band is weaker than between the other two bands. Many authors, e.g. Wiley and di Domenico (1970) and Lee and Look (1983), have neglected the split-off holes in their analyses;

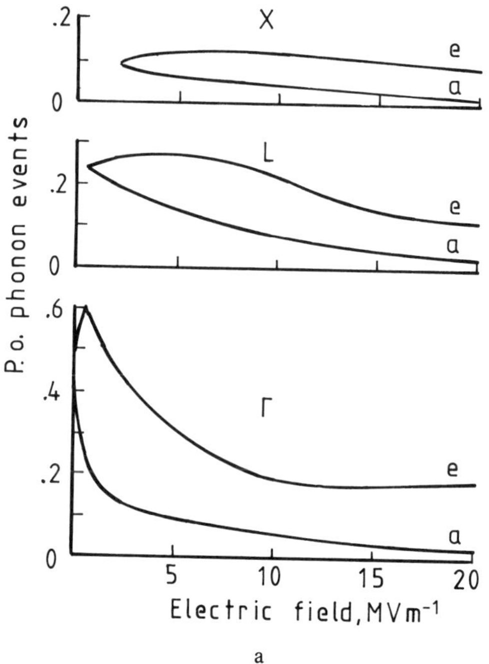
X
e
a
L
e
a
Γ
e
a
P. o. phonon events
.2
0
.2
0
.6
.4
.2
0
5
10
15
20
Electric field, MVm⁻¹

a

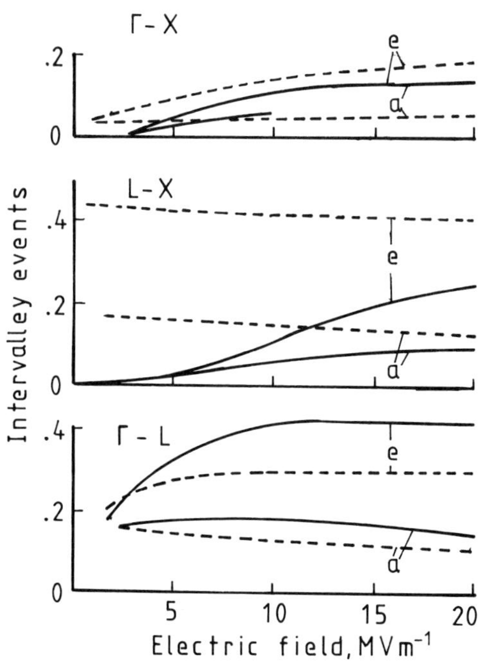
Γ-X
e
a
L-X
e
a
Γ-L
e
a
Intervalley events
.2
0
.4
.2
0
.4
.2
0
5
10
15
20
Electric field, MVm⁻¹

b

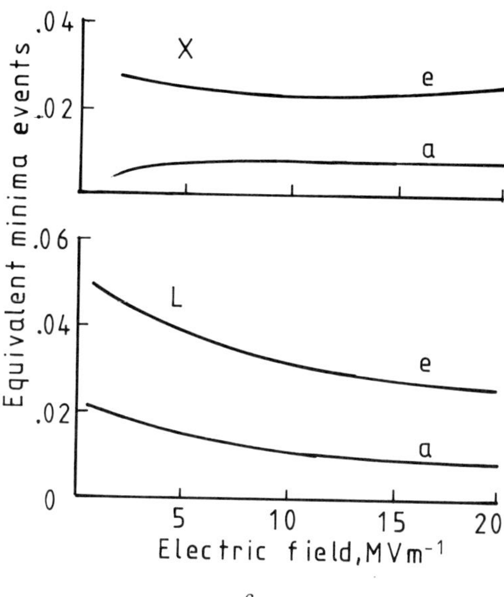

c

Figure 6.4 *Scattering events during steady state in the conduction band of GaAs at room temperature.*

a *Fraction of all scattering events caused by polar optical phonons against the electric field for each separate minimum.*
e = phonon emission
a = absorption

b *Fraction of all scattering events during steady state causing transfer between conduction band minima in GaAs against the electric field.*
—— *transfer* $\Gamma \rightarrow X$, $L \rightarrow X$ *and* $\Gamma \rightarrow L$
--- *transfer in the opposite direction, e.g.* $X \rightarrow \Gamma$, $X \rightarrow L$ *and* $L \rightarrow \Gamma$
e = phonon emission
a = absorption

c *Fraction of scattering events during steady state causing transfer within the equivalent X and L minima in GaAs.*
e = phonon emission
a = absorption

Brennan and Hess (1984), however, did include split-off holes in their work. Neglecting split-off holes is not recommended, especially at higher electric fields for which the contribution from the split-off holes becomes noticeable. It is fairly easy to include it in the model and split-off hole transport should be permitted even if such events are rare. It is a good principle to generally include certain effects in the model which are weak except under certain conditions, e.g. the communication with the split-off band. Another effect which is normally weak is the impact ionisation, which plays a significant role only when there are many energetic carriers. The inclusion of the possibility that it may take place simplifies the housekeeping because we need not introduce any criteria for when it should be considered. The model should automatically take care that rare events do not happen more often than they should through the expressions for the occurrence rate.

Figure 6.5 shows that the drift velocity of holes in GaAs increases steadily with

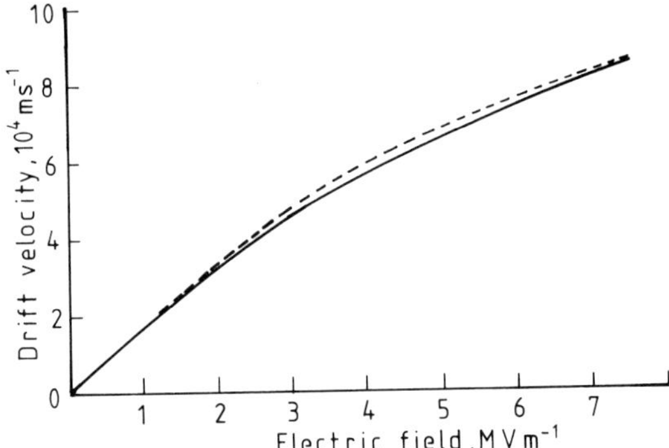

Figure 6.5 *The steady state drift velocity of holes in bulk GaAs at room temperature against the electric field. Full curve: calculated by means of Monte Carlo simulation by the author. Dashed curve: experimental data from Dalal (1970)*

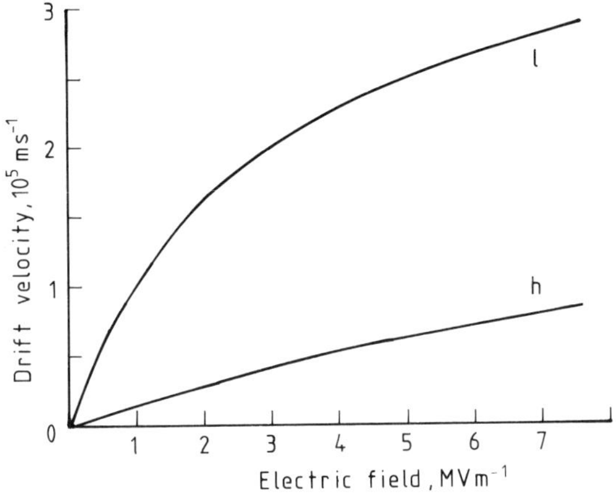

Figure 6.6 *Calculated average steady state drift velocity for light and heavy holes in bulk GaAs at room temperature during steady state against the electric field*

the electric field strength, although the increase shows signs of abating. The calculated drift velocity is in excellent agreement with the experimental results published by Dalal (1970). Figure 6.6 shows the drift velocity of the light and heavy holes separately. The curve for the heavy holes follows that of the average curve in Fig. 6.5, but lies slightly below it. Figure 6.7 shows that only between 2–3% of the entire hole population consists of light holes – the heavy holes are in overwhelming majority, and therefore dominate the transport almost entirely. Similar results will also be obtained for silicon. The split-off band starts to get populated above 5 MV m^{-1}. Even at the highest field for which the simulation has been carried out, 10 MV m^{-1}, the split-off hole population does not exceed that of the heavy hole band. The relatively insensitive reaction of the distribution of the holes over the three bands of the field explains the steady increase of the

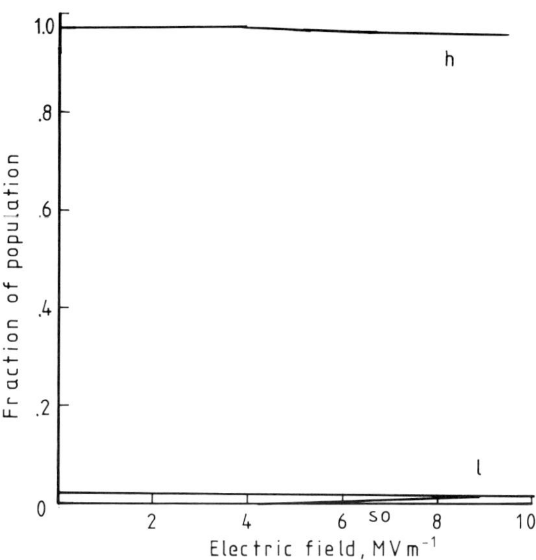

Figure 6.7 *Calculated population ratio for the three valence bands of GaAs at room temperature during steady state against the electric field. l: light holes. h: heavy holes. s.o.: split-off holes*

drift velocity. Some authors have therefore been tempted to treat only the heavy hole band. In spite of the small amount of light holes such an approach is physically unsound. The light and heavy holes should be considered separately because the scattering rates are quantitatively different for the two species of carriers. Figure 6.8 shows the relative occurrence of the various scattering events against the field. The relative weight of these events is responsible for the almost constant population of the various valence bands. The amount of events involving the split-off band has been omitted from the figure due to their relative infrequency.

The six X minima have the lowest energy of the conduction band of silicon. The band structure is anisotropic, Equation (3.65), so that we shall prefer the anisotropic model. The X minima are situated at the points

$$\mathbf{k}_\alpha^z = \begin{pmatrix} 0 \\ 0 \\ \xi\, k_{Br} \end{pmatrix}, \ \mathbf{k}_\alpha^y = \begin{pmatrix} 0 \\ \xi k_{Br} \\ 0 \end{pmatrix}, \ \text{and } \mathbf{k}_\alpha^x = \begin{pmatrix} \xi\, k_{Br} \\ 0 \\ 0 \end{pmatrix} \tag{6.4}$$

and at the three additional points $-\mathbf{k}_\alpha^z$, $-\mathbf{k}_\alpha^y$ and $-\mathbf{k}_\alpha^x$ where k_{Br} represents the distance from the centre of the Brillouin to the (100) face and the value of ξ is as given in Table 6.2. In particular, the band structure in the vicinity of the minimum at $\mathbf{k}_\alpha^i$ is:

$$E(1 + \alpha E) = \tfrac{1}{2}\hbar^2(\mathbf{k} - \mathbf{k}_\alpha^i) \bullet \mathbf{m}^{-1} \bullet (\mathbf{k} - \mathbf{k}_\alpha^i) \tag{6.5}$$

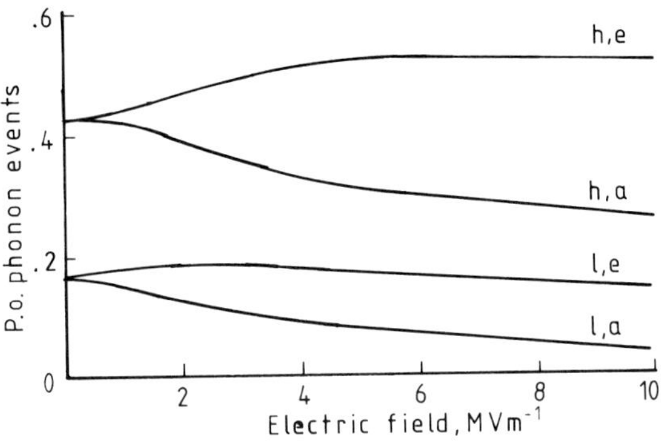
P.o. phonon events
.6
.4
.2
0
h,e
h,a
l,e
l,a
2
4
6
8
10
Electric field, MVm⁻¹

a

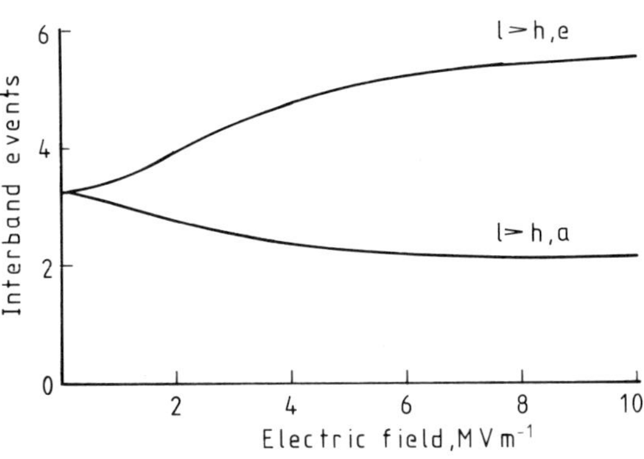
Interband events
6
4
2
0
l>h,e
l>h,a
2
4
6
8
10
Electric field, MVm⁻¹

b i

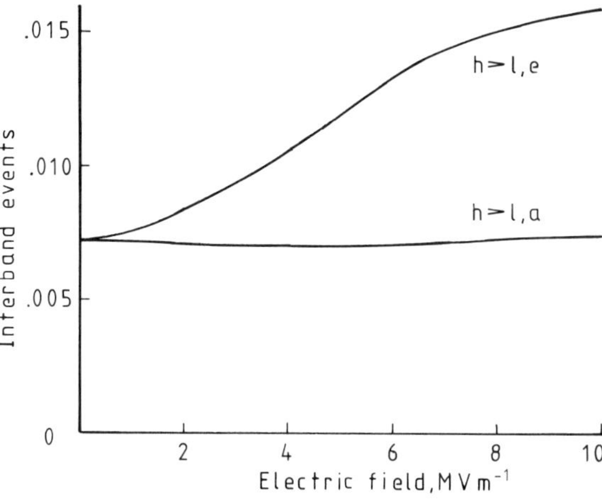
Interband events
.015
.010
.005
0
h>l,e
h>l,a
2
4
6
8
10
Electric field, MVm⁻¹

b ii

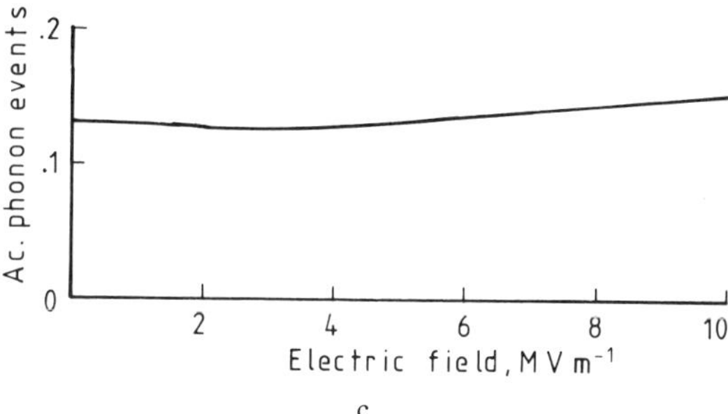

c

Figure 6.8 *Scattering events during steady state in the valence bands of GaAs at room temperature:*
a *Fraction of all scattering events caused by polar optical phonons against the electric field in the (l) light hole and (h) the heavy hole band. e = phonon emission; a = absorption*
b *i.ii Fraction of all scattering events causing transfer between the light and heavy hole bands. e = phonon emission; a = absorption*
c *Fraction of all scattering events in the heavy hole band caused by acoustic phonons*

with k_α^i representing either of the vectors given by Equation (6.4) or the negative of it and

$$\mathbf{m}^{-1} = \begin{pmatrix} m_t^{-1} & 0 & 0 \\ 0 & m_t^{-1} & 0 \\ 0 & 0 & m_l^{-1} \end{pmatrix}, \begin{pmatrix} m_t^{-1} & 0 & 0 \\ 0 & m_l^{-1} & 0 \\ 0 & 0 & m_t^{-1} \end{pmatrix}$$

and

$$\begin{pmatrix} m_l^{-1} & 0 & 0 \\ 0 & m_t^{-1} & 0 \\ 0 & 0 & m_t^{-1}. \end{pmatrix} \tag{6.6}$$

for $\mathbf{k} = \pm\, \mathbf{k}_\alpha^z$, $\mathbf{k} = \pm\, \mathbf{k}_\alpha^y$ and $\mathbf{k} = \pm\, \mathbf{k}_\alpha^x$ respectively. Here m_t and m_l represent the transversal and the longitudinal effective mass of the electrons respectively. Equation (6.5) describes families of constant energy ellipsoids in wave vector space with main axes along either of the Cartesian coordinate axes and the $\langle 111 \rangle$ directions. Figure 6.9 shows the distribution of the carriers over the different ellipsoids in the Brillouin zone for GaAs at an ambient electric field of 5 MV m^{-1}.

The eight L minima of silicon lie 0.93 eV above the X ones (Chelikowsky and Cohen, 1974). These can only be reached by the very hottest or most energetic electrons. No one has ever included them in their model because the necessary physical data about them has been unavailable and interaction with them is unlikely. We shall therefore disregard them in our present discussion. The Γ minima lie far beyond reach by transport in the local electric or magnetic field. In our anisotropic model of the band structure it is necessary to keep track of which of the six X minima the electron resides in. We find that they are not equally

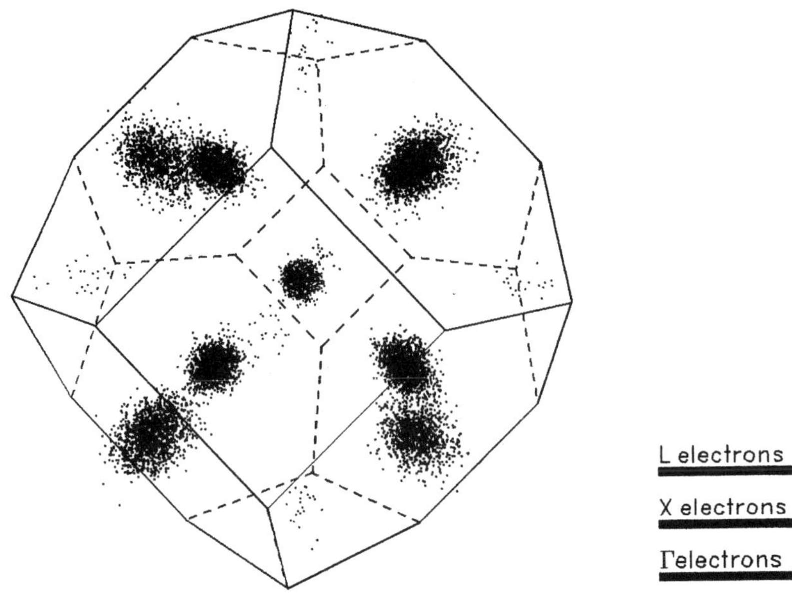

Figure 6.9 *View of the electron distribution in k-space during steady state at room temperature or in the Brillouin zone for gallium arsenide at an ambient electric field of 5 MV m^{-1}. Cyan: Γ electrons. Blue: L electrons. Green: X electrons. Note, this figure is reproduced in colour as Plate 2*

populated under a strong ambient electric field (Stark effect). This is also true for uniaxially strained crystals, but in this case because the six-fold degeneracy of the energy band minima is lifted.

The phonon spectrum of silicon has been determined by Brockhouse (1959). There is some inconsistency in the literature about which phonons participate in electron scattering. From group theoretical considerations Asche and Sarbei (1981) state that zeroth order intravalley scattering is forbidden. Jørgensen (1978) says that zeroth order optical intravalley phonon scattering is not possible. He included quasielastic acoustic phonon scattering in his model. He also stated that the only possible zeroth order intervalley scattering processes involve transversal optical and longitudinal acoustic *f* phonons, and longitudinal optical *g* phonons.

Canali *et al.* (1975) found that the rate of scattering from the three other intervalley phonons *f*(TA), *f*(LO) and *g*(TA) were an order of magnitude lower than the three phonons *f*(LA), *f*(LT) and *g*(TA), and therefore did not contribute significantly to the transport. Higher order processes and multiphonon scattering likewise contribute little to the calculated drift velocity or electrical conductivity.

Figure 6.10 shows that the calculated drift velocity versus the electric field rises steadily and saturates around 5 MV m^{-1}. The drift velocity is about the same for ⟨110⟩ oriented and ⟨100⟩oriented material, but is up to 10% higher in the ⟨111⟩ direction. At the highest fields the crystal orientation with respect to the direction of the electrostatic field is immaterial. Our calculations agree well with the

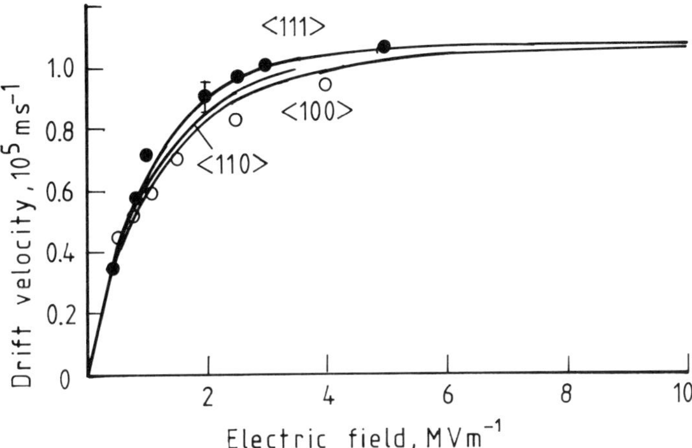

Figure 6.10 *Calculated room temperature average steady state drift velocity at room temperature against the electric field for the ⟨100⟩ and ⟨11⟩ directions in crystalline silicon. The points refer to experimental data of Canali* et al. *(1975).* ●*: ⟨111⟩.* ○*: ⟨100⟩. The bar represents the experimental error in the velocity*

experimental data of Canali *et al.* (1975). Ottaviani *et al.* (1975) have measured the drift velocity up to 5 MV m^{-1} for temperatures up to that of the room and confirmed that it depends on the crystal orientation. Reggiani *et al.* (1981) calculated it from an iterative solution of Boltzman's transport equation.

When the electric field is directed along the ⟨100⟩ crystal axis, the occupation of the four ellipsoids with major axis perpendicular to the field is lower than that of the two with major axis along it. (Figure 6.11.) At zero field the occupancy of all six X minima is equal; at higher field the anisotropy of the carrier distribution over them tends to reduce due to the increased scattering between the X minima.

Polycrystalline silicon consists of tiny crystallites of random orientation. To

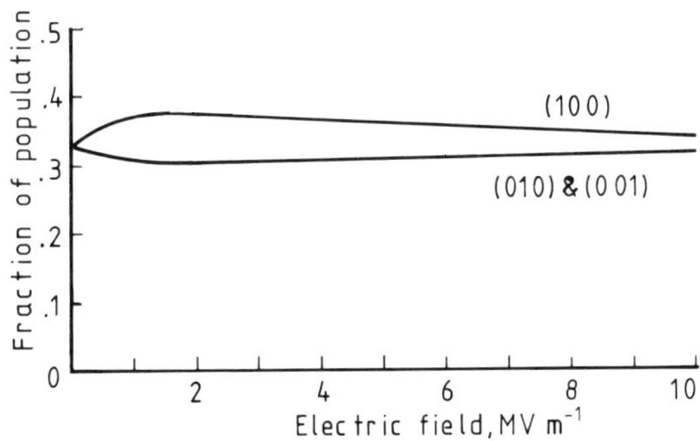

Figure 6.11 *Occupation of conduction band minima, or ellipsoids in silicon during steady state at room temperature against the electric field oriented in the ⟨100⟩ direction. The two ellipsoids along the ⟨100⟩ direction have their major axis along the field; those with the major axis along the ⟨010⟩ and ⟨001⟩ directions are oriented perpendicular to it*

simulate the transport in such a material we could use the anisotropic band structure. The orientation of the crystallite through which the electron travels should be selected by means of random numbers; a corresponding coordinate transformation into a local coordinate system where the band structure is given by Equation (6.5) should also be carried out. Alternatively the material could be treated as isotropic with the density of states effective mass representing the electronic mass. *Amorphous material*, however, exhibits a short range crystalline order: the correlation between the structure at two different places vanishes gradually with the distance between them. The best model for amorphous semiconductors is probably the isotropic one.

6.3 THE CARRIER DISTRIBUTION

The carrier distribution can be obtained by counting the particles against energy at a given time. We have, of course, to follow a large number of particles to gain sufficient statistics. Both the transient development of the distribution and the steady state one can be obtained in this way. The latter can also be calculated by sampling a smaller population of particles several times.

Figure 6.12 shows the particle distribution in silicon calculated in this manner (•). The *Maxwell–Boltzmann distribution* for $T_f = 300\,\mathrm{K}$,

$$N(E) = cE \exp\left(-\frac{E}{k_B T_f}\right), \tag{6.7}$$

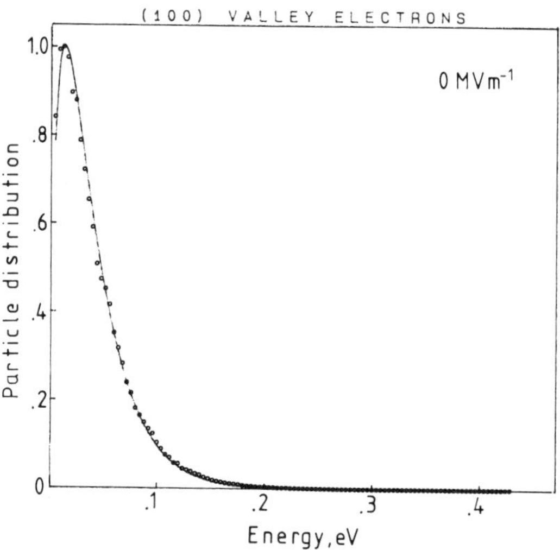

Figure 6.12 *Calculated number of particles against energy during steady state at room temperature in silicon for zero field (•). The number of particles has been recorded by counting the particles in 4 meV wide energy intervals. The full line represents the room temperature distribution calculated from Equation (6.7)*

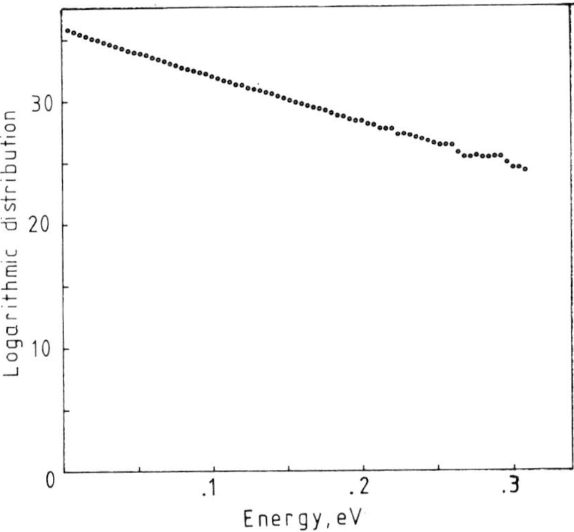

Figure 6.13 *Plot of log [N(E)/E] against the energy in Si at room temperature of the distribution obtained from the Monte Carlo simulated distribution of Fig. 6.12*

has also been traced as a full curve for comparison. Here E represents the energy of the particle, k_B Boltzmann's constant and c a normalisation constant, the value for which is not essential to give here. Figure 6.13 shows that log $[N(E)/E]$ for the same distribution is a straight line with a slope corresponding to $T_f = 300$. This proves that the electrons are Maxwell–Boltzmann distributed in the absence of an electric field.

At higher fields, at least up to $10\,\mathrm{MV\,m^{-1}}$, log $[N(E)/E]$ still proves to be a straight line, but with an electron temperature as shown in Fig. 6.14.

Considering the velocity distribution in one particular ellipsoid reveals a few interesting properties: At zero field, Fig. 6.15a, the velocity distribution along the length of the ellipsoid is narrower than that in the other two directions orthogonal to it. The velocity distribution in the longitudinal direction has to be narrower

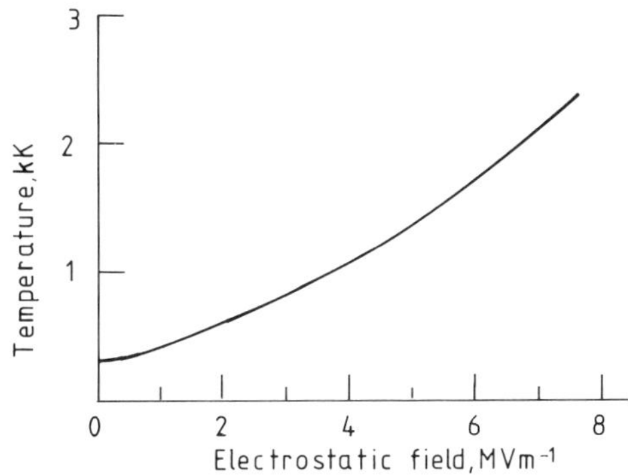

Figure 6.14 *The steady state electron temperature against the electric field for silicon*

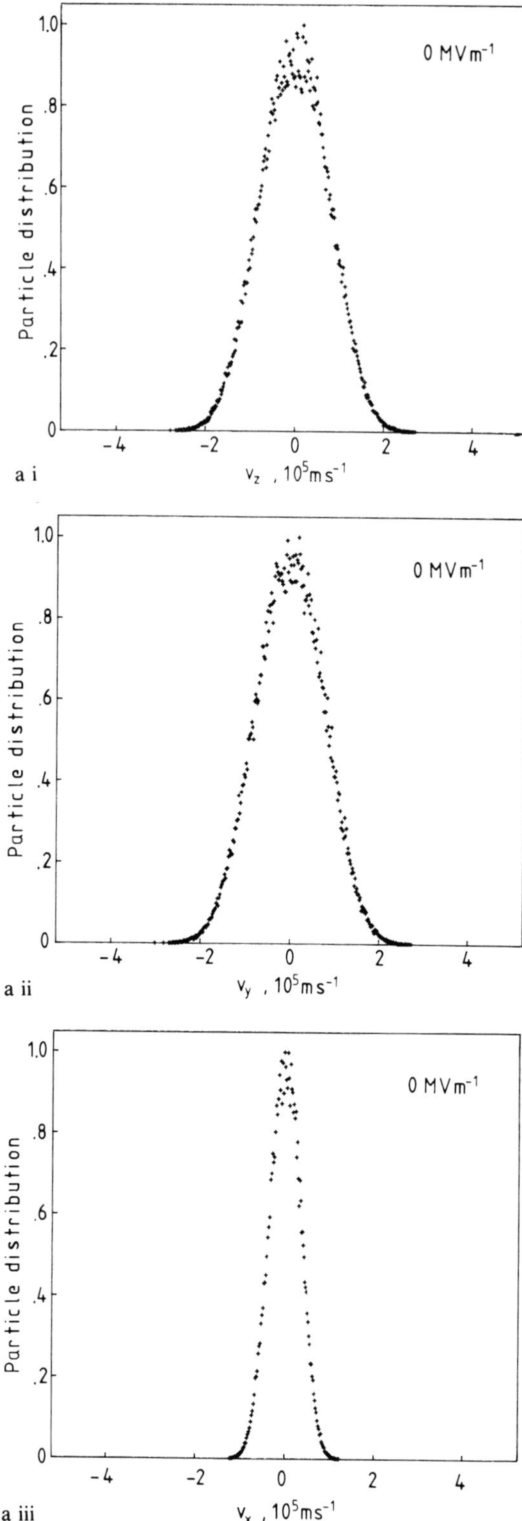

Figure 6.15 *Steady state transversal, v_y and v_z, and longitudinal, v_x, velocity distribution at room temperature in silicon at: a i,ii,iii Zero field; b i,ii,iii 1 MV m^{-1}*

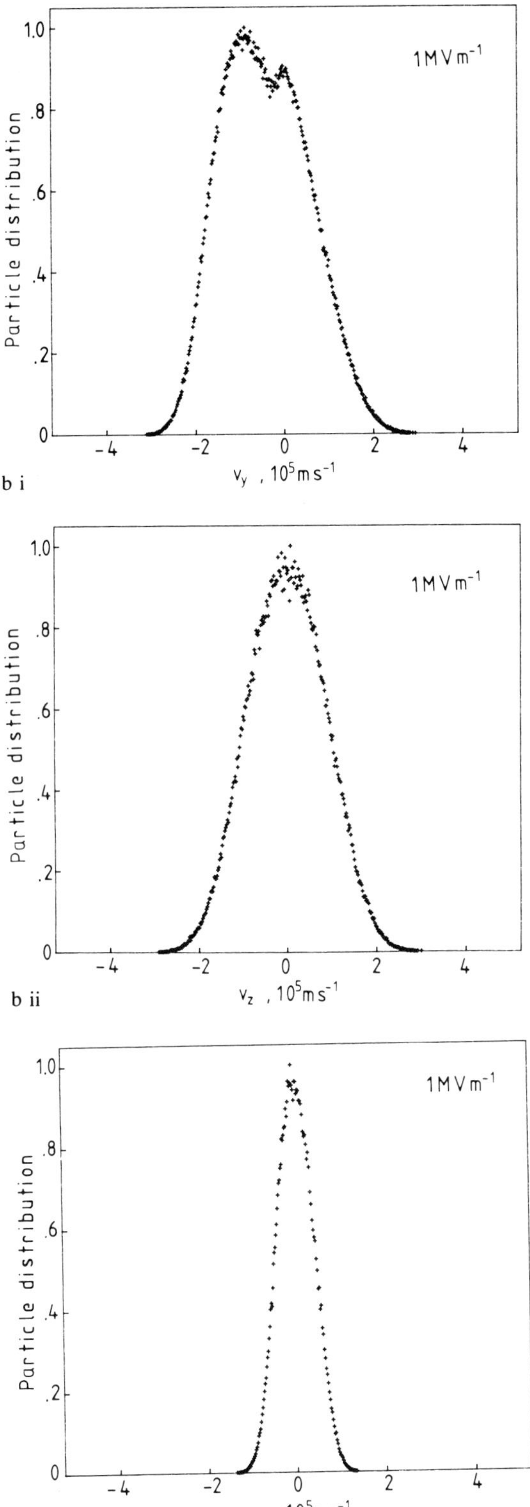

Figure 6.15 *contd.*

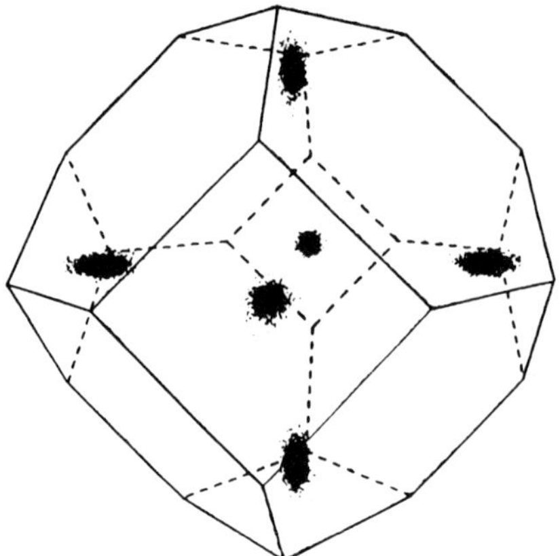

Figure 6.16 *View of the steady state electron distribution in k-space or in the Brillouin zone for silicon at zero field at room temperature*

because the effective mass is larger in this direction. Figure 6.16 represents a view of the electrons in k-space at zero field. The population shown here consists only of a small fraction of the simulated one, but the elongated ellipsoids are clearly visible.

Imposing an electric field along one of the major axes of an ellipsoid results in a slight broadening of the velocity distribution in the other two directions. The velocity distribution along the field splits in two maxima, the higher one oriented along it. The second maximum at zero velocity is due to electrons arriving by scattering from the other four ellipsoids oriented at right angles to it. (Figure 6.15b.)

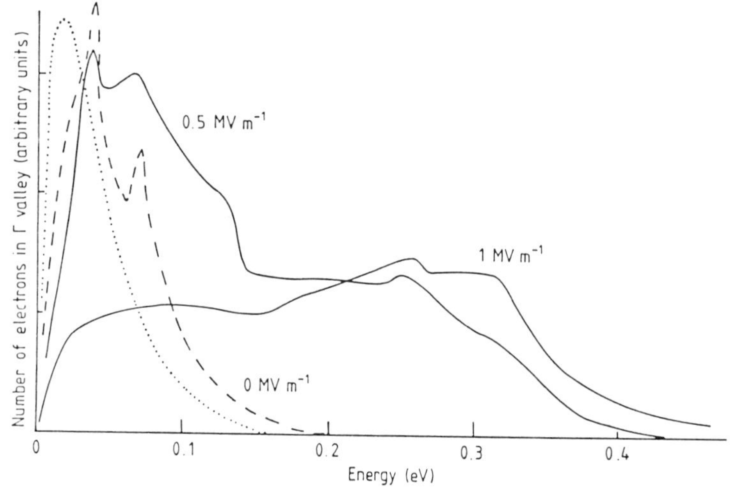

Figure 6.17 *Number of particles in GaAs against energy for different uniform electric fields at room temperature. The parameter indicates the fields in* $MV\,m^{-1}$

The energy distribution in gallium arsenide is not Maxwellian even at zero electric field. (Figure 6.17.) It peaks at 35 meV, which agrees well with the expected value of $3/2\ k_B T_f = 39$ meV at $T_{Lt} = 300$ K. The second distinct peak 30 meV higher in energy is due to the absorption of optical phonons at long wavelengths. The shoulder to the left of the main peak is due to the emission of energetic phonons. These shapes have been observed experimentally by Takenaka *et al.* (1979, 1980), Inoue *et al.* (1978) and predicted theoretically by Maksym (1982) and Požela and Reklaitis (1980).

In the presence of an electric field, carriers are transferred to and from the higher energy minima of the conduction band. (Figure 6.2.) The optical phonon peak weakens but additional peaks start to appear in the distribution of Γ electrons due to exchange with the population of the various secondary energy minima of the conduction band. (Figure 6.17.)

6.4 ENERGY AND MOMENTUM RELAXATION

The electron gas responds to a change in the electric field by redistributing itself. This takes time; in macroscopic theory it is expressed through the momentum and energy relaxation times, which are taken to be constants. Monte Carlo modelling, however, reveals that the energy relaxation time depends on the initial and final strengths of the field (Constant, 1985) and on the time the field takes to change. It is therefore necessary to make a clear definition of the relaxation times. The relaxation times are macroscopic quantities that can be extracted from a simulation; they should never be used as input parameters for the particle model. Figures 6.18 and 6.19 show that a distinction has to be made between momentum and energy relaxation. The *momentum relaxation time* can be defined as (Constant, 1985)

$$\tau_p = v_s m^*/(eF) \tag{6.8}$$

and the *energy relaxation time* by

$$\tau_E = \left| \frac{E_f - E_i}{eFv_s} \right| \tag{6.9}$$

where v_s represents the saturation velocity, F the electric field strength, m^* the average effective mass of the carriers and e the elementary electronic charge. E_f and E_i represent the final and initial average energy respectively of the particles. The meaning of these expressions is vague because it is unclear whether v_s, F and m^* refer to the final or the initial value of the field – we may take them to represent the average of the two.

Figure 6.18 shows that the velocity relaxation in silicon is insensitive to whether the field decreases or increases, but does depend on the initial or final field. Increasing the field, the drift velocity increases too much initially, before relaxing to its steady state value. The reason for this is that the field first accelerates the particles; these faster particles are then subjected to increased scattering resulting in a net loss of energy to the lattice. The effect of the latter is to over-relax the velocity so that the drift velocity again increases by acceleration in the field. The

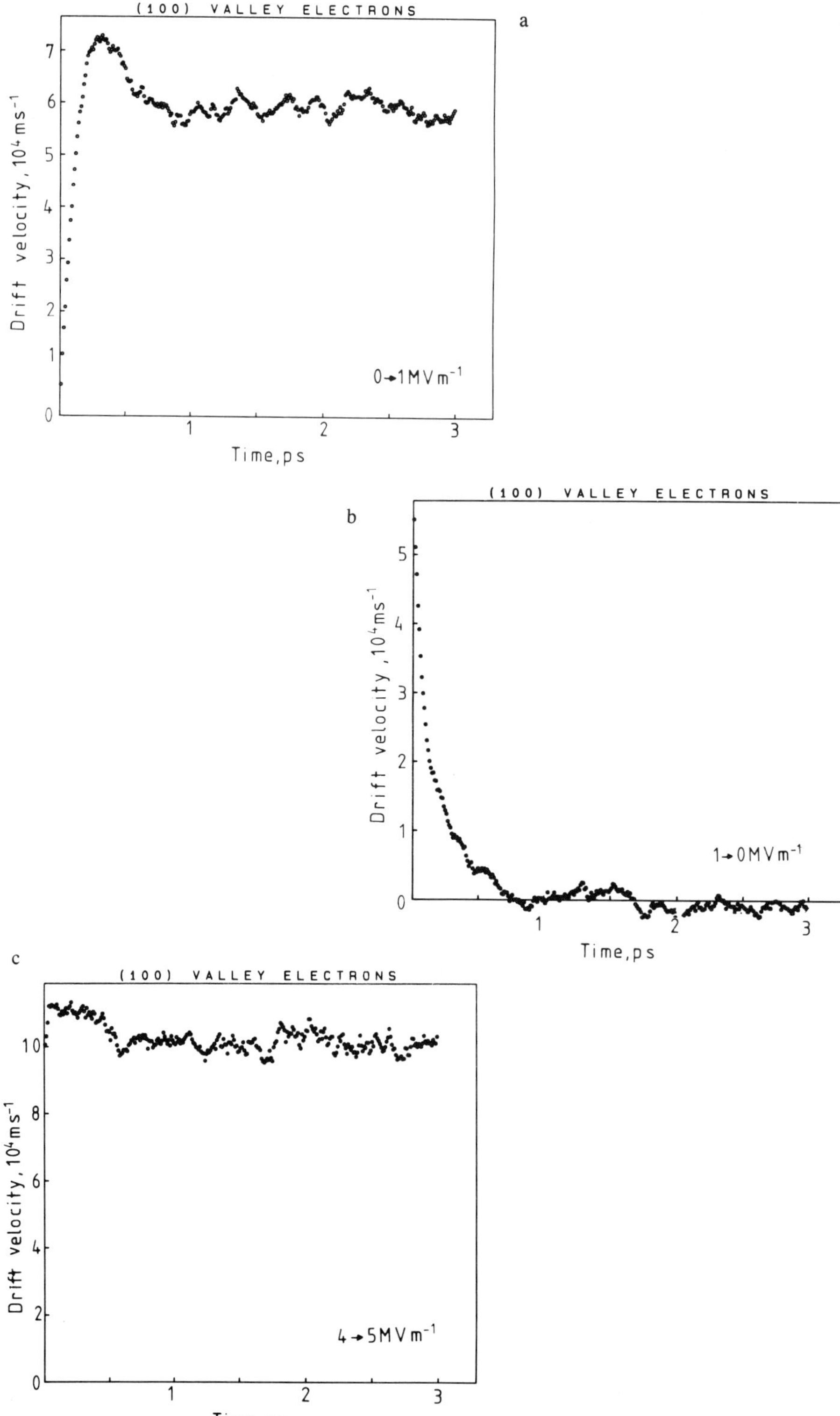

Figure 6.18 *The transient response of the drift velocity of the electrons in silicon to a 1 MV m^{-1} change in the ambient electric field: a, 0 to 1 MV m^{-1}; b, 1 to 0 MV m^{-1}; c, 4 to 5 MV m^{-1}*

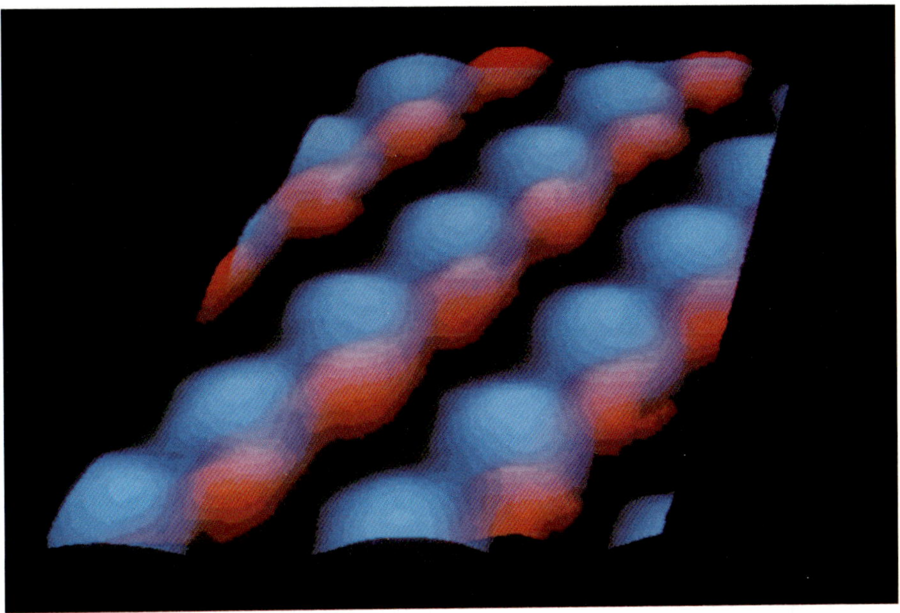

Plate 1. *Combined colour scanning tunnelling microscopic image of the GaAs (110) surface, showing the gallium atoms in blue and the arsenic atoms in red. The blue image was acquired by tunnelling into empty states which are localised around the gallium atoms, and the red one by tunnelling out of filled states localised around the surface arsenic atoms. Reproduced with permission of Feenstra* et al. *(1987). Note, this plate is reproduced as Figure 2.3*

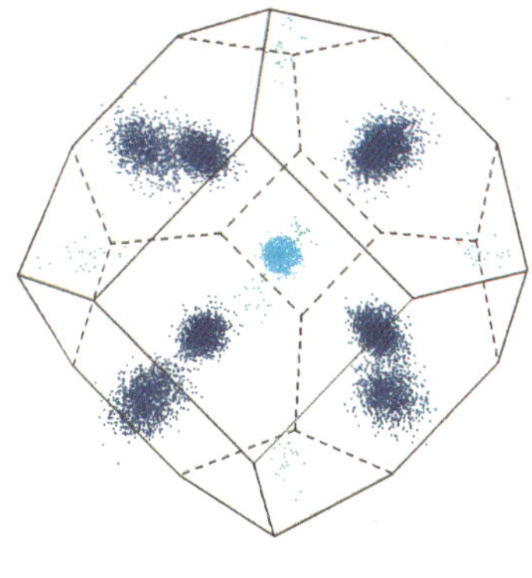

Plate 2. *View of the electron distribution in k-space during steady state at room temperature or in the Brillouin zone for gallium arsenide at an ambient electric field of 5 $MV\,m^{-1}$. Cyan: Γ electrons. Blue: L electrons. Green: X electrons. Note, this plate is reproduced as Figure 6.9*

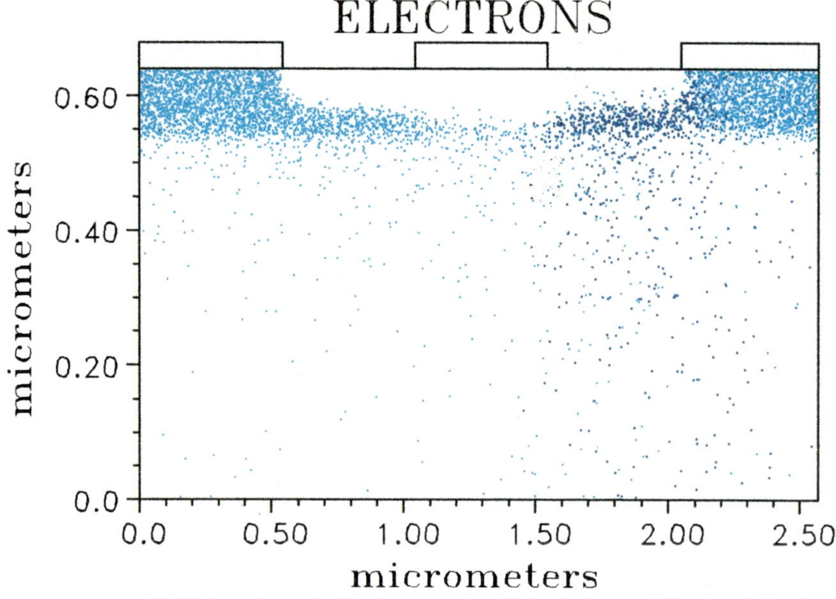

Plate 3. *Instantaneous distribution of electrons in the example transistor during steady state. Light blue points represent electrons in the Γ minimum of the Brillouin zone, dark blue L electrons. Hardly any X electrons (green) are present. Note the different linear scale along and perpendicular to the surface of the transistor. Drain bias: 2 V. Gate potential: −1.0 V (including the Schottky contact potential of −0.75 V). Field-adjusting time step: 25 fs. Note, this plate is reproduced as Figure 8.11*

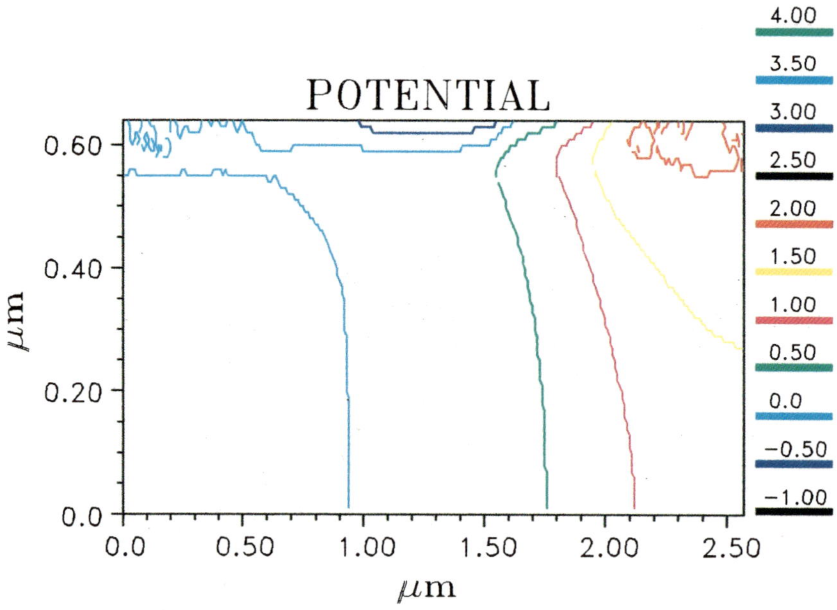

Plate 4. *Instantaneous equipotential lines for the same bias conditions as in Fig. 8.11. The geometrical scale is the same as in Fig. 8.11. Note, this plate is reproduced as Figure 8.13*

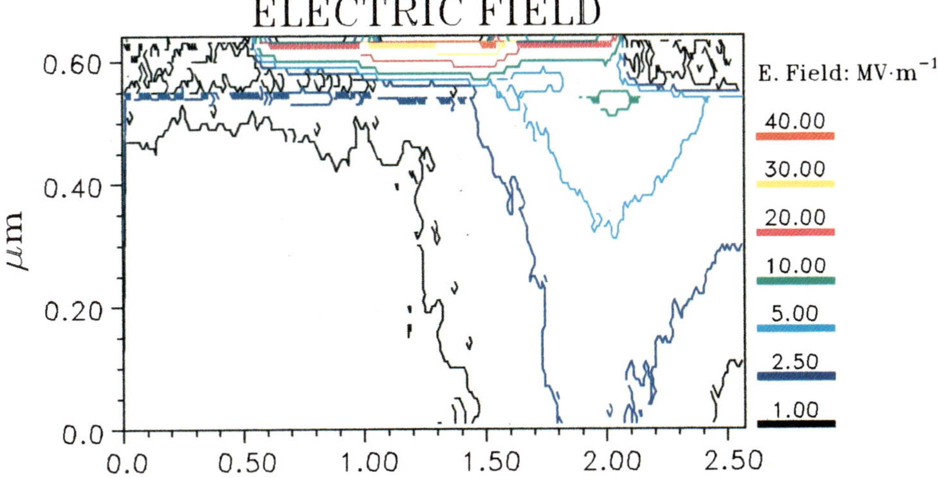

Plate 5. *Magnitude of electric field throughout the example transistor of the same bias as in Fig. 8.11. The units may be unfamiliar to the reader: 1 MV m^{-1} = 10 kV cm^{-1}. The geometrical scale is the same as in Fig. 8.11. Note, this plate is reproduced as Figure 8.14*

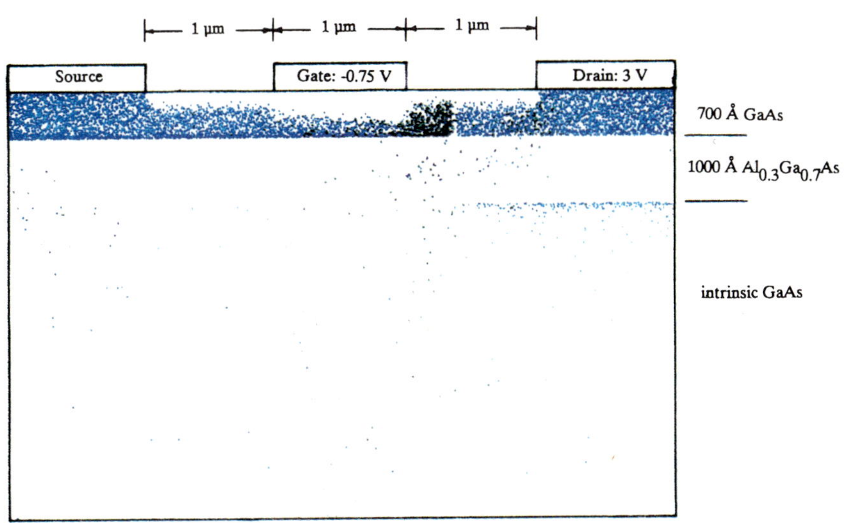

Plate 6. *Instantaneous electron distribution in the Gunn oscillating transistor of Fig. 8.17. Drain bias: 3 V, gate potential: −0.75, including the Schottky contact potential. The horizontal lines outside the left and right edge indicate the interface between the GaAs and the $Al_{0.3}Ga_{0.7}As$. Note the different linear vertical and horizontal geometrical scales. Note, this plate is reproduced as Figure 8.18*

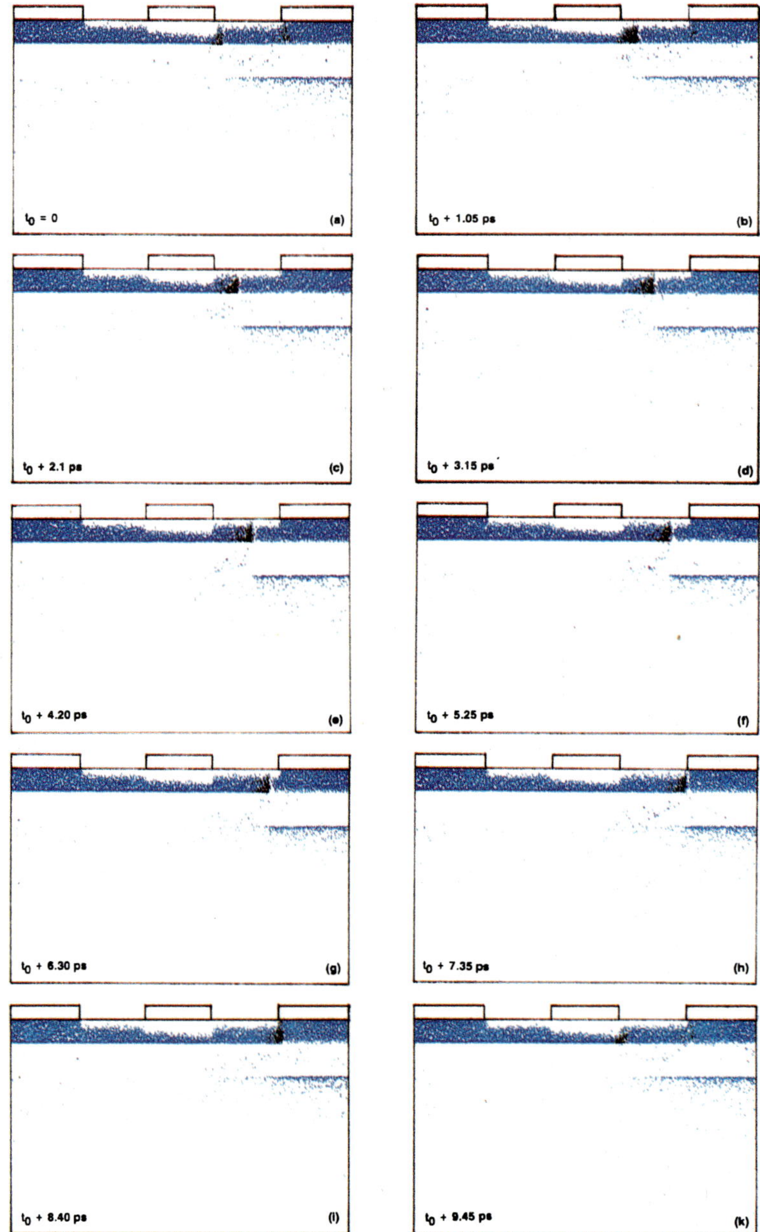

Plate 7. *The time evolution of the Gunn domain throughout the entire cycle. The domain moves towards the drain at the saturation drift velocity,* 10^5 *m s*$^{-1}$*. Light blue:* Γ *electrons, dark blue: L electrons. The instant shown as Fig. 8.16 corresponds to the time* $t = t_0 + 2.5$ *ps. Note, this plate is reproduced as Figure 8.19*

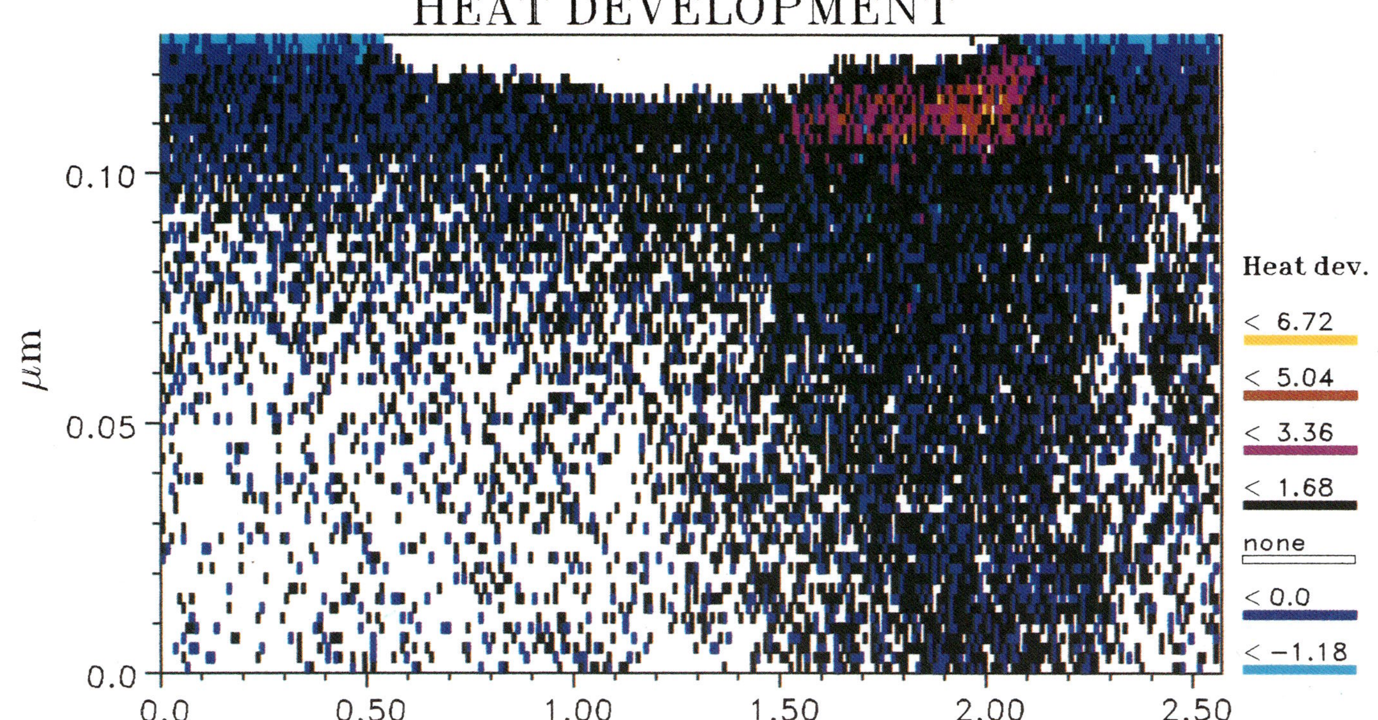

Plate 8. *The heating rate in the transistor of Fig. 8.4. The colour code gives ranges of heating rates in units of* 10^{15} *W* m^{-3}*, the blue hues represent cooling, the others heating. The strong cooling adjacent to the source and the drain is a result of injecting too-cold electrons. Drain bias: 2 V. Gate potential: −1 V, including the Schottky contact potential of −0.75 V. Note, this plate is reproduced as Figure 8.24*

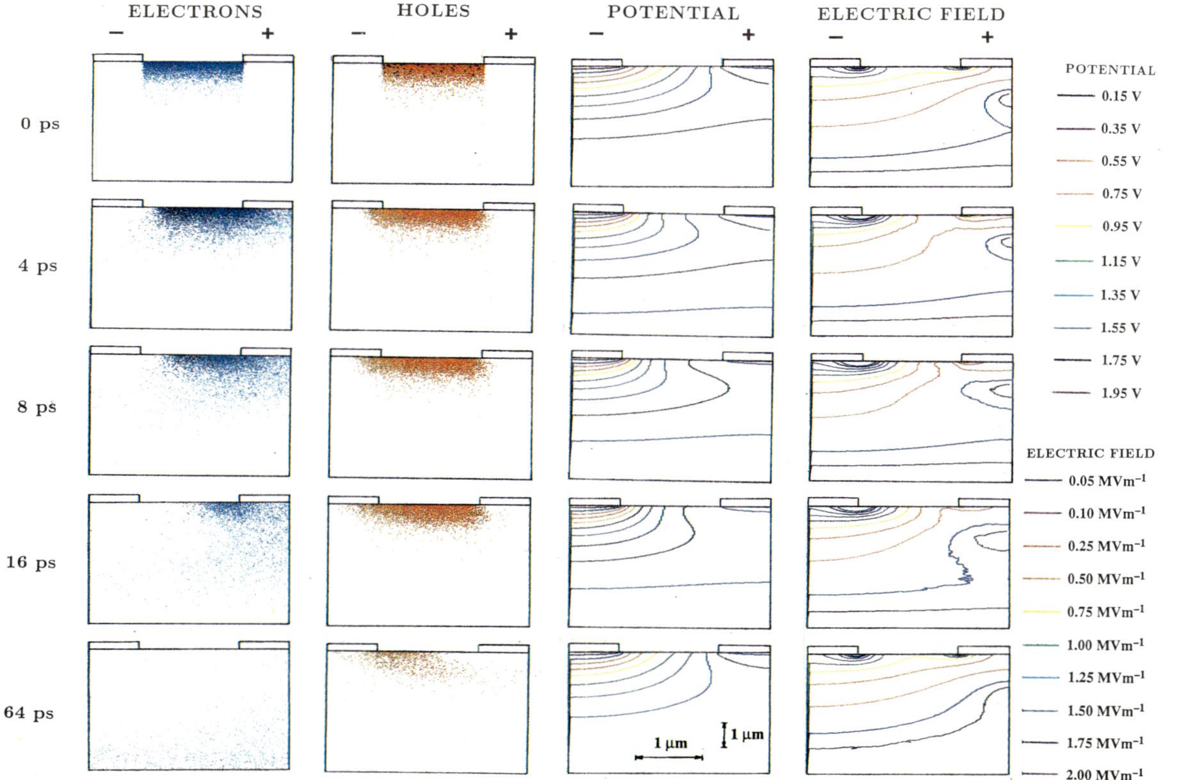

Plate 9. *Calculated distributions of electrons, holes, electric potential and magnitude of electric field at 0, 4, 8, 16 and 64 ps after the 70 fs light flash. The corresponding electron–hole pair density at the surface is* $I_0 = 5.1 \times 10^{21}\ m^{-3}$ *at time 0 (at light intensity of* $5.48 \times 10^9\ W^{-2}$*). The distance between the anode and the cathode is 1.5 μm. Anode bias: 2 V. The linear scale in the two orthogonal directions is indicated in the bottom panel. Note, this plate is reproduced as Figure 11.10*

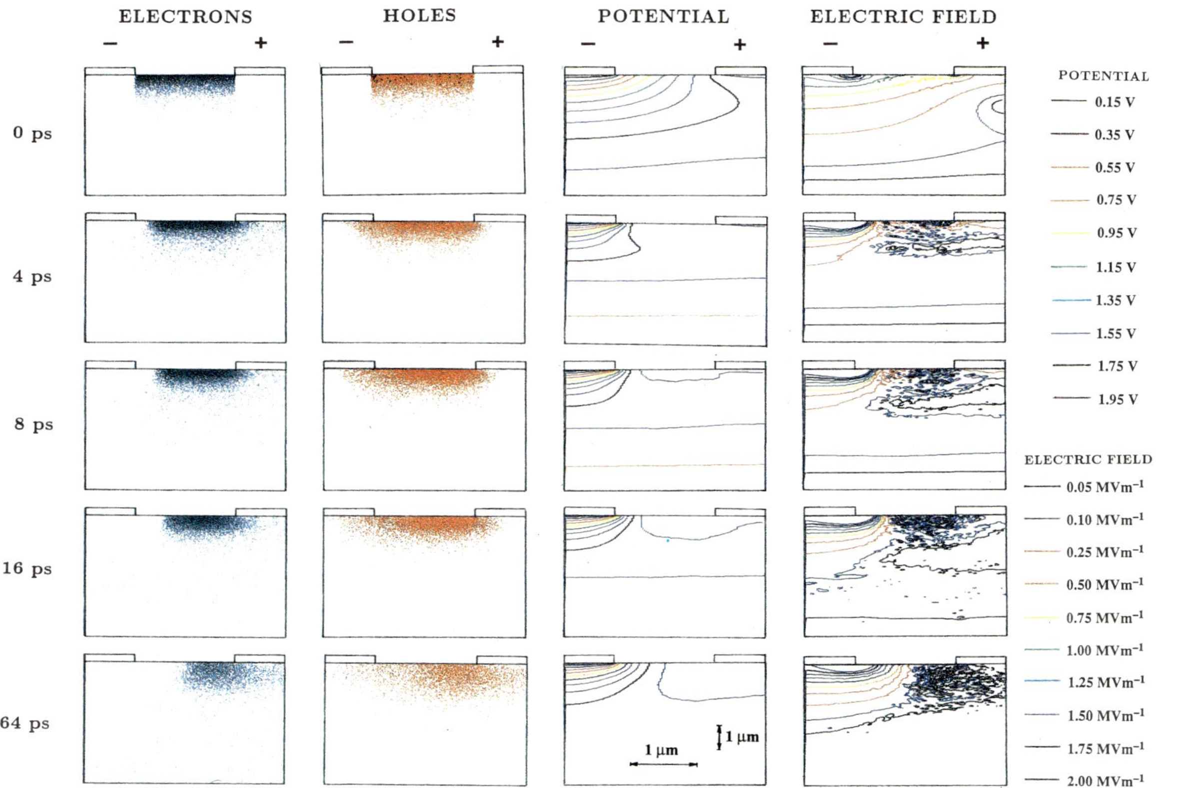

Plate 10. *Distribution of electrons, holes, potential and magnitude of electric field for the same diode as in Fig. 11.10. Anode bias: 2 V. Light intensity: 5.48×10^{10} W m^{-2}, ten times that of Fig. 11.10 or 11.9. Note, this plate is reproduced as Figure 11.12*

Plate 11. *The entrance to the casino in Monte Carlo (Photo: J. Rosenzweig). Note, this plate is reproduced as Figure 13.1*

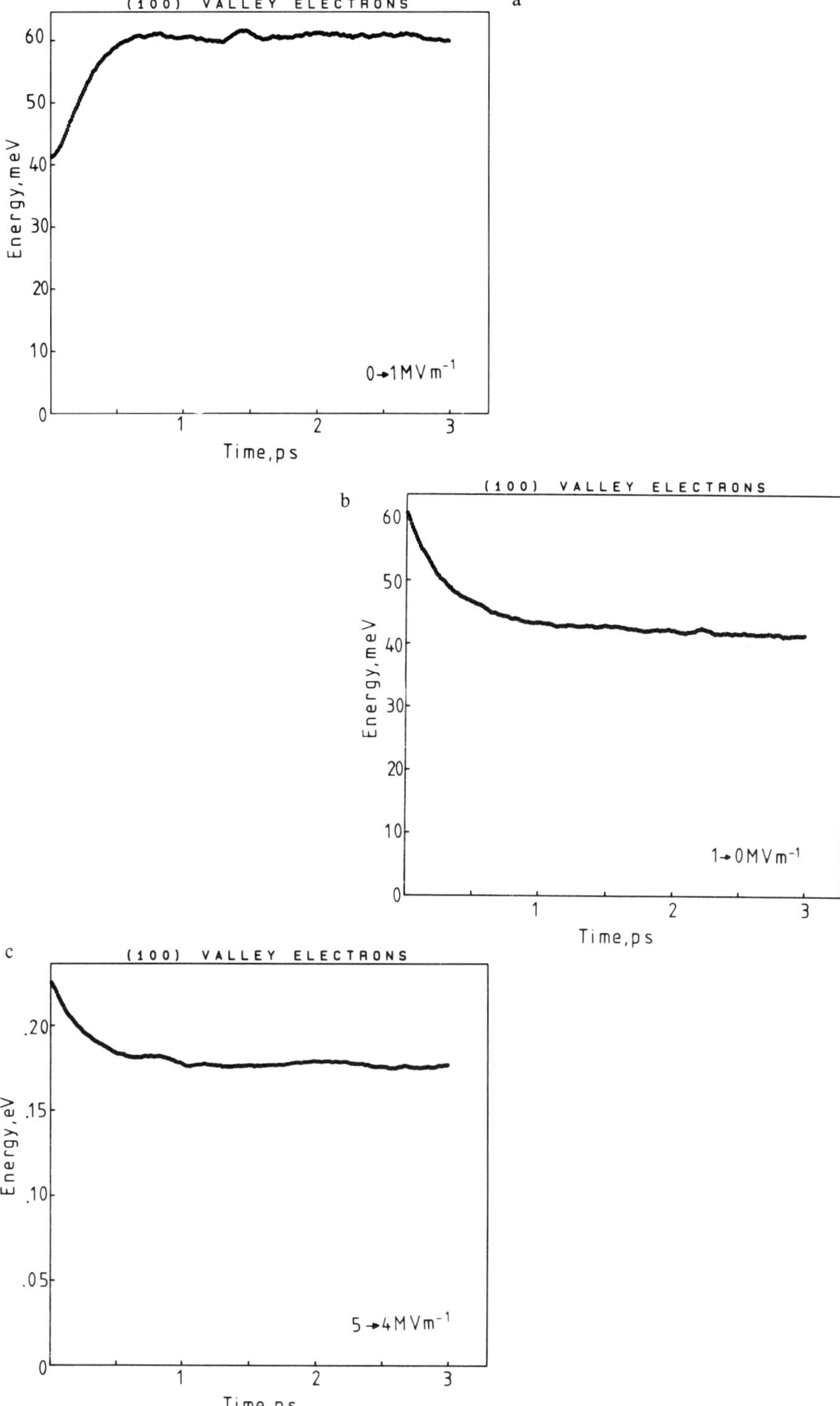

Figure 6.19 *The transient response of the average energy of the electrons in silicon to a change in the ambient electrostatic field: a, 0 to 1 MV* m^{-1}*; b, 1 to 0 MV* m^{-1}*; c, 5 to 4 MV* m^{-1}

average drift velocity thus exhibits a dampened oscillatory evolution, which gives rise to noise. Eventually equilibrium in the exchange of energy between the carriers and the lattice will be reached.

Figure 6.19 shows the response of the energy to the same change in the field ($1\,\mathrm{MV\,m^{-1}}$). The energy relaxation is the same whether the field changes from 0 to $1\,\mathrm{MV\,m^{-1}}$ and vice versa, but different if the initial field is 0 or $5\,\mathrm{MV\,m^{-1}}$. The oscillation in the energy is less pronounced than for the momentum. Figure 6.20 shows approximate energy and relaxation times for a change from zero field, as obtained from the Monte Carlo simulation. More accurate values can be obtained using more particles.

Gallium arsenide shows similar relaxation behaviour. (Figure 6.21.) The initial rise in the drift velocity is followed by scattering into the higher energy minima of the conduction band, whereby their velocity decreases due to the gain of the effective mass. This is an additional property of gallium arsenide which silicon does not possess. The overshoot effect is most pronounced around $2\,\mathrm{MV\,m^{-1}}$. For fields above $5\,\mathrm{MV\,m^{-1}}$ the overshoot effect becomes insignificant (Warriner, 1977; Moglestue, 1986a).

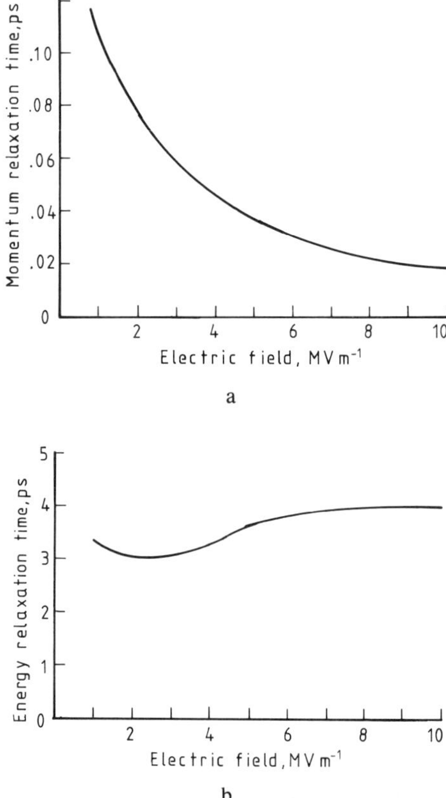

Figure 6.20 *Momentum (a) and energy (b) relaxation times in silicon at room temperature according to Equations (6.8) and (6.9) respectively*

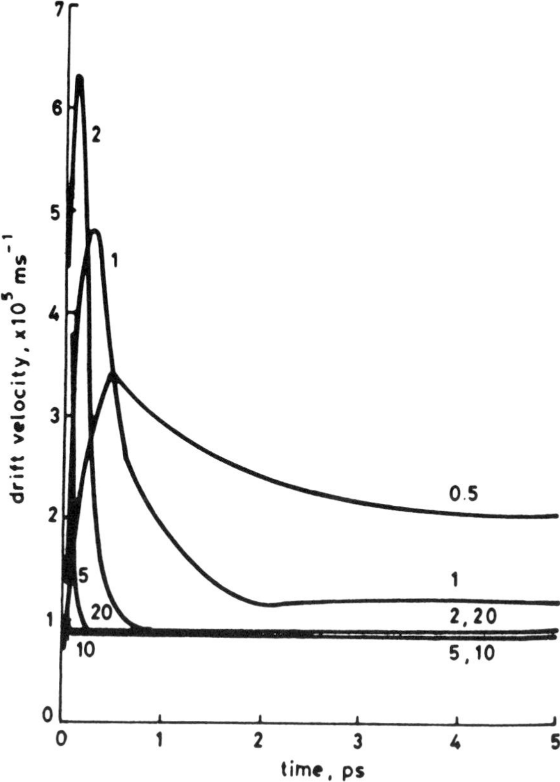

Figure 6.21 *Response of the drift velocity in GaAs at room temperature to a sudden change in the ambient electric field from zero. The parameter represents the magnitude of the uniform electric field in MV m^{-1}*

6.5 FREE FLIGHT PATH

Naively, the relaxation time can be associated with the inverse total scattering rate - they are of the same order of magnitude. The concept of average drift velocity, which enters the drift-diffusion and the hydrodynamic model, may tempt us to define the free flight path as

$$s_p = v_s \tau_p . \tag{6.10}$$

Although this gives values for s_p of the right order of magnitude it is far from being a physical reality. Although Equation (6.10) suggests an orderly motion in the direction of the field, in reality the particles' motion is rather chaotic. Equation (5.19) defines the duration of a free flight, while its length is obtained from Equation (5.43). An extensive discussion of the path lengths of free flights has been published by Moglestue (1986a). Figure 6.22 shows that the average flight path of the electrons in GaAs decreases drastically from 70 nm at zero field to about

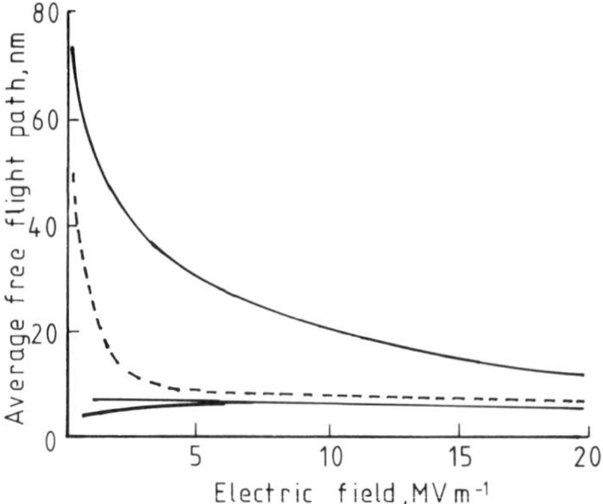

Figure 6.22 *Free flight path of electrons at room temperature in gallium arsenide against the uniform electric field for three types of equivalent minima of the conduction band (Γ, X, L) and all carriers (A)*

8 nm at 4 MV m^{-1}, above this it remains practically insensitive to the field strength. This statement, however, should not be taken at face value. In GaAs the electrons are divided over three kinds of conduction band minima; the mean free path in the L and X minima lies around 7 nm and is rather independent of the field. The mean flight path of Γ electrons is always larger than the average one for the entire electron population but decreases with increasing field. However the average path of free flight may be considered a guide to the feasibility of ballistic transport.

6.6 TWO-DIMENSIONAL TRANSPORT AT INTERFACES

A strong electric field perpendicular to the interface between two different materials can create a quantum well in which the transport is confined to a plane – it is *two-dimensional*. This means that the position of a particle in the direction perpendicular to the interface cannot be defined. The situation of inversion under the gate of a MOSFET is an example of such a system. Two-dimensional transport in silicon has been reviewed thoroughly by Ando *et al.* (1982).

A layer of silicon dioxide separates the metal gate contact from the p-type silicon semiconductor. (Figure 6.23.) The positive gate bias causes the conduction and valence bands to bend downwards. (Figure 6.24.) If the gate bias is large enough, the conduction band edge dives below the Fermi level (Fig. 6.24b) where

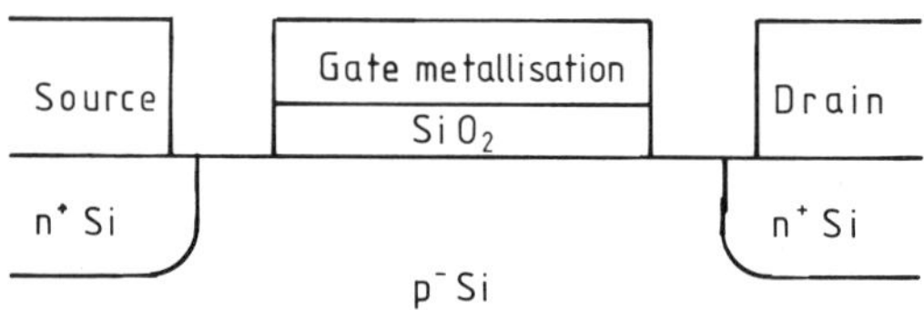

Figure 6.23 *Cross-section through a typical MOSFET. The gate is 1 μm long, which has been taken to be sufficient to be considered infinitely long*

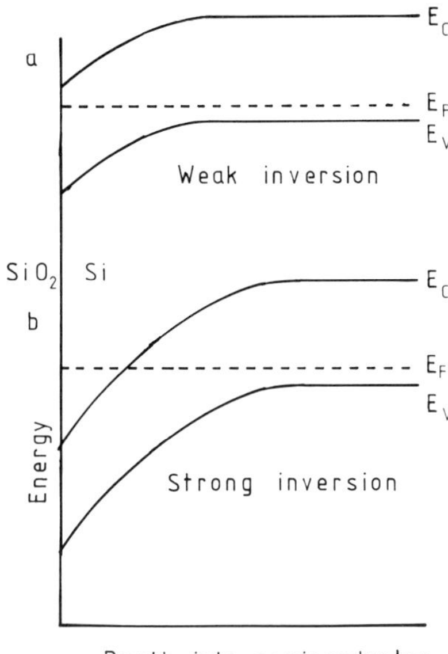

Figure 6.24 *Band structure of silicon against depth reckoned from the Si–SiO$_2$ interface:*

a *weak inversion: the conduction band edge lies above the Fermi level*
b *strong inversion: the conduction band edge submerges below it*

the semiconductor joins the oxide, so that electrons can flow along it. This is known as *inversion charge*. A quantum well is formed, where the conduction band is divided into sub-bands. It is therefore necessary to consider quantised transport. Siggia and Kwok (1970) gave experimental evidence of such transport, however they considered only the three lowest sub-bands; at very high electric fields oriented along the quantum well more sub-bands will be occupied. Krowne (1983) studied relaxation times from inter-sub-band scattering.

The electric field at the interface is

$$F_{\text{int}} = \frac{e(n_{\text{inv}} + n_{\text{depl}})}{\varepsilon_0 \varepsilon} \tag{6.11}$$

where ε represents the dielectric constant of the semiconductor, ε_0 the permittivity of vacuum and e the magnitude of the elementary electronic charge. n_{depl} represents the *depletion* sheet *charge*, i.e. the number density of ionised acceptors integrated from the interface downwards into the interior of the silicon, and n_{inv} the number density of accumulated electrons likewise integrated downwards. The electric field inside the oxide, which is constant when no net charges have been trapped in it, is given by Gauss' theorem:

$$F_{\text{ox}} = \frac{e(n_{\text{inv}} + n_{\text{depl}})}{\varepsilon_0 \varepsilon_{\text{ox}}} \tag{6.12}$$

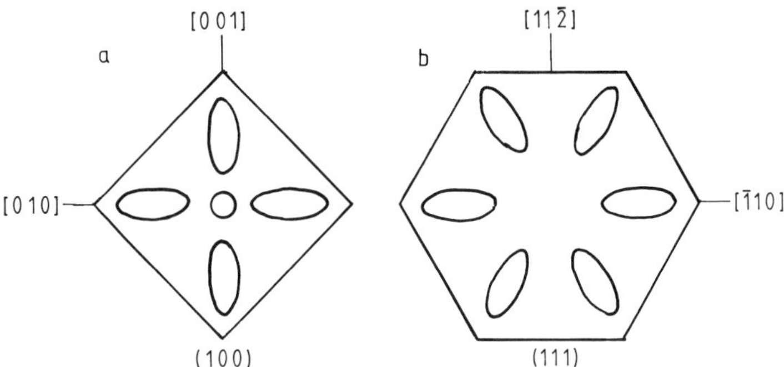

Figure 6.25 *Schematic diagram of the constant energy surfaces for the (a) (100) and (b) (111) surfaces in silicon. After Eisele (1978)*

where ε_{ox} denotes the dielectric constant of the oxide.

Assuming that the single-crystalline semiconductor is (i, j, k) oriented, the normal to the surface points in the direction $\mathbf{S} = (i, j, k)$ relative to the crystal Cartesian coordinate system defined by the cubic crystallographic cell. The most common orientations are in the $\langle 100 \rangle$, $\langle 110 \rangle$ and $\langle 111 \rangle$ directions. The plane representing the surface of the crystal passes through the origin of this Cartesian system. The corresponding plane in contravariant k-space also passes through the centre of the Brillouin zone. From Equations (2.3) and (2.7) it follows that it is spanned by the reciprocal vectors (1,0,0), (1,1,0) and (1,1,1) respectively.

A universal coordinate system is chosen in which the vector $\mathbf{S}_k$ coincides with the z axis. A transformation from the universal to the crystal coordinate system can according to Goldstein (1959) be obtained from the Eulerian relations. The plane through the centre of the Brillouin zone, representing the contravariant image of the crystal surface in k-space, will also be spanned by the (k_x, k_y) plane. The projection of the constant energy surfaces on to this plane represents the two-dimensional electronic band structures of the interface (Chu-Hao *et al.*, 1985). Figure 6.25 shows the two-dimensional Brillouin zones for (100) and (111) oriented diamond or zinc blende structured semiconductors. The projection in the $\langle 100 \rangle$ direction, Figure 6.25a, shows four ellipses with their major axes along the coordinate axes. The two ellipsoids with major axis along the k_z axis project on to one circle with centre in the origin. For the (111) orientation we have six ellipsoids with their major axes arranged like the spokes of a wheel. (Figure 6.25b.) In the effective mass approximation, the Schrödinger equation of an electron in the quantum well,

$$\tfrac{1}{2}\hbar^2 \nabla \mathbf{m}^{*-1} \nabla \Psi + V_t(z)\Psi + V_1(x, y)\Psi = (E_1 + E_t)\Psi, \tag{6.13}$$

can be separated into two parts, one describing the motion parallel, one the motion perpendicular to the interface. E_1 and E_t denote, respectively, the energy associated with the motion parallel and perpendicular to the interface. The parallel part of the separable Schrödinger equation reads

$$-\frac{1}{2}\hbar^2\left(\frac{1}{m_x^*}\frac{d^2\Psi_1}{d^2x}+\frac{1}{m_y^*}\frac{d^2\Psi_1}{d^2y}\right)+V_1(x,y)\Psi_1=E_1\Psi_1 \tag{6.14}$$

yielding the band structure

$$E_1(k_x,k_y)=\tfrac{1}{2}\hbar^2(k_x^2/m_t^*+k_y^2/m_y^*) \tag{6.15}$$

and the wave function

$$\Psi_1=\frac{1}{\sqrt{A}}u_k(x,y)\exp(-ik_xx-ik_yy) \tag{6.16}$$

where $u_k(x,y)$ represents the Bloch function reflecting the periodicity of the crystal in the plane and A the area of the interface. Usually this area can be considered extending to infinity, even if it is only of the order of square micrometres. Equation (6.16) resembles the wave function of free electrons; in fact, the carriers can be considered as free along the interface.

In the perpendicular direction the Schrödinger equation reads

$$-\frac{\hbar^2}{2m^*}\frac{d^2\Psi_{nt}}{dz^2}+V_t(z)_n\Psi_{nt}=E_{nt}\Psi_{nt}. \tag{6.17}$$

The total wave function is

$$\Psi=\Psi_1\Psi_{nt}. \tag{6.18}$$

The energy E_{nt} forms distinct sets of levels which have been indicated by the second label t. When

$$V_t(z)=eF_{\rm int}z \tag{6.19}$$

Equation (6.17) has the solution (Stern, 1972)

$$\Psi_{nt}(z)=Ai\{[2m_t^*eF_{\rm int}/\hbar^2]^{1/3}[z-E_n/(eF_{\rm int})]\} \tag{6.20}$$

The second solution of Equation (6.17), the associated Airy function $Bi(\zeta)$, is unphysical because it diverges for positive arguments, i.e. when the kinetic energy, E_{nt}, is less than the potential one. The regular Airy function, $Ai(\zeta)$, approaches zero for positive arguments, describing the evanescent tail of the distribution into the potential barrier $V_t(z)$. The two Airy functions are shown in Fig. 3.15. The corresponding energy levels are

$$E_{nt}=\left(\frac{\hbar^2}{2m_t^*}\right)^{\frac{1}{3}}\left[\frac{2}{3}\pi eF_{\rm int}\left(n+\frac{3}{4}\right)\right]^{\frac{2}{3}} \tag{6.21}$$

with $n = 0,1,2,\ldots$.

The two-dimensional Brillouin zone for the $\langle 100 \rangle$ oriented surface contains four constant energy ellipses, two along each of the coordinate axes and two coinciding circles with centre in the origin. (Figure 6.25.) The effective mass perpendicular to the plane of the four ellipses is the same for all of them, viz. m_t^*, the transversal bulk mass. The perpendicular mass for the two circles is m_l^*, the longitudinal bulk mass. Equation (6.21) therefore yields two separate sets of energy levels according to whether $m^* = m_l^*$ or $m^* = m_t^*$. We have two *energy ladders*. Figure 6.26 shows the energy levels for the two ladders in the (100) oriented silicon–silicon dioxide interface. Stern (1972) distinguishes between the two ladders by marking the symbol for one of them with a prime. Generally there are three different effective masses perpendicular to the plane of motion. With a uniaxial stress we may even have six ladders. For GaAs the situation becomes even more complex. We therefore suggest giving E_n an additional label to indicate which ladder we refer to: E_n becomes E_{nt}. When the crystal is $\langle 111 \rangle$ oriented, Fig. 6.25b, the perpendicular mass takes only one value, the energy levels are six-fold degenerate.

Figure 6.27 shows the wave function of the first three levels; the wave function of the nth level has $n - 1$ zeros. The square of the amplitude, $|\Psi_{nt}(z)|^2$, represents the probability density for the position of the particle against the distance from the interface to the oxide.

The potential $V_t(z)$ is generally not of the form described by Equation (6.19). The quantum well contains charge which modifies the potential according to Poisson's equation:

$$\frac{d^2V_t}{dz^2} = \frac{1}{\varepsilon_0 \varepsilon_s} \left[\rho_{\text{depl}} - e \sum_n n_{nt} \Psi_{nt}^2(z) \right]. \tag{6.22}$$

where $\rho_{\text{depl}} = n_{\text{depl}} e$ represents the sheet depletion charge and the sum the inversion charge with n_{nt} the sheet density of electrons of level n of ladder t.

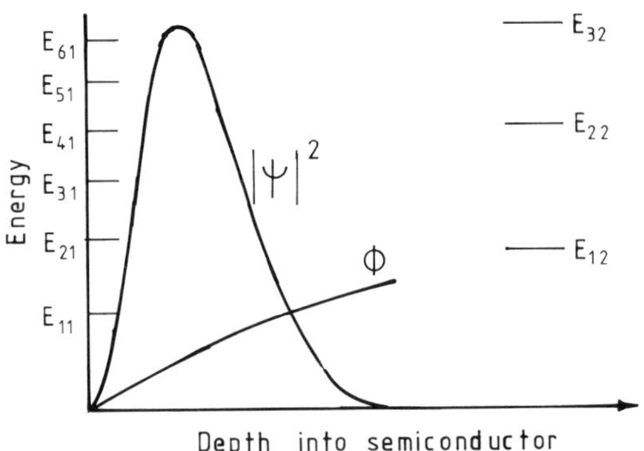

Figure 6.26 *The energy levels of the two ladders at the (100) oriented interface between silicon and silicon dioxide. $|\Psi|^2$ represents the probability distribution for the position of the electron in the lowest energy level; Φ the electrostatic potential which has been calculated self-consistently assuming a Fermi–Dirac distribution of the electrons over the energy levels*

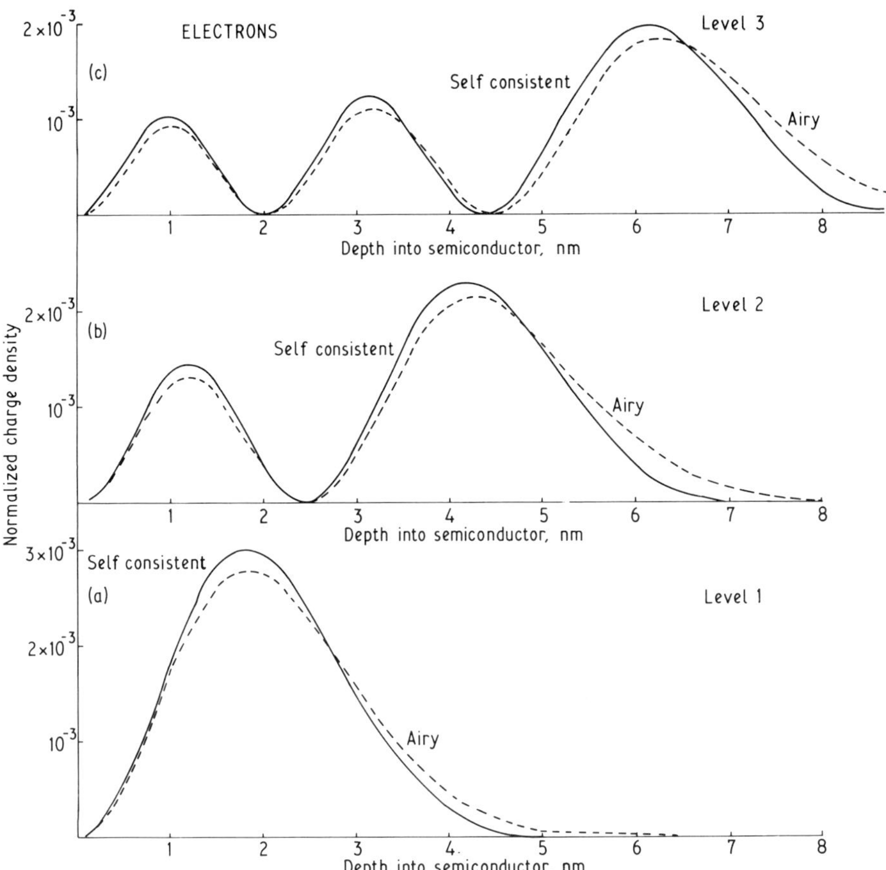

Figure 6.27 *The envelope wave function for the first three energy levels at the (100) oriented interface between silicon and silicon dioxide. The full curve represents self-consistent calculations; the dashed curve the Airy function approximation*

A self-consistent solution of the coupled Equations (6.22) and (6.17) has to be carried out. Siggia and Kwok (1970) have done it by calculating the dielectric function approximately using Green's functions. Stern (1972) used variational calculation to obtain the second energy level. Researchers interested in mobility calculations have assumed that higher energy levels did not need to be considered as they are most likely to be empty if the electrons obey Fermi–Dirac statistics. However, this is not true for transport in a strong longitudinal electric field. Since many sub-bands are occupied, it is necessary to know the shape and occupation of each to gain an accurate estimate of the charge distribution and the electrostatic potential. A fully self-consistent calculation applied to an arbitrary number of sub-bands has been published by Moglestue (1986b). His calculations also apply to holes.

Figure 6.27 shows that when $n_{inv} \leqslant n_{depl}$, the result of self-consistent calculation does not differ significantly from the Airy function solution, Equation (6.20). When $n_{inv} > n_{depl}$, the results *are* significantly different.

In Monte Carlo particle modelling of transport it is necessary to use scattering rates for two-dimensional gases (see Chapter 4). As with bulk silicon, optical phonons do not contribute to scattering; only quasielastic acoustic phonons and phonons causing electron transfer between the various energy minima in the two-dimensional band structure can cause scattering. According to Equation (4.60) a transfer between energy levels within the same ladder is impossible because of the orthogonality of the wave functions $\Psi_{nl}(z)$. The only way would be via another ladder or via a phonon with a transversal wave vector component matching the energy difference between two levels.

Monte Carlo modelling of the electron mobility has been carried out several times in the past. The first was probably published by Basu (1977), whose calculated drift velocity against the electric field showed the correct trend, but did not agree fully with the experimental results. Terashima *et al.* (1985) achieved a good agreement with experiments for high electric fields, but their estimate was too low for lower fields. Chu-Hao *et al.* (1985) achieved an overall good agreement when they included roughness scattering. Roughness scattering is poorly understood; today it is possible to make such smooth silicon–silicon dioxide interfaces that this scattering can be virtually eliminated. In this case the introduction of such a scattering mechanism is not justified. Moglestue (1983a), omitting roughness scattering altogether, obtained good low-field agreement by postulating a higher value for the acoustic and optical deformation potentials than in bulk material, with the justification that the oxide causes the lattice at the interface to be strained.

The author's unpublished calculations of the drift velocity, using the thus modified values for the deformation potentials, agree well with the experimental data of Canali *et al.* (1975). (Figure 6.10.) Furthermore, the occupation at the higher sub-bands increases with the strength of the longitudinal electric field. The sub-bands get closer with increasing energy until eventually they form a continuum where the scattering is bulk-like. The two-dimensional scattering rates are generally lower than in bulk and reduce with higher energy levels, as in Equation (4.52). For very narrow channels, however, they will be higher. An electron in a moderately high energy sub-band can attain a kinetic energy of 3 eV because of the scarcity of scattering, which is sufficient for it to enter into the oxide; this is a possible mechanism of leakage through oxides too thick for tunnelling.

7

The Field Equation

7.1 INTRODUCTION

Poisson's Equation (1.9) reads

$$\nabla\left(\varepsilon\frac{d\Phi}{d\mathbf{r}}\right) = -\frac{\rho_e(\mathbf{r}, t)}{\varepsilon_0} \tag{7.1}$$

where Φ represents the potential, ε the dielectric function which is material, and possibly also position dependent, ε_0 the permittivity of vacuum and ρ_e the charge density. The charge consists of:

i) the stationary charged donors and acceptors
ii) the mobile current carriers
iii) carriers which are permanently or temporarily trapped.

An analytical solution of Equation (7.1) is impossible except in trivial cases. Several different numerical methods exist which have been especially developed for computers, the most popular being one based on finite elements. All numerical approaches have one thing in common, i.e. that the solution area has to be divided into a mesh. The finite difference scheme makes use of a grid that is finest where the gradient in the charge density or in the electrostatic field is greatest. Various ways of obtaining the solution have been reviewed by Jacoboni and Lugli (1989).

There are two good reasons, however, for introducing a *uniform* mesh in the Monte Carlo particle model: one, it permits the use of fast Fourier transforms; two, the noise inherent in the model, which is of a physical origin, generates local fluctuations in the mobile charge distribution and in the local electric field everywhere the current flows. This means that there are no 'large areas' with small, uninteresting changes in the electric field or in the charge distribution.

We shall discuss this grid in Section 7.2.

The algorithm described in this chapter to solve the field equation is based on Hockney's POT4 method (1965). Different ways of assigning the charges to the mesh will be presented in the following section. The actual algorithm for the solution of the field equation makes use of Green's functions and a fast Fourier transform in $n - 1$ dimensions where n represents the dimensionality of the problem ($n = 2, 3$), and a cyclic reduction scheme in the direction where the Fourier transform has not been applied. This algorithm used to be the world's fastest for many years. The Fourier transform requires that the potential, Φ, is treated as unknown everywhere. However this conflicts with the boundary conditions of the problem. In Section 7.4 we shall introduce a way around this by calculating the capacity matrix which is based on a Green's function approach to the solution. The capacity matrix expresses the correction in the charges that has to be applied in order to satisfy the boundary conditions.

The discretisation of the problem requires that the boundary lines or faces be divided into small areas, or electrodelets, as will be explained in Section 7.5.

A semiconductor device consists of one or more different materials. Within each material we assume the dielectric function to be constant. This approximation is good enough even for alloys of a variable spatial stoichiometric composition. The normal component of the electric field is discontinuous across an interface between two dielectrics. This can be accounted for by placing adequate charges at the interfaces.

The solution obtained here is only spatial. A time-dependent solution is required because of the dynamics of the charge. This can be taken care of by repeating the solution at regular time intervals. We shall return to this point in Section 8.2.

7.2 CHOICE OF MESH

The simulated area has the shape of a rectangular box. We define a *global coordinate system* with its x axis along the length, the y axis along the height and z axis along the width. (Figure 7.1.) The length, the height and the width are designated by L_x, L_y and L_z respectively. Performing a Fourier transform in the x and z directions the problem decomposes into a set of one-dimensional Poisson equations. The application of the Fourier transform dictates that the mesh has to be chosen uniform in these directions. Experience has shown that the maximum processing speed will be achieved when we use a cyclic reduction scheme in the y direction instead of a Fourier transform. This algorithm is most optimal when the mesh is uniform in this direction too.

With N_x cells along the length, N_y along the height and N_z cells along the

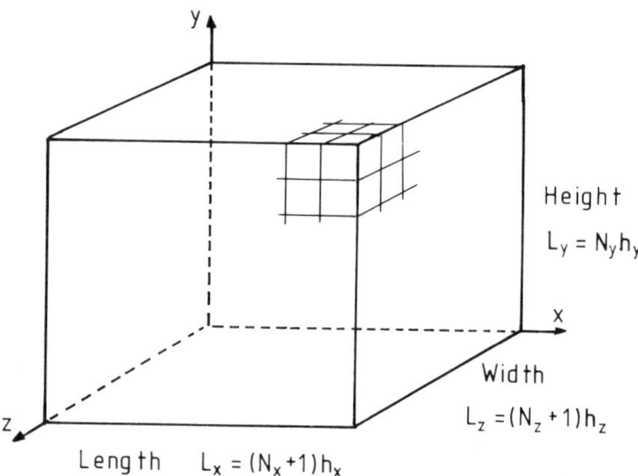

Figure 7.1 *The area over which Poisson's equation has to be solved. The global coordinate system referred to in the text has been drawn and the cell structure indicated in the upper right hand corner*

width, the cell size becomes h_x, h_y and h_z respectively in these three directions:

$$L_i = h_i N_i \tag{7.2}$$

for $i = x, y, z$. The Fourier transform is strictly applicable to variables ranging from minus to plus infinity. It is therefore necessary to extend the simulation in the directions of the transform. To comply, the box is extended by mirroring in the (y, z) and in the (x, y) planes. The cyclic reduction is easier to handle when a similar reflection in the (x, z) plane is also carried out. This means:

$$\Phi(-x) = \Phi(x) \quad \text{and} \quad \Phi(L_x + x) = \Phi(L_x - x), \tag{7.3a}$$

$$\Phi(-y) = \Phi(y) \quad \text{and} \quad \Phi(L_y + y) = \Phi(L_y - y) \tag{7.3b}$$

and

$$\Phi(-z) = \Phi(z) \quad \text{and} \quad \Phi(L_z + z) = \Phi(L_z - z) \tag{7.3c}$$

and similar expressions for ρ_e. Consequently

$$\rho_e(i + 2nL_i) = \rho_e(i) \tag{7.4a}$$

and

$$\Phi(i + 2nL_i) = \Phi(i). \tag{7.4b}$$

We have met a similar periodic extension already in Section 2.2.

Our problem has been transformed into a periodic one with periods $2L_x$, $2L_y$ and $2L_z$, but this does not affect our solution because we are only interested in Φ inside the original simulation area. The Fourier transforms dictate us to choose

$$N_x + 1 = 2^{ix} + 1 \tag{7.5a}$$

cells along the length and

$$N_z + 1 = 2^{iz} + 1 \tag{7.5b}$$

cells along the width, where i_x and i_z are positive integers. The computed values of the potential refer to the centre of the mesh cells. The resolution of our solution is of course defined by the size of the mesh cell.

The Fourier transform renders $N_x N_z$ independent one-dimensional equations, each with N_y variables representing the N_y values of the potential on the original mesh in the y direction. When these equations are solved by means of cyclic reduction, which implies that the unknown values are found pairwise, then

$$N_y = 2^{iy} \tag{7.5c}$$

where i_y represents a positive integer.

In the case of a two-dimensional geometrical model, Poisson's equation has to be solved in two dimensions, i.e. the z direction no longer counts. This corresponds to a choice of infinitely long cells in the z direction, $h_z \to \infty$, and $N_z = 1$ or $i_z = 0$.

7.3 ASSIGNMENT OF CHARGE TO THE MESH

It is necessary to attribute both the stationary and the mobile charges to the centre of the mesh cell. There are several ways of doing this. The simplest, the *Nearest Grid Point Scheme* (Birdsall and Fuss, 1969), attributes all the charge of the cell to its centre. With other methods such as the *Cloud-In-Cell Scheme* of Hockney *et al.* (1974a) the charge is spread out uniformly over a box which does not necessarily have to be congruent with the mesh cell. Charge is assigned to each of the mesh cells it covers according to coverage as illustrated in Fig. 7.2.

With the *Triangular-Shaped Cloud Scheme* of Eastwood and Hockney (1974) the charge extends over a cloud of density

$$\rho_e = \frac{e_s}{\alpha^3_s h_x h_{yz}} \left(1 - \frac{x}{\alpha_s h_x}\right) \left(1 - \frac{y}{\alpha_s h_y}\right) \left(1 - \frac{z}{\alpha_s h_z}\right) \qquad (7.6)$$

where x, y, z represent a position in a local charge coordinate system which has its origin at the centre of the charge cloud and with axes parallel to those of the global coordinate system. The charge, e_s, occupies a box of volume $\alpha_s h_x h_y h_z$ where α_s represents a freely chosen factor which is 1 when the volume of the cloud equals that of the mesh cell. The charge is attributed to the mesh cells it covers according to the fractional charge in each. Figure 7.3 shows the charge density along one Cartesian coordinate.

The *Gaussian-Shaped Cloud Scheme*, Fig. 7.4, also attributed to Eastwood and Hockney, spreads the charge over a Gaussian distribution of the shape:

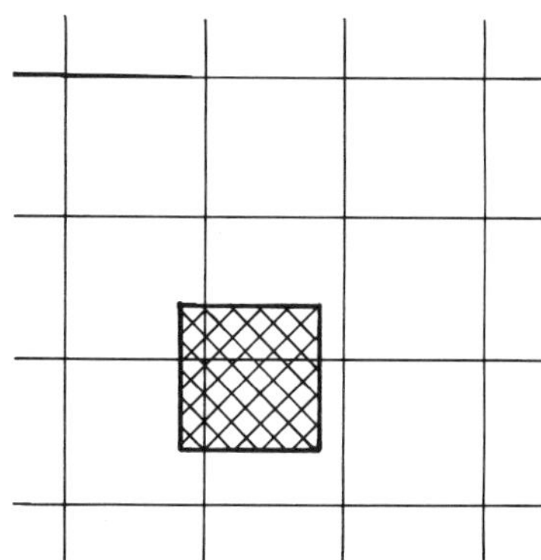

Figure 7.2 *Two-dimensional illustration of the Cloud-in-Cloud assignment scheme to attribute charge over the adjacent cells. In this figure the charge cloud is congruent with the mesh cell, though this need not be true in general*

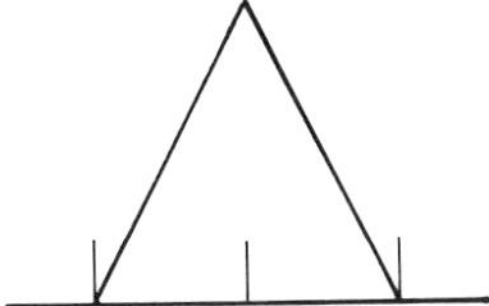

Figure 7.3 *One-dimensional illustration of the Triangular-Shaped Cloud scheme to distribute charge over neighbouring mesh cells. The bars represent mesh cells*

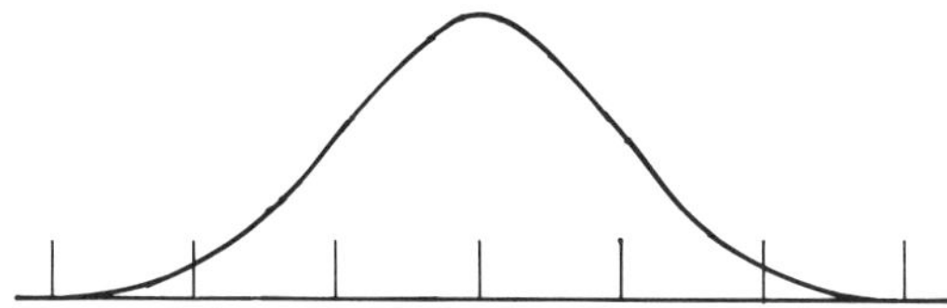

Figure 7.4 *One-dimensional illustration of the Gaussian-Cloud scheme to distribute charge over neighbouring mesh cells. The standard deviation parameter of the Gauss function has been chosen to be one mesh cell in this illustration. The bars represent mesh cells*

$$\rho_e = \frac{e_s}{\pi^{3/2}\alpha_s^3 h_x h_y h_z} \exp\{[(x/h_x)^2 + (y/h_y)^2 + (z/h_z)^2]/\alpha_s\} \tag{7.7}$$

which is also of a rectangular shape. The coordinates x, y and z are relative to the local charge coordinate system defined above.

The Nearest Grid Point Scheme is a limiting case of all the other assignment schemes and is the simplest to manage. The reader may object to the idea of rectangular-shaped charges, arguing that they are unphysical and may depend on the choice of mesh. We are free to choose a spherical cloud, however:

$$\rho_e = \frac{3e_s}{\pi R^3}\left(1 - \frac{r}{R}\right) \tag{7.8a}$$

where r represents the distance from the centre of the cloud and R its radius, $r \leqslant R$, or a Gaussian spherical cloud:

$$\rho_e = \frac{e_s}{2\pi^{3/2}R^3} \exp[-(r/R)^2]. \tag{7.8b}$$

The latter would match the charge density of a quantum mechanical wave packet. Although Equations (7.8a, b) describe more realistic charge shapes than the rectangular ones, these assignment schemes require a large computational effort. Readers will have to judge for themselves whether the improvement in accuracy is sufficient to justify this extra expenditure in processing time.

An extended assignment scheme is less noisy than the Nearest Grid Point one, i.e. it produces less noise in the charge distribution. However, great care should be taken when applying it in the presence of sharp potential barriers, e.g. heterojunctions, to ensure that no charge is incorrectly placed. We should also take care that charge is not lost when a part of it extends outside the simulated area.

7.4 THE FAST FOURIER TRANSFORM

Whatever assignment scheme is employed, the charge attributed to each cell is placed in its centre. The potential and the charge of the ith mesh cell in the x direction, the jth cell in the y direction and the kth cell in the z direction are

$$\Phi_{k,j,i} = \Phi[(i+\tfrac{1}{2})h_x,\ (j+\tfrac{1}{2})h_y,\ (k+\tfrac{1}{2})h_z] \tag{7.9}$$

and

$$\rho_{l,k,j,i} = \rho_e[(i+\tfrac{1}{2})h_x,\ (j+\tfrac{1}{2})h_y,\ (k+\tfrac{1}{2})h_z] \tag{7.10}$$

where the indices i, j and k run from 0 to N_x, $N_y - 1$ and N_z respectively.

In this notation the discretised form of Equation (7.1) reads

$$\frac{\Phi_{k+1,j,i} + \Phi_{k-1,j,i} - 2\Phi_{k,j,i}}{h_z^2} + \frac{\Phi_{k,j+1,i} + \Phi_{k,j-1,i} - 2\Phi_{k,j,i}}{h_y^2}$$

$$+ \frac{\Phi_{k,j,i+1} + \Phi_{k,j,i-1} - 2\Phi_{k,j,i}}{h_x^2} = \rho_{ek,j,i}/\varepsilon\varepsilon_0. \tag{7.11}$$

Here the following simplification has been made: the dependence of the dielectric function ε on position has been suppressed by omitting the spatial derivative in it, but its local variation has still to be considered by using the local value everywhere. We shall return to this point later. Note that there are $N_x N_y N_z$ such equations of the form (7.11).

Hockney's POT4 method, which forms the basis for our method of solution, was developed for the two-dimensional case. This is regained after a Fourier transform in the z direction:

$$2\sum_{k=0}^{N_z-1}\left\{\left[\mu'\left(\cos\frac{\zeta\pi}{N_z} - 1\right) - \lambda' - 1\right]\Phi_{\zeta,j,i}\right.$$

$$\left. + \lambda'(\Phi_{\zeta,j,i+1} + \Phi_{\zeta,j,i-1}) + \Phi_{\zeta,j+1,i} + \Phi_{\zeta,j-1,i}\right\}\varepsilon_k \cos\frac{k\zeta\pi}{N_z}$$

$$= h_y^2 \sum_{k=0}^{N_z} \rho_{e\zeta,j,i}\varepsilon_k \cos\frac{k\zeta\pi}{N_z} \tag{7.12}$$

where

$$\varepsilon_k = \begin{cases} \tfrac{1}{2} & \text{for } k = 0 \text{ and } N_z \\ 1 & \text{otherwise.} \end{cases} \tag{7.13}$$

Expressed in terms of the Fourier transform

$$\chi_{\zeta,j,i} = \frac{1}{N_z}\sum_{k=0}^{N_z-1}{}' \varepsilon_k \chi_{k,j,i} \cos\frac{\zeta k\pi}{N_z} \tag{7.14a}$$

with χ representing ρ_e or Φ and ζ is an integer. We shall use Greek characters to indicate integers in Fourier space. Here

$$\left.\begin{aligned}\lambda' &= (h_y/h_x)^2\\ \mu' &= (h_y/h_z)^2\end{aligned}\right\}. \tag{7.14b}$$

As each term $\cos(k\zeta\pi/N_z)$ is linearly independent we have $N_z + 1$ decoupled equations:

$$\begin{aligned}\mu'[\cos(\zeta\pi/N_z) - 1]\Phi_{\zeta,j,i} &+ [\Phi_{\zeta,j+1,i} + \Phi_{\zeta,j-1,i} - 2\Phi_{\zeta,j,i}]\\ &+ [\Phi_{\zeta,j,i+1} + \Phi_{\zeta,j,i-1} - 2\Phi_{\zeta,j,i}]\lambda' = \rho_{e\zeta,j,i}\end{aligned} \tag{7.15}$$

with $\zeta = 0, 1, 2, \ldots, N_z - 1$.

When the problem is two-dimensional this step is unnecessary and the term containing h_z in Equation (7.11) is omitted. Equation (7.15) will be of the same form as Equation (7.11) when it has been amended likewise. The effect of the Fourier transform has, as shown, been to reduce the three-dimensional problem to a two-dimensional one. There are $N_z + 1$ equations (7.15) which can be solved independently of one another.

Defining the vectors

$$\left.\begin{aligned}\boldsymbol{\Phi}^{\mathrm{T}}_{\zeta,j} &= \{\Phi_{\zeta,j,0}, \Phi_{\zeta,j,1}, \ldots, \Phi_{\zeta,j,Nx}\}\\ &\text{and}\\ \boldsymbol{\rho}^{\mathrm{T}}_{e\zeta,j} &= \{\rho_{e\zeta,j,0}, \rho_{e\zeta,j,1}, \ldots, \rho_{e\zeta,j,Nx}\}\end{aligned}\right\} \tag{7.16}$$

the set of Equations (7.15) can be written

$$\boldsymbol{\Phi}_{\zeta,j-1} + \Lambda^0_{\zeta,j} + \boldsymbol{\Phi}_{\zeta,j+1} = h_y^2\,\boldsymbol{\rho}_{e\zeta,j} \tag{7.17}$$

where row i of the tridiagonal matrix Λ^0 reads

$$\ldots\ \lambda',\ g,\ \lambda',\ \ldots \tag{7.18}$$

with λ' occurring next to the diagonal, and the diagonal elements read

$$g = -2[\lambda' + 1 + \mu'(1 - \cos\zeta\pi/N_z)]. \tag{7.19}$$

Confined to the original area, the top left corner of the original simulation area Λ^0 is

$$\begin{bmatrix} g & 2\lambda & 0 & 0 & \dots \\ \lambda & g & \lambda & 0 & \dots \\ \cdot & & & & \\ \cdot & & & & \\ \cdot & & & & \end{bmatrix}. \tag{7.20}$$

When the original problem is two-dimensional the term in μ' of Equation (7.19) vanishes.

Writing Equation (7.17) for three successive values $j-1$, j and $j+1$, then multiplying the middle one by $-\Lambda^0$ and adding them eliminates $\Phi_{\zeta,j-1}$ and $\Phi_{\zeta,j+1}$:

$$\Phi_{\zeta,j-2} + (2\mathbf{I} - \Lambda^{02})\Phi_{\zeta,j} + \Phi_{\zeta,j+2} = h_y^2(\rho_{e\zeta,j-1} - \Lambda^0\rho_{e\zeta,j} + \rho_{e\zeta,j+1}) \tag{7.21}$$

where $\mathbf{I}$ represents the identity matrix and the ith row of the pentadiagonal matrix $2\mathbf{I} - \Lambda^{02}$ reads:

$$\dots 0, -\lambda'^2, -2\lambda' g, 2 - 2\lambda'^2 - g^2, -2\lambda' g, -\lambda'^2, 0, \dots \tag{7.22}$$

thus with $-\lambda'^2$ in columns 1 ± 2 and $-2\lambda' g$ in columns $i \pm 1$.

This step could be repeated until only one matrix equation remains, but it has been shown (Hockney, 1980) that the optimal computational speed will be obtained by applying a Fourier transform in the x direction instead to get the $(N_x + 1)(N_z + 1)$ sets

$$\Phi_{\zeta,j-2,\xi} - A^0\Phi_{\zeta,j,\xi} + \Phi_{\zeta,j+2,\xi} = \rho^0_{e\zeta,j,\xi} \tag{7.23}$$

each with $N_y/2$ unknowns, $\Phi_{\zeta,j,\xi}$, with even label in j and with

$$\rho^0_{e\zeta,j,\xi} = \rho_{e\zeta,j-1,\xi} + B^0\rho_{e\zeta,j,\xi} + \rho_{e\zeta,j+1,\xi} \tag{7.24}$$

where

$$A^0 = 2\{\lambda'^2 \cos\xi\pi/N_x - 4\lambda'[\lambda' + 1 + \mu'(1 - \cos\zeta\pi/N_z)] -$$
$$- 3\lambda'^2 + 4\lambda' + 1 + 2\mu'(1 - \cos\zeta\pi/N_z)[(1 - \cos\zeta\pi/N_z)\mu' + 4(\lambda' + 1)] \tag{7.25}$$

and

$$B^0 = 2[1 - \lambda'(1 - \cos\xi\pi/N_x) + \mu'(1 - \cos\zeta\pi/N_z)]. \tag{7.26}$$

However, there are more than $N_y/2$ equations of the form (7.23) because of the periodic extension in the y direction introduced in Section 7.2, but only $N_y/2$ different ones. We are going to make use of this in order to eliminate the unknowns.

The result of the last Fourier transform has been to decouple Poisson's equation into one-dimensional ones. The quickest method of solving them is by the successive elimination of every second $\Phi_{\zeta,j,\xi}$. Write down Equation (7.23) for the three successive indices $j-2$, j and $j+2$, multiply the middle one by $-A^0$ and add them to get

$$\Phi_{\zeta,j-4,\xi} + A^1\Phi_{\zeta,j,\xi} + \Phi_{\zeta,j+4,\xi} = h_y^2\rho^l_{e\zeta,j,\xi} \tag{7.27}$$

with

$$A^1 = 2 - A^0 \tag{7.28}$$

and

$$\rho^1_{e\zeta,j,\xi} = \rho^0_{e\zeta,j-2,\xi} - A^0\rho^0_{e\zeta,j,\xi} + \rho^0_{e\zeta,j+2,\xi} \tag{7.29}$$

This procedure is repeated writing down three equations of the form (7.27). The next time the middle index takes the values $j-8$, j and $j+8$; next time $j-16$, j and $j+16$; and so on, terminating with $-2N_x$, 0, $2N_x$. The coefficient of the middle term is substituted recursively every time by

$$A^l = 2 - A^{l-1} \tag{7.28$'$}$$

and the right hand side by

$$\rho^l_{e\zeta,j,\xi} = \rho^{l-1}_{e\zeta,j-2^l,\xi} - A^{l-1}\rho^{l-1}_{e\zeta,j,\xi} + \rho^{l-1}_{e\zeta,j-2^l,\xi}. \tag{7.29$'$}$$

When this procedure stops we have obtained the equation

$$\Phi_{\zeta,2Ny,\xi} - A^n\Phi_{\zeta,0,\xi} + \Phi_{\zeta,-2Ny,\xi} = \rho^n_{e\zeta,j,\xi}. \tag{7.30}$$

The periodic extension in the y direction permits us to write

$$\Phi_{\zeta,2Ny,\xi} = \Phi_{\zeta,-2Ny,\xi} = \Phi_{\zeta,0,\xi} \tag{7.31}$$

which gives

$$\Phi_{\zeta,0,\xi} = \frac{\rho^n_{e\zeta,0,\xi}}{2 - A^n} \tag{7.32}$$

The other values for $\Phi_{\zeta,j,\xi}$ are then obtained by back substitution and the inverse Fourier transform gives the potential distribution over the original simulation area due to the given charges.

There are alternative ways of solving the one-dimensional Poisson Equations (7.23) but the method presented here is the fastest. It has one disadvantage, however, namely that the rounding-off errors accumulate as we proceed. If N_y is sufficiently large (i.e. in the hundreds) this error may overwhelm the calculated values for Φ. The calculation should therefore be performed with an extended length mantissa.

7.5 THE BOUNDARY CONDITIONS

So far we have solved Poisson's equation for a given charge distribution and paid no attention to the boundary conditions imposed upon the problem. The solution we have obtained is not unique, but a realisation of one of the many possible ones. Defining the boundary conditions we can select a solution that satisfies these as well.

Adding charges to the cells where boundary conditions have been defined we can, at least in principle, find a solution by the method just described which also satisfies the imposed boundary conditions. This can be done in a systematic manner based on the use of Green's functions.

There are two types of boundary conditions. The first requires that the potential assumes given values on specific faces which may lie on the surface of the original simulation area or in its interior. For two-dimensional problems the faces become line segments. The second condition is due to dielectric interfaces. From Gauss' theorem

$$\varepsilon_1 F_{1t} - \varepsilon_2 F_{2t} = \rho_{e,sf}/\varepsilon_0 \tag{7.33}$$

where ε_1 and ε_2 represent the dielectric constant at each side of the interface, and F_{1t} and F_{2t} the components of the electrostatic field perpendicular to it at both sides respectively. The induced polarisation generates interface charges $\rho_{e,sf}$, and ε_0 represents the permittivity of vacuum. The electric field can be expressed in terms of the potential:

$$F_{it} = \frac{\partial \Phi}{\partial \mathbf{r}_t} \tag{7.34}$$

where the position vector $\mathbf{r}_t$ stands perpendicular to the interface.

The solution satisfying the boundary conditions now has to be found by means of the capacity matrix method (Hockney, 1968) which is based on the Green's function technique. The *Green's function*, $G(\mathbf{r}, \mathbf{r}^0)$, satisfying our boundary conditions, is defined by

$$\frac{d^2}{d\mathbf{r}^2} G(\mathbf{r}, \mathbf{r}_0) = \frac{1}{4\pi} \delta(\mathbf{r} - \mathbf{r}_0) \tag{7.35}$$

which is Poisson's Equation (7.1) for a δ-function charge at the point $\mathbf{r}^0$. The δ-function has the property

$$\frac{1}{4\pi} \int \delta(\mathbf{r} - \mathbf{r}_0) d^3\mathbf{r} = 1. \tag{7.36}$$

In the discretised scheme this equation has been integrated only over the one mesh cell containing the δ-function point charge. This is permissible because the other areas do not contribute to the integral. This means that, for the Green's function, we solve with a unit charge in one cell using the algorithm presented in the previous section.

When the Green's function of our problem is known, the solution of our original problem satisfying the boundary conditions is

$$\Phi(\mathbf{r}) = \int G(\mathbf{r}, \mathbf{r}_0)\rho_e(\mathbf{r})d^3\mathbf{r}_0 \tag{7.37}$$

where r_0 represents the position of the unit charge. This also applies with unit charge placed on each of the electrodelets.

We find the relevant Green's function by placing a unit charge in each of the mesh cells where the boundary conditions have been defined in turn. The boundary conditions have been imposed on segments of area or line. Usually the potential is defined constant over one such segment; we therefore refer to such a segment as an *electrode*. The electrodes can lie on the surface or in the interior of the original simulation area.

Without sacrificing significantly on accuracy, more mesh cells passing through an electrode can bc combined to share the unit charge to define the Green's function. The electrode is thus divided into smaller parts which range in size from one to just a few mesh cells - these parts are referred to as *electrodelets*. Use of this scheme can save a significant amount of computer time. It is a matter of experience how large these electrodelets should be chosen.

Dielectric interfaces can be treated likewise according to Equation (7.33) by placing unit charges on them. One such charge may cover more than one mesh cell along the interface, so that it too can be divided into segments like the electrodes. For simplicity we shall also refer to these segments as electrodelets.

The two types of boundary conditions can be combined into one matrix equation:

$$\Phi_B + \varepsilon_1 \frac{\partial \Phi_B}{\partial \mathbf{r}_t}\bigg|_1 - \varepsilon_2 \frac{\partial \Phi_B}{\partial \mathbf{r}_t}\bigg|_2 = \mathbf{c} \tag{7.38}$$

where Φ_B has N_{sf} components representing the electrodes and N_{ifs} representing the interface segments. The vector $\mathbf{c}$ has correspondingly N_{sf} components representing the imposed potential values and the rest of the right hand side of Equation (7.33). In our discretised solution the partial differentials of Equation (7.38) will be replaced by finite differences so that Equation (7.38) can be written in the matrix form

$$\mathbf{B}_{bc}\Phi = \mathbf{c} \tag{7.38$'$}$$

where $\mathbf{B}_{bc}$ represents a matrix with N_{bd}^2 elements, $N_{bd} = N_{sf} + N_{ifs}$. Most of them, however, are zero. The non-zero off-diagonal elements originate from the partial differentials. Equation (7.11) can also be written in tensor form:

$$\mathbf{P}_{sp}\Phi = \rho_e/\varepsilon\varepsilon_0 \tag{7.39}$$

where Φ represents a vector of the potentials $\Phi_{k,j,i}$ with $(N_x + 1)N_y(N_z + 1)$ components. When the problem is two-dimensional Φ will contain only $(N_x + 1)N_y$ components. Both $\mathbf{B}_{bc}$ and $\mathbf{P}_{sp}$ are linear operators on Φ and Φ_B respectively. Computing the potential at all electrodelets with a unit charge at electrodelet i yields N_{bd} values of the potential, one for each electrodelet. These values form a vector $\Phi^{(i)}$. Performing this for all electrodelets in turn we form the matrix

$$C_{cap}^{-1} = \{\Phi^{(1)}, \Phi^{(2)}, \ldots . \Phi^{(Nbd)}\} \tag{7.40}$$

which is the inverse of the *capacity matrix*. The capacity matrix can be determined once and for all for a given problem and is independent of the actual values of the potential imposed on the electrodes.

The solution satisfying the constraints can now be computed as follows: solve the equation

$$\mathbf{P}_{sp}\Phi^{(a)} = \mathbf{b} \tag{7.41}$$

with zero sources at the boundary points. Here **b** represents $\rho_e/\varepsilon\varepsilon_0$ of Equation (7.33) or (7.11) with zeros where the boundary conditions have been defined. Calculate

$$\mathbf{v} = \mathbf{B}_{bc}\Phi_B \tag{7.42}$$

and the error

$$\varepsilon_\Delta = \mathbf{v} - \mathbf{b}. \tag{7.43}$$

Compute the correction sources from

$$\mathbf{s}_{cor} = -\mathbf{C}_{cap}\varepsilon_\delta. \tag{7.44}$$

Re-solve the problem with

$$\mathbf{P}_{sp}\Phi = \mathbf{b} - \varepsilon_\Delta. \tag{7.45}$$

This solution, Φ, will satisfy our boundary conditions because the sources cancel the errors ε_Δ.

8

Steady State Simulation of Devices

8.1 INTRODUCTION

In Chapters 2 and 3 we presented the phonon spectra and the band structure respectively of the semiconductor; in Chapter 4 the interaction between them. The necessary basic transport physics should now be familiar to the reader. In Chapter 5 the Monte Carlo method was introduced; in Chapter 6 it was used to simulate the bulk properties of semiconductor material. However, the Monte Carlo particle model is aimed at semiconductor devices: its main purpose is to simulate the transport of confined particles under influence of an imposed external field. In the previous chapter an algorithm for a quasi time- and space-continuous solution of Poisson's field equation was introduced. Now our goal has been achieved and we are able to simulate devices of any geometry. Once the model has been primed it is ready to perform the following interesting task: to simulate devices.

Readers who have written or obtained their own software may decide to prime their model to simulate a simple resistor. This resistor may consist of a uniformly doped block of semiconductor material, one end grounded, a d.c. bias applied to the other end, and the user expecting a steady current to flow through it. However, we warn against this approach as this resistor will most likely behave like a Gunn diode when the average electric field is above that for the maximum electronic drift velocity of the material. This resistor will show Gunn oscillations and if the reader is not aware of what is happening they may be greatly confused by the results. This, however, is not that relevant to silicon.

Figure 8.1 summarises the organisation of the Monte Carlo simulator software. After solving Poisson's equation with the present position of the carriers, the transport history is calculated in turn for each carrier for a set time, the field-adjusting time-step. To start with the carrier is allowed to complete its present free flight before the selection of scatter and the duration of the next flight is carried out. It may be necessary to revise the duration of the remainder of the present flight in the light of the updated electric field. For each carrier the following information at the end of the last field-adjusting time-step is required: the position; the wave vector; the scattering rate in order to revise the duration of the remainder of the flight in case the local electric field has changed as a result of solving the field equation; the remaining time of flight; information about the band extremum in which the particle resides.

In this chapter we shall describe how to use the Monte Carlo particle model to simulate entire devices. Heuristically, the best approach is by way of an example in the form of a GaAs *metal-semiconductor field-effect transistor (MESFET)*. The discussion of more complex devices will be postponed until Chapters 10 and 11, where heterostructures, bipolar transistors and quantum transport will be treated.

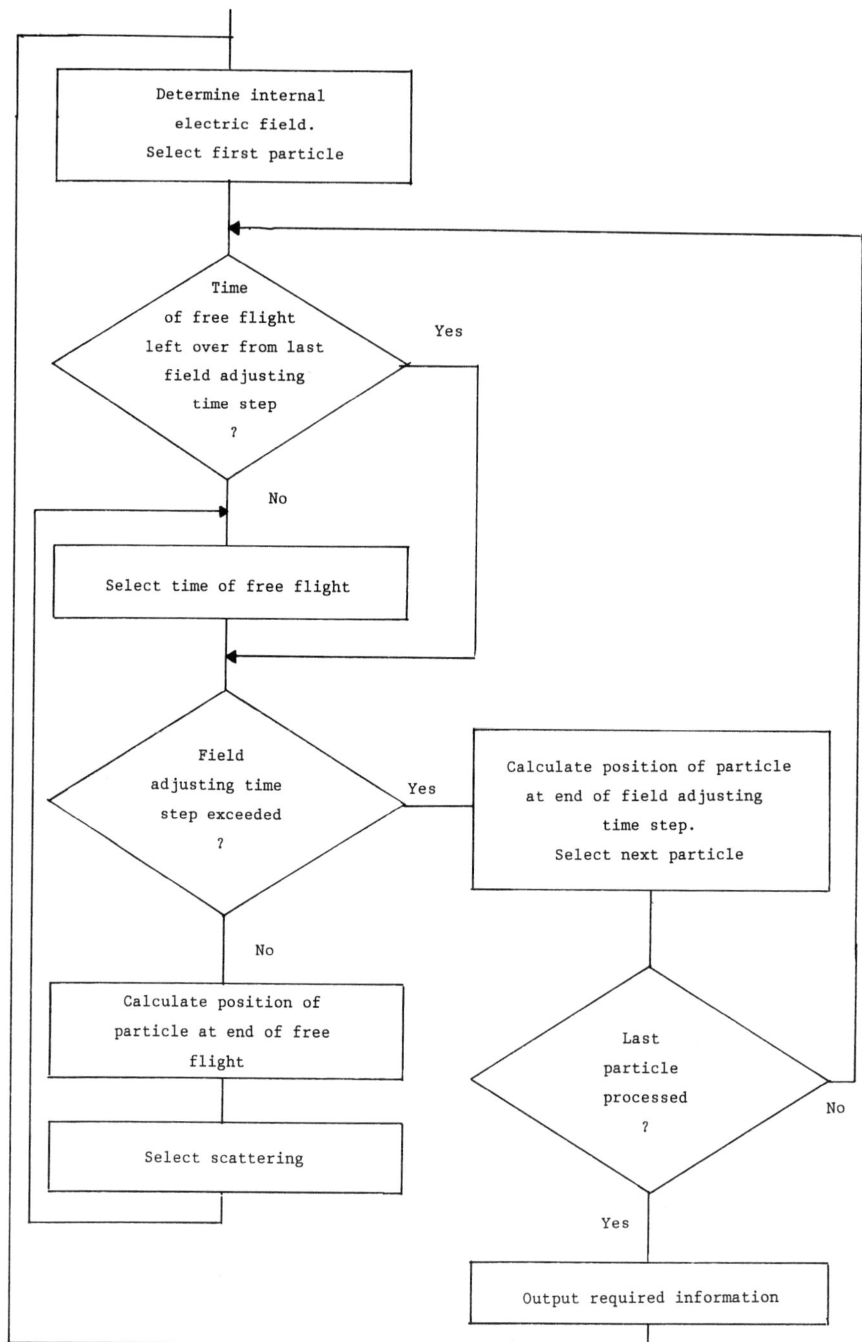

Figure 8.1 *Flow chart summarising the Monte Carlo particle model. Note that the process forms a closed loop. It is passed through as long as is found necessary*

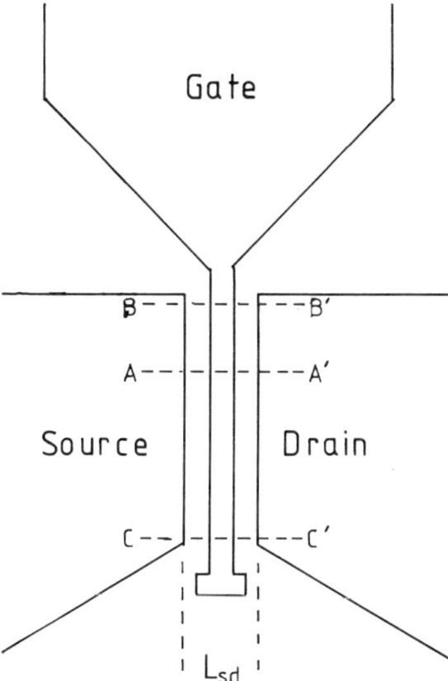

Figure 8.2 *Typical geometry of a single transistor which we have chosen as our example transistor. The distance between the source and gate and the gate and drain, and the length of the gate, are all 0.5 μm. B–B′ and C–C′ indicate the ends of the transistor, A–A′ a section chosen anywhere between*

Figure 8.2 shows the design of a typical microwave transistor manufactured on a wafer; Fig. 8.3 a photograph of an actual transistor. Figure 8.2 also represents our example transistor. The source, gate and drain contacts are coplanar. The actual transistor is that part between the parallel source and drain contact edges which is limited by the two lines B–B' and C–C'. The distance BC (or $B'C'$) is referred to as the *width*, W, *of the transistor*, L_{sd} represents the distance between the source and the drain, its *length*. If $W \gg L_{sd}$, in other words, if the effects of the corners and the end of the gate electrode can be neglected without making any significant error in calculating the transistor properties, and if it is immaterial where the section A–A' between B–B' and C–C' has been chosen, it is sufficient to use a model which is two-dimensional in the geometrical space; this means that the third dimension along the width of the transistor can be disregarded.

Figure 8.4 shows the cross-section through A–A' perpendicular to the surface of the transistor. This particular example transistor has an n-type epilayer of thickness 0.1 μm doped uniformly at $2 \times 10^{23}\,\mathrm{m}^{-3}$. (Throughout this book concentrations will be given in m^{-3} rather than in the more familiar cm^{-3} to comply with the SI system of units.) For modelling purposes the substrate has been assumed intrinsic, i.e. ideally it is free of any impurities. This can probably never be achieved in real life; the best we can hope to obtain is a semi-insulating substrate, that the material is compensated, in other words there are just as many donors as acceptors. Compensated material has been discussed by Walukiewicz *et al.* (1979). From a transport point of view, there is a subtle distinction between fully compensated material and intrinsic material in that the former contains an equal concentration of positively and negatively charged ions which all give rise to ionised impurity scattering.

In the next section the boundary conditions for determining the electric field

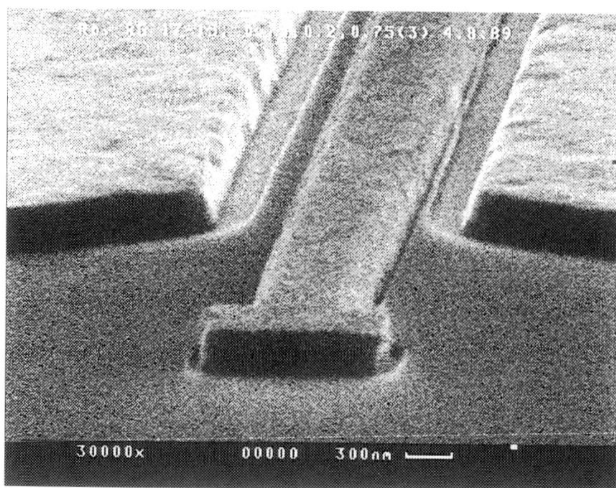

Figure 8.3 *Scanning electron microscopy photograph of a transistor of similar design to that shown in Fig. 8.2. It differs in detail from our example transistor in that it has a gate recess – which the picture does not clearly show – nor are the lateral dimensions exactly those chosen for our transistor. It was manufactured at the Fraunhofer Institute of Applied Solid State Physics in Freiburg im Breisgau, Germany in 1989*

inside the transistor will be defined. Then a discussion of the physical and modelling aspects of the contacts will be given in Section 8.3. The topic of free surfaces will also be treated here. The problem of initiating a simulation will be tackled in Section 8.4.

Typically a transistor contains millions of mobile particles; the size and speed of the computer, however, compel us to follow only a fraction of these. The implication of this limitation will be discussed when introducing the superparticle in Section 8.5. With this concept we are able to calculate the steady state current characteristics with today's computers, Section 8.6, starting from a chosen distribution of carriers. Here we shall discuss concepts like transconductance, transadmittance, gate capacitances etc. which are quite familiar to the electrical engineer. But we shall re-examine the meaning of these concepts in light of our experience with the Monte Carlo model. The feasibility of including the dynamics of trapping and the release of carriers will also be discussed here. Section 8.7 discusses the mechanism of negative differential resistivity, which is undesirable in amplifiers. The following section demonstrates wandering Gunn domains in transistors, an effect of turbulence in the transport which can occur even if the drain and the gate have been connected to a d.c. power supply. Modelling of luminescence and local heating will be introduced in the last two sections.

8.2 DEFINITION OF THE DEVICE GEOMETRY AND THE FIELD-ADJUSTING TIME STEP

The boundary conditions for solving Poisson's field equation are:

i) The potential at the electrodes as given. They may be constant, but can also be varying with time. We shall choose stationary potentials for our example simulation

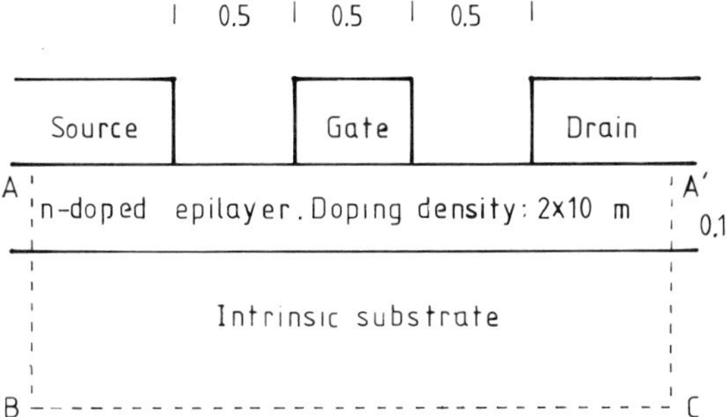

Figure 8.4 *Cross-section A–A′ through the simulated example transistor at A–A′ of Fig. 8.2 perpendicular to the surface of the semiconductor. The line AA′ follows the semiconductor surface which is assumed even. The doping of the epilayer is taken to be uniform throughout. The geometrical dimensions are given in micrometres*

ii) Neumann's boundary condition applies to the free surfaces and to the internal boundary $ABCA'$ (Fig. 8.4) of the simulation area. This states that the gradient of the potential is zero at the boundaries.

The application of Neumann's conditions along an, in principle, arbitrarily chosen boundary forces the electric field to be perpendicular to it. If the boundary is chosen so that the direction of the calculated field deviates considerably from what we would get choosing the boundary further away, the calculated field distribution will be incorrect. The border of the simulation area should therefore be chosen such that this error becomes insignificant. The choice is often a matter of good judgement and experience. A large simulation area certainly reduces the significance of this error but includes a larger number of particles, without adding any more information to the current transport and makes the simulation take longer to carry out. The choice of the boundary represents a compromise between these two opposing requirements: the simulation area should be as small as possible but the distortion of the electric field due to the choice of the boundary should be kept to a minimum. It is also possible to define devices with internal contacts, which means defined potentials at internal faces.

The use of fast Fourier transforms, as described in the previous chapter, requires that the simulation area be divided into a uniform grid with $2^n + 1$ cells (where n represents an integer) along the directions the transform is being applied. Choosing a uniform mesh along all the coordinate directions simplifies the housekeeping as we can work with particle coordinates in terms of mesh cells so that the position of the particle also gives the cell it is in. This manner has been preferred for all simulations carried out by the author of this book. Our way of solving Poisson's equation dictates that a rectangular simulation area be chosen.

Figure 8.5a shows a three-dimensional mesh that can be chosen for our sample transistor. A coordinate system with origin at B of Fig. 8.4 is chosen with the y axis along BA and x axis along BC. The z axis points perpendicularly out of the plane. When the choice of the location of the section A–A' of Fig. 8.2 between B–B' and C–C' is immaterial, the length of the cells in the z direction equals the width of the transistor. The position in this direction does not need to be considered. Differently formulated, only one cell in the z direction is defined, which

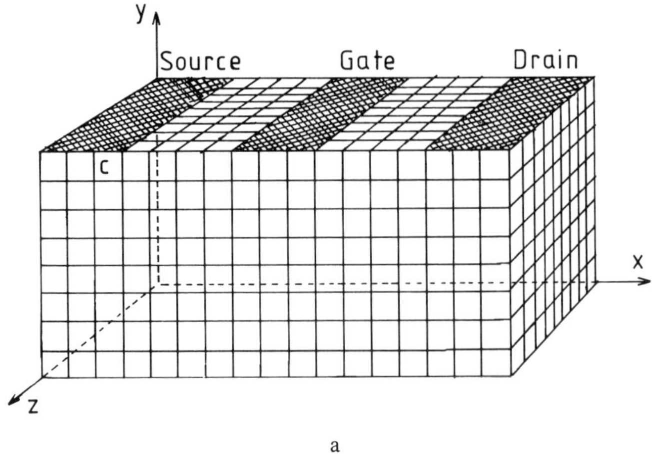

a

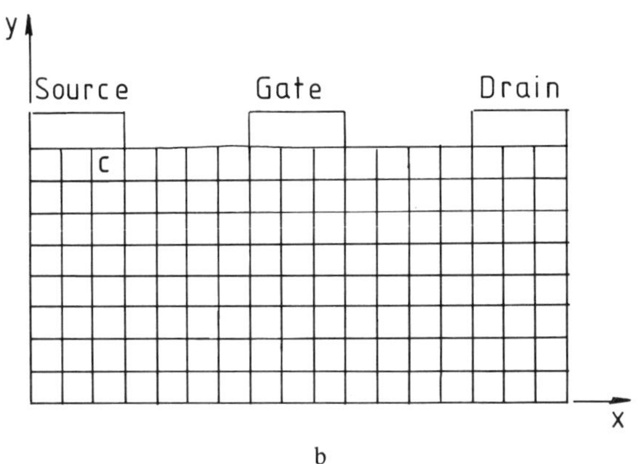

b

Figure 8.5 *A possible division of the simulation area ABCA′ of Fig. 8.4 into a three- (a) and two-dimensional (b) rectangular mesh. We have chosen to perform our simulation in two geometrical dimensions. Note that the mesh size is not the actual one chosen for the simulation. The shaded areas on top represent the contacts. C indicates a cell adjacent to an Ohmic contact. In Section 8.3.2 this cell serves as an example to illustrate the modelling of injection from the contact into the semiconductor*

extends over the entire width of the device. Then the model is two-dimensional in geometrical space. We therefore choose the mesh like the one represented in Fig. 8.5b. If there is a variation of the transistor properties along z, it is necessary to solve the field equations in three dimensions. This may be the case for wide transistors operating at terahertz frequencies. Then the number of cells in the z direction also has to be a power of two, plus one; the model is three-dimensional in geometrical space. In k-space we should always include three coordinates. A two-dimensional mesh of square cells of 10 by 10 nm^2 has been chosen for our example transistor, $2^8 + 1 = 257$ cells in the x direction and $2^6 = 64$ cells in the y direction. These, and other, model particulars are given in Table 8.1. Poisson's equation has been re-solved every 25 fs.

Table 8.1 Model particulars for the example transistor

Quantity	Symbol	Value
Size of mesh cell: in the x direction	h_x	10 nm
Size of mesh cell: in the y direction	h_y	10 nm
Size of mesh cell: in the z direction	h_z	1 m
Number of mesh cells: in the x direction		257
Number of mesh cells: in the y direction		64
Number of mesh cells: in the z direction		1
Field-adjusting time-step	δT	25 fs
Schottky contact potential	V_s	−0.75 V
Free surface potential	V_{sf}	−0.60 V

The source contact is 54 cells wide; the 0.5 μm long gate is situated between cells 104 and 154, the numbering starting from the left in Fig. 8.5b. The drain is 53 cells wide. The source contact extends to the left outside the simulated area, the drain to the right. Sufficiently large parts of these two contacts have to be included in the simulation to avoid the effect of the electric field not being in the direction it should be in, as discussed above. Including too little of the source will also result in it being unable to supply enough electrons: including too little of the drain may result in the electrons accumulating because they cannot get out fast enough. A scatter plot, like the one shown in Fig. 8.11, will reveal whether the source or the drain contact are sufficiently large to supply or absorb enough carriers. Inability of the source to supply enough carriers will manifest itself as a void in the electron distribution near the contact; the lack of capacity of the drain to absorb enough carriers will show as an excess of carriers accumulating there. This defect can also be remedied by selecting a shorter field-adjusting time step.

The epilayer of our example transistor is uniformly doped at $2 \times 10^{23}\,\mathrm{m}^{-3}$. The transition to the substrate is therefore sharp. Any other doping profile can equally well be chosen; Monte Carlo particle simulation of ion implanted field-effect transistors have been performed (Moglestue, 1981a, 1982). The effect of a p-type buffer has also been studied by Sanghera *et al.* (1980).

The choice of the *field-adjusting time-step*, δT, the frequency with which the field equation is being solved, is somewhat related to the size of the mesh. As a rule of thumb the field-adjusting time step is approximately equal to the time most of the carriers need to travel a distance less than that corresponding to the shortest side of the mesh cell. When the particle crosses into the next cell we do not update its acceleration until the next time it is scattered. The error committed thus is negligible if the time step is not too large. The optimum choice of time step is (Hockney, 1971)

$$\delta T \simeq \min\left\{\tfrac{1}{2}h_c/\lambda_D, 1\right\}/\omega_p \tag{8.1}$$

where h_c represents the smallest cell dimension, λ_D the Debye length, Equation (8.9), and ω_p the plasma frequency given by Equation (3.117). This choice is not rigid, there is still some freedom; a good choice depends on experience and judgement. Too large a time step can lead to numerical instability or stochastic heating. Following this criterion, Fig. 8.6, should yield a safe choice. Note that the scattering time has *nothing* to do with the field-adjusting time step.

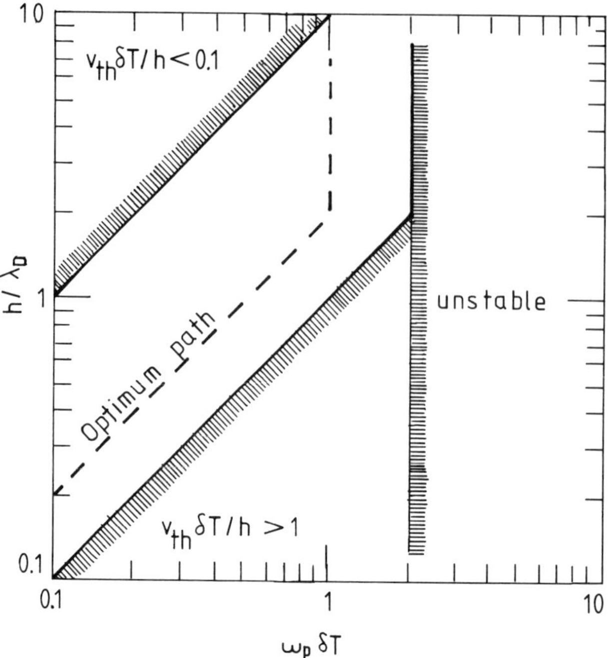

Figure 8.6 *The parameter plane to determine the correlation between the field-adjusting time step δT and the shortest side of the cell of the mesh over which Poisson's equation is to be solved. ω_p represents the plasma frequency given by Equation (3.117) and λ_D the Debye length, Equation (8.9). The relation between δT and the shortest cell size, h_c, is given by Equation (8.1). From Hockney (1971)*

8.3 CONTACTS AND SURFACE CHARGES

Modelling contacts is not straightforward because the physics of injection and absorption is rather complex and our knowledge of it not well established. As mentioned in the introduction, there are two types of contacts, the *Ohmic* ones which allow unhindered passage of carriers through it, and the *Schottky* ones where the carriers experience a potential barrier. All contacts can be considered as being Schottky; the Ohmic contact represents a special ideal case with a zero barrier.

Ohmic contacts are usually deposited by evaporating layers of metal (e.g. gold, germanium and nickel) on to the semiconductor surface. The wafer containing the transistors is then heated for a short while to allow alloys to form between the metals so as to eliminate the contact barrier. Metal atoms diffuse into the semiconductor in the process. The Schottky contact of the gate is formed by depositing gold, titanium and platinum during a separate step in the manufacturing process afterwards without annealing so that no diffusion of metal is expected to take place.

8.3.1 The Schottky contact

The *work function*, the energy required to liberate an electron at the Fermi surface, ϕ_m, is a material property. This energy is the same everywhere in the material, and if no surface states form, the work function is constant right up to the surface, Figs. 8.7a–d. It is 5.2 eV for gold and 4.7 eV for GaAs. When gold and GaAs come in contact, we should expect a barrier or contact potential of (4.7–5.2) V = −0.5 V between the two materials due to the difference in the work function.

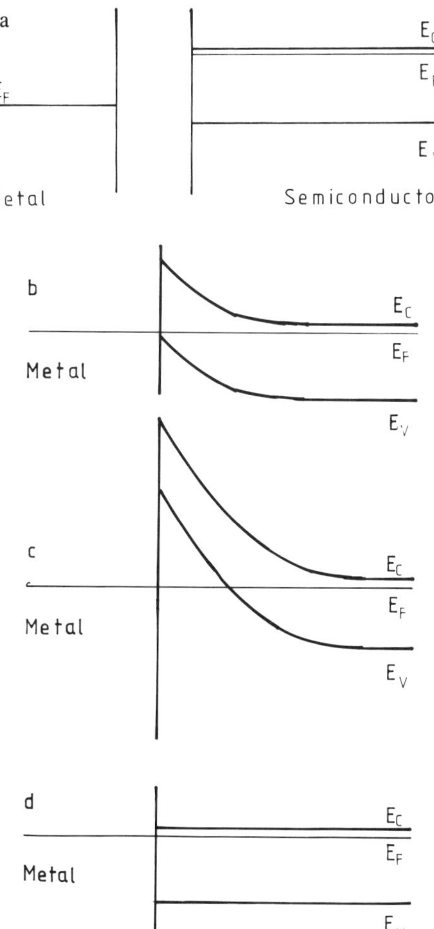

Figure 8.7 *A Schottky contact under various biases. The semiconductor is doped such that the Fermi level lies below the conduction band edge:*
a Metal and semiconductor not in contact with each other
b Non-biased Schottky contact
c Strongly negatively biased contact. At sufficient bias the conduction band edge rises above the Fermi level in the metal so that the contact starts to inject holes
d Positively biased contact to compensate exactly for the contact potential. Electrons are now allowed to enter from the metal into the semiconductor. The positive bias exactly compensates the Schottky contact potential, giving rise to the flat-band condition

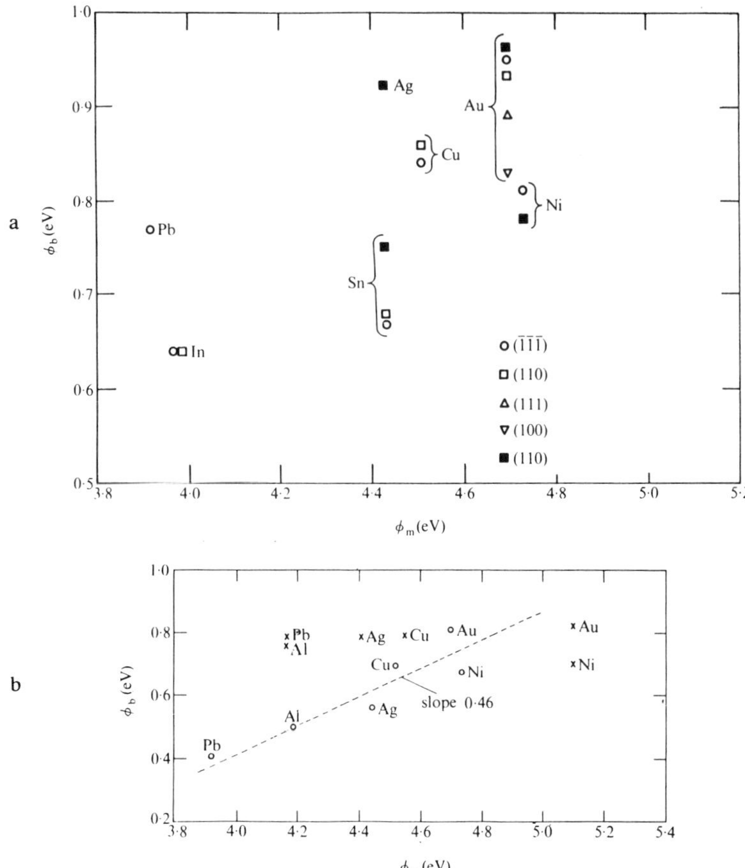

Figure 8.8 *a Barrier heights on etched (○) and cleaved (×) silicon surfaces. From Rhoderick (1978). In the cases of Ni and Au, the lower values of the work function ϕ_m have been used for the etched surfaces, and the higher for the cleaved (ultra high vacuum) surfaces*
b Barrier heights of chemically etched gallium arsenide. From Rhoderick (1978)

Unfortunately this is not correct as shown below; this approach to contacts is too simplistic. The atoms at the surface which are to become the contact will most likely rearrange (the surface reconstructs itself) because of unsatisfied chemical bonds. This implies a change in the electronic configuration at the surface, whereby the number of electron states changes. Electrons can be trapped here. Most likely additional atoms will also settle on it (oxidation and adsorption) forming additional electronic states. Experience shows that the contact barrier, V_s, settles between about -0.4 and -0.9 eV. Figure 8.8a and b show that the contact potential indeed depends on how the contact has been manufactured and not on the difference in the work function of the materials. Look *et al.* (1990) have published an extreme case of a Schottky barrier of -0.17 eV for GaAs grown by molecular beam epitaxy at 200°C.

When the two materials come in contact with each other the Fermi levels will align when no current is flowing. This implies that the conduction and valence band edges bend. (Figure 8.7b.) The alignment of the Fermi energy results in a carrier depletion which for uniformly doped semiconductors extends a distance, d, into it. This can be calculated from Poisson's equation in one dimension:

$$d = \sqrt{2\varepsilon\varepsilon_0 V_s/\rho_e} \tag{8.2}$$

where ε_0 represents the permittivity of vacuum, ε the dielectric constant of the semiconductor, V_s the contact potential and ρ_e the charge density of the ionised dopants. Usually V_s is negative for metal semiconductor contacts, seen from the metal side. (Many authors use the symbol Φ_B for this potential.)

When the contact has been *reverse biased* (given a negative bias, V_{gs}) the barrier V_s in Equation (8.2) increases (negatively) to $V_s + V_{gs}$. (Figure 8.7c.) The contact is *forward biased* when $V_{gs} > 0$. When $V_s = -V_{gs}$ then $V_s + V_{gs} = 0$ so that the effect of the Schottky contact potential has just been cancelled. This situation, known as *flat band* is depicted in Fig. 8.7d. The number of electrons that can overcome the Schottky barrier is

$$N = \int_{V_s}^{\infty} \frac{D_S(E)}{1 + \exp[E/(k_B T_f)]} \, dE \tag{8.3}$$

where $D_S(E)$ represents the density of states, k_B Boltzmann's constant and T_f the temperature of the electron gas. It is assumed that the electrons in the metal are in thermal equilibrium. The upper limit of the integral should be the valence band edge of the metal rather than infinity but this extension makes no significant effect on the numerical result. This expression also applies to holes when their energy is reckoned downward from the top of the valence band. (Section 3.8.)

The injection into a Schottky contact is modelled according to Equation (8.3). The contact should be allowed to absorb any particle that enters it.

The gate contact of a metal semiconductor field-effect transistor is of Schottky type. To model this transistor the contact potential *including* the Schottky contact potential has to be used as a boundary condition for Poisson's equation. We shall refer to the gate bias applied in the laboratory as the *gate bias* or *applied gate bias*, the sum of the gate bias and the Schottky contact potential as the *gate potential*. For example, when the Schottky contact potential is -0.75 V, and the gate bias is $+0.25$ V applied the gate potential is -0.50 V. It is very important that this distinction between these two concepts is made, because the carriers sense the gate potential rather than the gate bias applied in the external circuit.

8.3.2 The Ohmic contact

As with the Schottky contacts it is in general necessary to distinguish between the *applied Ohmic bias* and the *Ohmic contact potential*. One contact (e.g. the source or emitter of a transistor) serves as a *reference potential*, the other contact potentials are defined in relation to this. When the Ohmic contacts are made with the same kind of material with the same doping density as the reference contact the distinction between Ohmic contact potential and applied bias does not need to be made. When the material for the contacts is the same but with different doping as e.g. for the bipolar transistor or the p–n diode the difference in the bulk Fermi level has to be added to the contact potential. When the materials are different,

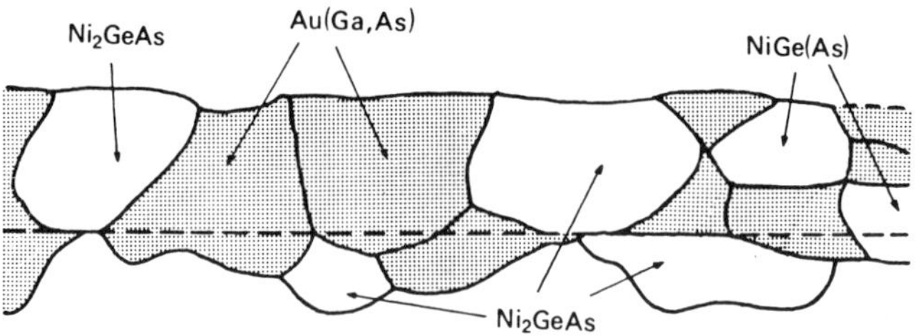

Figure 8.9 *Sketch showing a section through an Ohmic gold–germanium contact to gallium arsenide. The horizontal dashed line represents the surface of the semiconductor prior to the deposition of the contact. From Kuan* et al. *(1983)*

as for e.g. the Si/Ge diode, the difference in the work function of the two materials also has to be considered.

The physics of Ohmic contacts is very complex indeed. Actually an ideal Ohmic contact is a Schottky contact with a zero barrier. Figure 8.9 shows a typical structure of a gold–germanium–nickel contact. As this picture tells, metal and semiconductor molecules diffuse on manufacture across the original surface on which the metal has been deposited to form islands or pockets of different alloys. The band structure of such a system is complicated and deviates additionally from bulk because of interface states forming at the grain boundaries. The injection takes place partly by overcoming the barriers forming between the grains, partly by tunnelling through them. As the topography of individual contacts varies from one to another, and is additionally dependent on how the contact has been deposited, it is impossible to form a general picture of Ohmic contacts until the technology of contact making is sufficiently controlled to make reproducible structures. The criterion for Ohmic contacts used today is that the contact resistance should be reduced to a minimum. Interest in investigating the nature of Ohmic contacts is now emerging (Kim *et al.* 1990).

Boudville and McGill (1985) have suggested that the Ohmic contact consists of a triangular barrier of height equal to the Schottky contact potential and base given by the depletion length (Equation (8.2)). The electrons enter the semiconductor mainly by tunnelling through this barrier. This is indeed a simple model which should be considered. Lacking any proposal of a better model, we have modelled the Ohmic contacts by allowing the free exchange of carriers (electrons or holes) through it. For each field-adjusting time step, the number of electrons (holes) in each mesh cell adjacent to the contact is noted. If this cell has a net positive (negative) charge in it, *electrons (holes) are added until it is as charge neutral as practically possible*. The injected particles may have a Maxwellian or any other velocity distribution, with a resultant velocity component pointing into the semiconductor. A contact should either inject holes or electrons, but usually not both. Which type of particles are to be injected depends on the band structure of the contact.

In our example transistor we have not considered that the contact areas have a higher doping than the rest of the epilayer due to metal diffusing into it during manufacture. This, however, is probably a more realistic model for e.g. gold–germanium–nickel contacts. We shall return to this in Section 10.7.

For example, if the cell marked C in Fig. 8.5a and b has a net positive charge, more electrons have to be added to make it charge neutral; however, if it has a net negative charge the surplus carriers will *not* be removed from the cell because these particles may be scattered into the interior again.

As the reader may have gathered, the contact semiconductor material has to be doped in order for the contact to be effective in the model. This is also the case for the real material. A metal would not normally be able to inject current into the pure semiconductor material, because the pure material is a perfect insulator. When a carrier has entered the semiconductor, it may be scattered back into the contact by a phonon or an impurity. In the model this particle will be absorbed and not returned at another place in the device. A particle can only be injected according to the procedure just described. In a real system the particle will enter the contact again at a later time. Both the source and the drain inject and absorb particles. The source injects more particles than it absorbs; at the drain side the absorption rate exceeds that of the injection. Particles that are absorbed at one contact should *not be* re-entered at another one. They should, as has already been said, only be injected according to the scheme described above. As a consequence of this, the total number of particles in the simulated device does fluctuate with time; this is analogous to a grand canonical ensemble of statistical mechanics. This is also true for real devices. There is thus no explicit correlation between the source and the drain.

8.3.3 Free surfaces

As mentioned when discussing Schottky contacts, the free surfaces also reconstruct to create bound electrons. These represent a surface charge, giving rise to a *surface potential* and local depletion. These surface charges can partly be eliminated by passivation. For the GaAs free surface the surface potential is $V_{sf} = 0.6$ V. Surface charges should be modelled by imposing a *surface charge* density of

$$\rho_{sf} = \sqrt{2\varepsilon\varepsilon_0 V_{sf}\rho_e} \tag{8.4}$$

at the mesh cells adjacent to the surface. Here ρ_e represents the density of ionised dopants adjacent to the surface; the other symbols have the same meaning as in Equation (8.2). Imposing a potential V_{sf} as a boundary condition for the field equation at the free surfaces is incorrect.

8.4 INITIALISATION OF A SIMULATION

Having defined the device geometry, doping and bias of the electrodes, the simulation can start. The very first problem appearing now is the choice of an initial distribution of the carriers, both in geometrical and in momentum space. To start with it is most likely to be unknown and therefore a guess has to be made. Fortunately this is a fairly easy task because even when this guess is wrong, the correct distribution will eventually emerge. When the correct distribution has been obtained, it will stay correct even when the biases change during the simulation. The carriers will react to a change in the bias in the form of the correct transient response.

The time it takes to obtain the correct carrier distribution depends on how well our first guess has been. Usually it is of the order of a few hundred time steps. The time it takes to obtain a new steady state from a correct distribution after changing the bias, however, depends on the geometry of the device. In this case the evolution of the new steady state represents the genuine transient response of the device.

A possible suggestion for the initial distribution is to assume charge neutrality throughout the device as far as practically possible, and a Maxwellian or Fermi–Dirac distribution of the particles' energy.

During simulation regular security dumps of the particle coordinates should be made. This reduces the loss of computer time should the simulation stop prematurely. Such a security dump contains the geometrical and momentum space distribution of the carriers and can also serve as an initialisation of a continuation simulation.

8.5 THE SUPERPARTICLE

The length of the simulated area AA' of Fig. 8.4 is 2.57 μm for the example transistor, the depth of the epilayer is 0.1 μm, and, assuming a reasonable length of 100 μm, we find that the simulated transistor under flat-band gate bias would contain $N = 5.14 \times 10^6$ electrons. This is too large a number of electrons to follow even with today's modern supercomputers. The time such a simulation takes would be too long. It is feasible to follow the transport histories of only N_s ($\leqslant N$) carriers. Sampling theory allows us to get a meaningful picture of the transport provided the selection of the N_s particles out of the N is random: this means that no bias towards certain properties has been applied in the selection procedure. Each of the selected particles thus represents N/N_s real particles; these representatives are referred to as *superparticles*.

Solving the field equations and calculating the currents passing through the electrodes, each particle has to be attributed a charge

$$e_s = eN/N_s \tag{8.5}$$

where e represents the elementary charge of a particle. *During transport, however, the particle carries the elementary charge, e, only*. The charge attributed to the superparticle must, once it has been chosen or defined, *never change during the course of the simulation*. A change is dangerous because it yields unphysical results. We shall find that the number of superparticles changes during transient states of the calculation; this reflects what happens in true devices which also experience fluctuations in the number of electrons and holes. In other words, the charge of the device changes when the biases change. It has an electrical capacitance. During steady state the number of carriers remains, apart from small fluctuations, on the average constant.

In our *two-dimensional* simulation the selection of the section A–A', Fig. 8.2, is immaterial. The superparticles may be considered as rods extending throughout the entire width of the transistor. The supercharge carried by the superparticles is proportional to the width of the transistor. The currents are also proportional to the width of transistors of identical lateral geometry. As the width does not enter the simulator explicitly, the calculated currents and superparticle charges can be normalised to the unit length, here chosen to be the metre to comply with the SI units. Many authors prefer to quote the currents per millimetre width.

In our example transistor the epilayer covers $257 \times 10 = 2570$ mesh cells (the epilayer is 10 cells thick). If we choose the superparticle factor $\nu_s = 4$ superparticles to neutralise the positive background charge due to the ionised donors for each cell in the uniformly doped epilayer, our simulation of the example transistor will start with 10 280 superparticles when starting from an entirely charge neutral initial condition.

One cell of the epilayer contains a doping charge of

$$Q_c = h_x h_y \rho_e \tag{8.6a}$$

where ρ_s represents the density of ionised dopants, h_x and h_y the cell size in the x and y directions respectively. Q_c has dimension charge per unit length which is in line with the calculated currents we discussed above. Q_c can be given in C m^{-1}. For three-dimensional simulations

$$Q_c = h_x h_y h_z \rho_e \tag{8.6b}$$

where h_z represents the cell size in the z direction. Now Q_c has dimension Coulomb. If ν_s superparticles are required to neutralise the doping charge of the mesh cell, the charge per superparticle is chosen to be

$$e_s = Q_c / \nu_s \tag{8.7}$$

The *superparticle factor*, ν_s, does not have to be an integer, although the choice $\nu_s < 1$ may lead to problems in obtaining an accurate solution of Poisson's equation.

When the active layer has not been uniformly doped, the choice of e_s is not that straightforward as above. In this case the superparticle charge may be chosen to neutralise only the layer of cells immediately adjacent to the injecting contacts.

The usual number of superparticles lies between 5000 and 100 000. As the devices become smaller and computers progressively faster, it is only a matter of time before we can follow all particles, i.e. let one superparticle represent one real particle. We should never have more superparticles than real ones because the result is unphysical.

8.6 THE STEADY STATE CHARACTERISTICS

Even from a wrong initial distribution of carriers in the device the correct one will eventually evolve. At constant drain and gate bias (the source has been grounded) the steady state of our example transistor will evolve. Apart from fluctuations, the average number of superparticles and the drain current remain constant and the currents in and out of the device balance. The reader may think that they could relax the requirement of maintaining the regular solution of the field equation once the steady state has been achieved believing that the field distribution does not change any further. *It is strongly advised to continue solving the field equation regularly* during steady state, because there are always noisy fluctuations of physical origin. Ceasing to solve it means that the local fluctuations in the field are being frozen, which can lead to unphysical results. In Section 8.8 we shall see an example where regular oscillations, e.g. Gunn oscillations, take place in a transistor even when both the gate and the drain are kept at constant potential. The Gunn oscillations would not have been revealed without maintaining the regular re-solution of the field equation.

Figure 8.10 shows the number of superparticles and the accumulative count of particles leaving the drain against time. The slope of the latter curve yields the drain current. The curves show clear fluctuations, which are due to noise which is of physical origin. This noise will be discussed in Chapter 12. The regularities in the fluctuations are due to plasma oscillations which, according to Equation (3.117), have a frequency of 9.2×10^{13} Hz. The oscillating period is shorter than our field-adjusting time step but shows because of the frequency modulation with the frequency with which we solve the field equation.

As stated, we obtain the current by counting the net charge passing through the contacts. This works well for steady state transport; when we turn to transient analysis, however, we also have to include the displacement current - we shall return to this in Section 9.2. Recently Gružinskis *et al.* (1991) suggested a more accurate method to calculate the currents based on integrating over the charges in the regions between the electrodes. The method, however, does not apply to transient studies. An estimate can also be obtained from Ramo–Shockley's theorem (Kim *et al.*, 1991) which for constant biases estimates the current at electrode or port $N°$ i to be:

$$I_i = \sum_{j=1}^{N_S} e_j \mathbf{v}_j \bullet \mathbf{F}_i(\mathbf{r}_j) \tag{8.8}$$

where the summation j runs over all particles, $\mathbf{v}_j$ represents the velocity of carrier $N°$ j which has charge e_j (accounting for the possibility of two species of carriers, namely electrons and holes). $\mathbf{F}_i(\mathbf{r}_j)$ represents the field from electrode $N°$ i when it is biased at unit potential and all other electrodes have been connected to ground and there are no charges, neither stationary nor mobile, in the device.

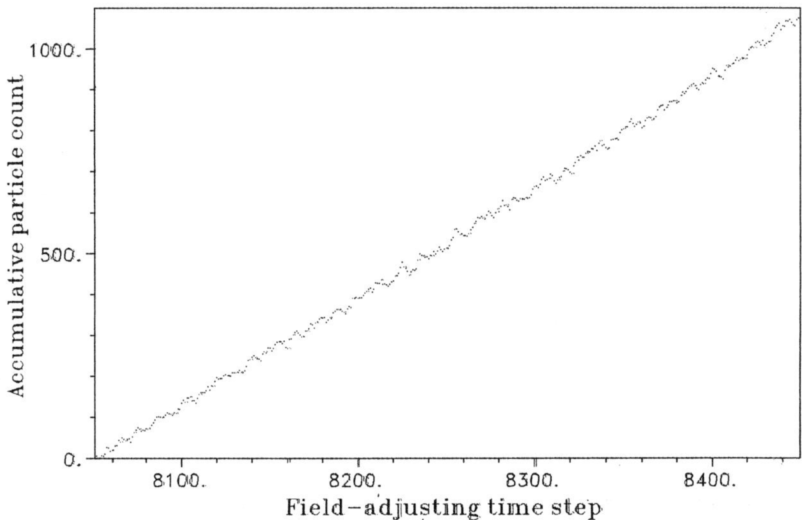

Figure 8.10 *The net accumulative count of the number of superparticles leaving the example transistor through the drain during steady state*

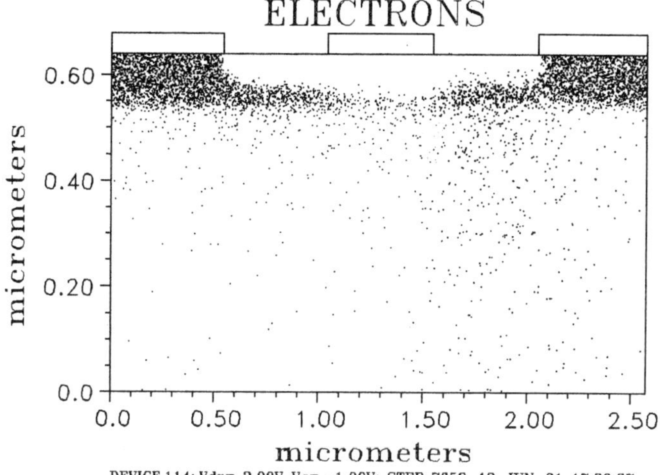

Figure 8.11 *Instantaneous distribution of electrons in the example transistor during steady state. Light blue points represent electrons in the Γ minimum of the Brillouin zone, dark blue L electrons. Hardly any X electrons (green) are present. Note the different linear scale along and perpendicular to the surface of the transistor. Drain bias: 2 V. Gate potential: −1.0 V (including the Schottky contact potential of −0.75 V). Field-adjusting time step: 25 fs. Note, this figure is reproduced in colour as Plate 3*

The map of the electric field from which $\mathbf{F}_i(\mathbf{r}_j)$ is derived is obtained by taking the gradient which has to be calculated from Poisson's equation once and for all for the device in study. In none of the examples presented here, however, has this equation been used.

Plate 3 shows, in colour, the instantaneous distribution of the electrons throughout the device during steady state. The light blue points represent electrons near the Γ minimum of the Brillouin zone, the dark blue points electrons near any of the L minima. The same band structure and the same scattering mechanisms have been included in the model as described in Section 6.3. We also consider ionised impurity scattering in the epilayer.

The negative gate potential and the trapped electrons at the free surfaces of the transistor both repel electrons, creating a positively space-charged *depletion zone*. The transition between this depletion zone and the charge neutral bulk is far from sharp as assumed in the gradual channel approximation. The thickness of the transition layer between the fully depleted material and the charge neutral one is about one *Debye length* λ_D:

$$\lambda_D = \sqrt{\frac{\varepsilon\varepsilon_0 k_B T_f}{e^2 n}} \tag{8.9}$$

with n representing the carrier density, ε_0 the permittivity of vacuum, e the elementary electronic charge, ε the dielectric constant, k_B Boltzmann's constant and T_f the temperature of the electron gas.

The Debye length reflects the success of the particles of different kinetic energy

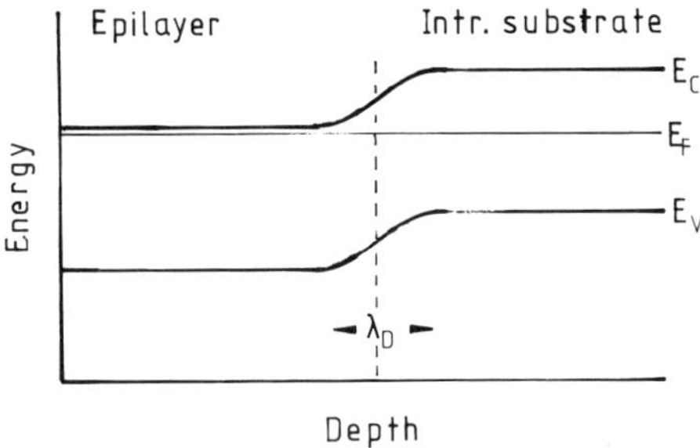

Figure 8.12 *Band diagram of the epilayer–substrate interface. The width of the transistor is about one Debye length, λ_D. E_C = conduction band edge. E_V = valence band edge energy. E_F = Fermi energy level. The energy and depth scales are both linear*

entering the depletion zone: for our example transistor the Debye length is about 10 nm. A transistor with a gate length of about one Debye length would therefore not function.

Furthermore, the transition in carrier concentration at the border between the epilayer and the substrate is also diffuse. Electrons from the epilayer enter the substrate leaving a positive space charge behind which tends to pull them back again. The electrons in the substrate represent a negative space charge, so that a dipole is established along the interface. More accurately, the charge distribution can be discussed in view of band bending: when no net current flows across the interface, the Fermi level remains constant in the direction perpendicular to it. It is near the conduction band edge in the epilayer, and near the centre of the band gap in the ideally intrinsic substrate. (Figure 8.12.) This results in band bending at the interface, which in turn establishes a potential barrier. The more energetic carriers moving parallel to the interface can, after scattering, head towards it and

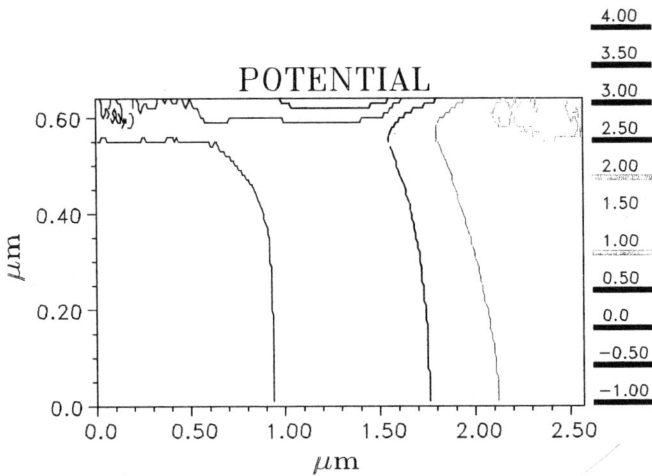

Figure 8.13 *Instantaneous equipotential lines for the same bias conditions as in Fig. 8.11. The geometrical scale is the same as in Fig. 8.11. Note, this figure is reproduced in colour as Plate 4*

overcome it if they have sufficient energy. In this way *substrate currents* evolve. The most likely place of passage over the barrier is in the high-field region between the gate and the drain. The existence of substrate currents was first discussed by Hockney *et al.* (1974) and their existence confirmed by simulation on several occasions afterwards.

Figure 8.13 shows an instantaneous distribution of the electrostatic field. The irregularities in the 0 and 2 V equipotential lines at respectively the source and the drain are due to noise. In the gate region the lines tend to follow the depletion edge and the interface between the epilayer and the substrate. Between the gate and the drain the lines run almost perpendicular to this interface. The electric field is the gradient of the equipotential lines, Equation (1.9). Figure 8.14 shows the magnitude of the electric field throughout the example transistor. This figure has been derived from Fig. 8.13. The noise manifests itself in the irregular features of the lines of constant magnitude.

The ridge in the electric field at the epilayer substrate interface is clearly visible in the left part of Fig. 8.14. The high-field region in the depleted area of the epilayer between the source and the drain is due to the positive space charge created by the ionised donors there, and is of no importance to the transport. However, too large a field causes dielectric breakdown. The *high-field region* between the gate and the drain peaks near the epilayer substrate interface under the edge of the drain contact. Here the electrons attain enough energy to be scattered into an L minimum by means of intervalley scattering. These electrons gain herewith a factor of $5\frac{1}{2}$ in effective mass, whereby they slow down and congregate, forming a negative space charge. These electrons are represented by dark blue spots in Fig. 8.11. Scrutinising Figs. 8.11 and 8.14 the reader will find that the distribution of the heavy electrons extends towards the drain beyond the peak of the high-field region. From Fig. 6.2 we might expect the majority of the carriers to return to the central minimum of the Brillouin zone. The cause of the delay in returning to it is that the electron has to wait for the right phonon to ferry it back to the Γ valley. This point illustrates an important difference between the

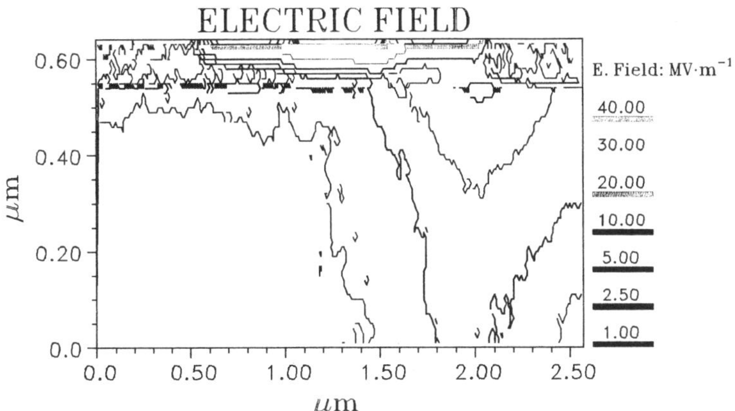

Figure 8.14 *Magnitude of electric field throughout the example transistor of the same bias as in Fig. 8.11. The units may be unfamiliar to the reader: 1 MV m^{-1} = 10 kV cm^{-1}. The geometrical scale is the same as in Fig. 8.11. Note, this figure is reproduced in colour as Plate 5*

drift-diffusion approach and our physical particle model. Electrons have mass, and therefore inertia, and depend on phonons to travel from one minimum of the conduction band to another. Both these effects have been neglected in the drift-diffusion model. In the hydrodynamic model various macroscopic relaxation times have to be introduced in order to describe it.

The current is obtained by counting the net flow of charge passing through the contact over a time sufficiently long to average over the fluctuations. For our example transistor this time has been 10 ps consisting of 400 field-adjusting time steps of 25 fs. We could just as well simulate over shorter intervals of time and average over the runs. Runs of a couple of picoseconds or less show relatively large fluctuations in the drain current. These are due to noise of terahertz frequencies. The longer the simulation time chosen, the more accurate the estimate of the currents will be. A simulation over a long time is more economical in computer time than increasing the number of superparticles, because the computer time needed to achieve a steady state after initialisation or change of bias also increases with the size of the ensemble.

The next point of the steady state characteristic is calculated changing the bias of one of the contacts and proceeding as above, starting from the previous steady state carrier distribution. It was found that the transients had decayed to insignificance after about 10 ps. We may generally find that the transients are oscillating. The change in the bias causes a shock causing the charge to flow back and forth like a liquid in a container being struck. In this case we have to be sure these oscillations have decayed before we can estimate the value for the steady state current.

Figure 8.15 shows the calculated *transfer characteristics* of the example transistor at 2 V drain bias; Figure 8.16 shows a *forward characteristic*. The drain current depends on the width of the transistor. As mentioned before, we have chosen to normalise to the width of one metre. The currents displayed in Figs 8.15 and 8.16 have been normalised thus. The actual current is obtained by multiplying the

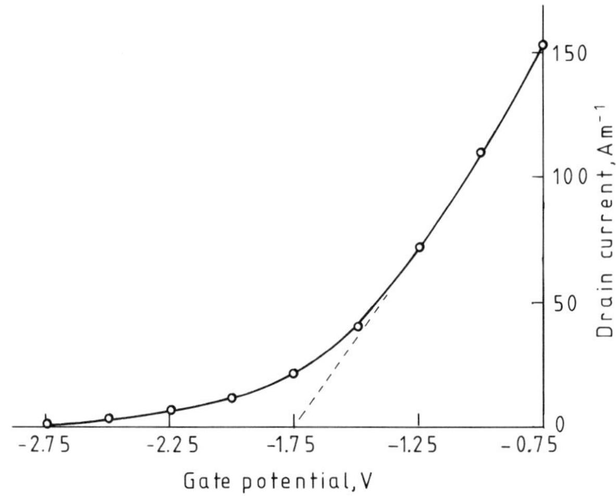

Figure 8.15 *The transfer characteristic against the gate potential for 2 V drain bias. The dashed line represents an extrapolation of the current to zero in order to define the gate threshold voltage.*

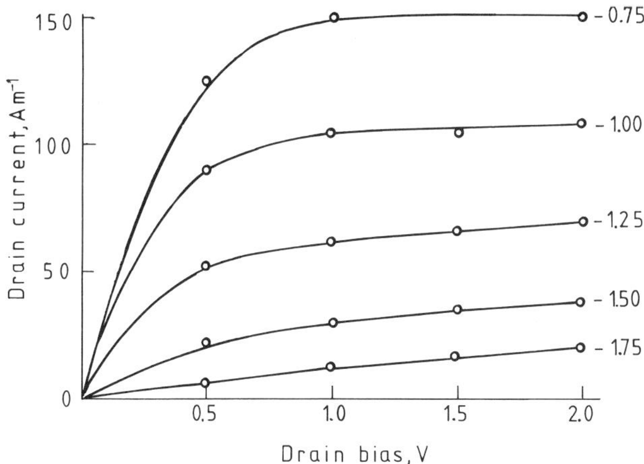

Figure 8.16 *Forward characteristic of the example transistor; the drain current against drain bias for different gate biases. The values of the latter include the Schottky contact potential. To obtain the corresponding laboratory potential or applied bias add 0.75 V*

calculated values for the currents by the actual width.

Figure 8.15 shows that the drain current reduces with increasing negative drain bias. Eventually the current reaches zero - the transistor has been *pinched off*, or is in the *off-state*. The transition to this off-state is smooth, the pinch-off or *gate threshold voltage* is therefore not clearly defined. An exact definition is only possible when the current abates linearly down to zero. Usually we determine the threshold by extrapolating the linear part of the transfer characteristic to zero as indicated by the dashed line in Fig. 8.15. The tail is caused by electrons entering the substrate from the source side and passing to the drain via the interior. Increasing the drain bias (negatively) these electrons will have to penetrate deeper; eventually this is no longer possible so the current passage closes. The slope of the transfer characteristic yields the *transconductance* against the gate bias. It has a maximum near -0.75 V gate potential. (Figure 8.15.)

The forward characteristics demonstrate the usual features: a sharp increase in the drain current for small drain bias, followed by near saturation for a larger bias. The initial slope is an expression of the contact resistance - the transistor is in its Ohmic range. The forward characteristics have been calculated for different drain biases, as indicated. The parameter of Fig. 8.16 represents the gate potential, which includes the Schottky contact potential. The saturated part of the curve for -0.75 V gate bias, i.e. of the curve corresponding to zero applied gate bias, represents the *saturation current* of the transistor. The slope of the slowly varying part of the forward characteristic is known as the *drain admittance*. This should ideally be zero - a small transadmittance together with a large transconductance means a large voltage gain, Equation (9.21). Usually the current characteristic slopes in the saturated region, so that the concept of saturation is ill-defined. Unless the transadmittance is zero, we also have to state the value of the drain current when quoting a value for the saturation current.

Scattering from optical and acoustic phonons, and from phonons causing electrons to transfer between the various minima of the conduction band, has been considered in the simulation of the transistor characteristics. Also scattering from the ionised donors has been included in the model. These scattering mechanisms have rates of the order of $10^{12}\ s^{-1}$, and above (Figure 5.4); the scattering times are thus of the order of a picosecond or less. The trapping and release of carriers has not been considered because their rates are of the order of microseconds, which require very much longer computer time waiting for something to happen. The most significant effect of trapping and releasing carriers is to modify the local field distribution.

Transistor current characteristics are usually measured by repetitive scans at megahertz frequency which is too short for the dynamics of trapping and release to reach equilibrium. Such a measurement technique corresponds to our simulation of the d.c. characteristics of the example transistor. Measuring the characteristics during d.c. operation by observing the drain current point for point takes seconds, or longer, per point. This would yield different characteristics because then there would be sufficient time to allow the trapping and release to attain equilibrium.

A direct Monte Carlo simulation of the dynamics of trapping is not straightforward because of the time scale involved. Accelerating the process artificially leads to unphysical results. We should note that the trapping and release rates depend on the local Fermi level. A possible method to simulate the dynamics of the trapping and release of carriers is to introduce accelerated trapping and release rates for a short time and then use the physically correct model to obtain the amended carrier distribution for the new distribution of trapped charges. This could be applied alternatively until the distribution of the trapped charges no longer changes. This approach would, of course, be invalid for a.c. analysis.

8.7 NEGATIVE AND POSITIVE DIFFERENTIAL RESISTIVITY

Figure 8.16 shows that the drain current increases slightly with the drain bias in the saturation region. The slope gives the *transadmittance*, G_d, of the equivalent circuit, Fig. 9.6, of the transistor. The voltage gain, Equation (9.21), is inversely proportional to y_{22}, which should be as small as possible in order to obtain a large gain.

It does happen that the drain current starts to reduce with increasing drain bias. This phenomenon, known as negative differential resistivity, has, among others, been observed by Tsironis and Dekkers (1980) and Willig and de Santis (1977). A transistor exhibiting this property is undesirable as an amplifier because it leads to oscillations in the microwave circuit. Tsironis and Dekkers (1980) attributed the effect to the measuring apparatus. Ridley (1982) attributed the effect to plasmon–phonon scattering.

How can, however, a transistor made from a material which is intrinsically differential negative resistive exhibit a positive transadmittance? Figure 6.1 shows that the drift velocity of the electrons against the electric field first increases, then decreases, for fields above $0.3\ \mathrm{MV\ m^{-1}}$. A decreasing drift velocity implies a decreasing current when the number of electrons remains the same, so that GaAs is negative differential resistive. Moglestue (1983b) has shown, by means of Monte Carlo particle modelling, that the substrate currents are responsible for the positive differential resistivity: prohibiting the electrons from entering the substrate by replacing the epilayer substrate interface with a reflecting face, the transistor will become *negative differential resistive*.

As the drain bias increases, the high-field region between the gate and the drain extends. An increasing number of particles obtain sufficient kinetic energy to be scattered into one of the secondary (namely L or X) minima of the conduction band, whereby their effective mass increases and their drift velocity decreases accordingly. This results in a reduction of the drain currents flowing directly through the epilayer.

However, some particles can enter the substrate by heading towards it after appropriate scattering from an ionised impurity atom or a low momentum phonon. Once inside the substrate this particle has a great chance of avoiding the high-field region by travelling around it. In spite of the detour it makes, it will arrive at the drain earlier than the particles taking the direct route, delayed because of their temporary gain in effective mass. When the substrate current is sufficiently strong, as it is for our example transistor, the differential negative resistivity has been more than compensated for, making the transistor *positive differential resistive*.

Prohibiting the particles from entering the substrate produces a negative differential resistive transistor. It is possible to make such a transistor by etching the substrate away from the back side, but it is easier to reduce the substrate current by introducing a p-type buffer between the substrate and the epilayer. The same effect can also be obtained by backgating the transistor, i.e. placing an additional contact underneath the gate or besides the transistor (d'Avanzo, 1982). Placing a material with a wider bandgap between the epilayer and the substrate, however, does not necessarily produce negative differential resistivity: a potential well forms at the interface between the two materials which gradually empties with increasing (negative) gate bias resulting in a positive transadmittance (Moglestue, 1990).

Negative differential resistivity is not expected in silicon MOSFETs because the drift velocity of the electrons increases with increasing electric field. (Figure 6.10.)

8.8 WANDERING GUNN DOMAINS

In many transistors the Gunn domain that forms between the gate and the drain is stationary. If the doping of the active layer of a transistor of otherwise identical geometry had been increased, the Gunn domain will no longer be stationary; it will move towards the drain, and as it gets absorbed there a new one nucleates near the gate. Sze (1969) has given a simple rule to predict when wandering Gunn domains can be expected:

$$n_e L_{sd} > \varepsilon\varepsilon_0 v_s/(e\mu_e) \tag{8.10}$$

where n_e represents the carrier density, L_{sd} the length of the Gunn domain path e.g. between the source and the drain, ε_0 the permittivity of vacuum, ε the dielectric constant of the material, v_s the saturation velocity of the carriers and μ_e their mobility. For n-type GaAs, $\varepsilon\varepsilon_0 v_s/(e\mu_e)$ is about 10^{14} m^{-2}. Samples with $n_e L_{sd}$ smaller than this value would exhibit a stable field distribution – this is not the case with the transistor of Fig. 8.17. This rule is only approximate; a more accurate formulation would involve complex mathematics which is not considered

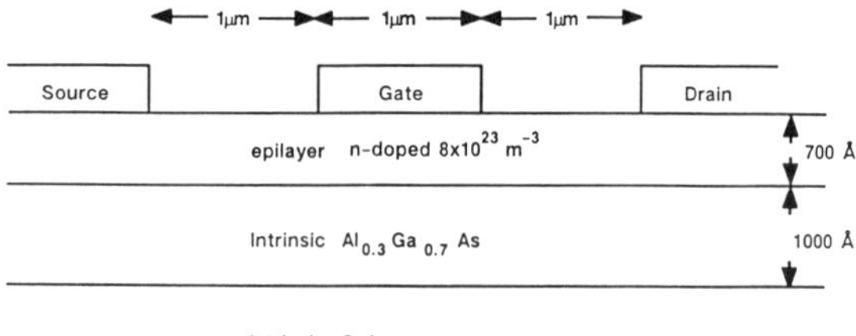

Figure 8.17 *Cross-section of a heterojunction field-effect transistor exhibiting wandering Gunn domains. All relevant transistor data can be read from the figure. The gate potential includes the Schottky contact potential*

necessary to introduce here, as will be explained below.

Gunn domains were first studied in one dimension by Lebwohl and Price (1971a) by means of Monte Carlo simulation. Recently Moglestue (1990) has simulated an *inverted heterojunction transistor (HEMT)*, Fig. 8.17, demonstrating the dynamics of wandering Gunn domains. The 0.07 μm epilayer is uniformly doped with a donor density of $8 \times 10^{23}\,\mathrm{m}^{-3}$. A 0.1 μm undoped $Al_{0.3}Ga_{0.7}As$ layer separating the epilayer from the substrate allows only a few electrons to enter the intrinsic substrate in a way which will be discussed in Chapter 10. Electrons which have succeeded in entering it prevent further entry of other carriers by electrostatic repulsion. Trapped electrons at the free surfaces between the contacts cause surface depletion, Fig. 8.18, corresponding to a surface charge potential of -0.6 V when no externally applied bias would exist (Chandra *et al.*, 1979).

The combination of doping density and epilayer thickness fulfils the condition for formation of moving Gunn domains, Equation (8.10). Although the problem is a steady state one, given that the drain and the gate have fixed biases and that no transient effects are present, the simulation clearly shows how the Gunn domains are generated, moved across to the drain at saturation drift velocity,

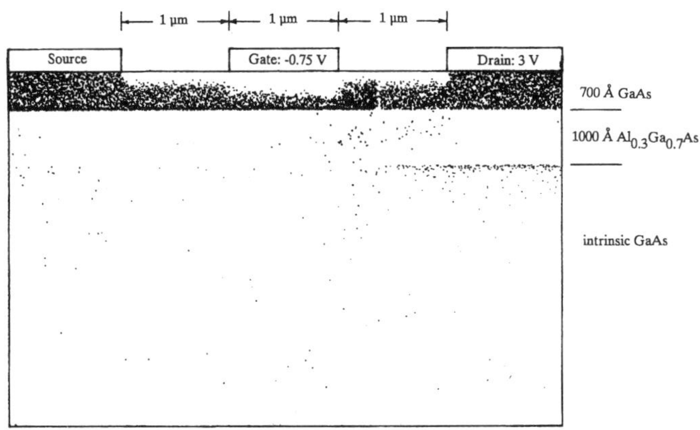

Figure 8.18 *Instantaneous electron distribution in the Gunn oscillating transistor of Fig. 8.17. Drain bias: 3 V, gate potential: −0.75, including the Schottky contact potential. The horizontal lines outside the left and right edge indicate the interface between the GaAs and the $Al_{0.3}Ga_{0.7}As$. Note the different linear vertical and horizontal geometrical scales. Note, this figure is reproduced in colour as Plate 6*

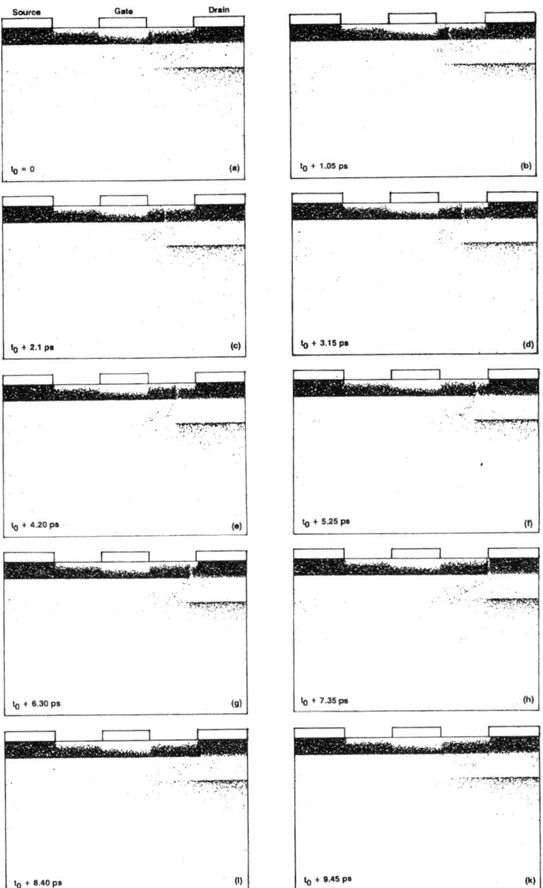

Figure 8.19 *The time evolution of the Gunn domain throughout the entire cycle. The domain moves towards the drain at the saturation drift velocity, 10^5 m s^{-1}. Light blue: Γ electrons, dark blue: L electrons. The instant shown as Fig. 8.16 corresponds to the time $t = t_0 + 2.5$ ps. Note, this figure is reproduced in colour as Plate 7*

about 10^5 m s^{-1}, and absorbed at the drain at the same time as the next one nucleates at the gate. (Figure 8.19.) The drain current shows a sharp peak every time a domain arrives at the drain. (Figure 8.20.) This result clearly demonstrates the necessity of solving the field equation regularly, even during steady state. If this important step has been omitted, the effect will not be seen.

The phenomenon has a hydrodynamic equivalent: depending on the speed or pressure with which a gas or liquid emerges from a nozzle into a wider pipe, the flow becomes laminar or turbulent. The wandering domains correspond to the turbulent flow. The problem of the electrical current is more complex than the corresponding hydrodynamic flow because of the mutual Coulombic repulsion between the particles. A complete picture of the flow can only be obtained by solving both Poisson's field and Boltzmann's transport equation. This is exactly what the Monte Carlo particle model is doing. It is therefore not necessary to discuss

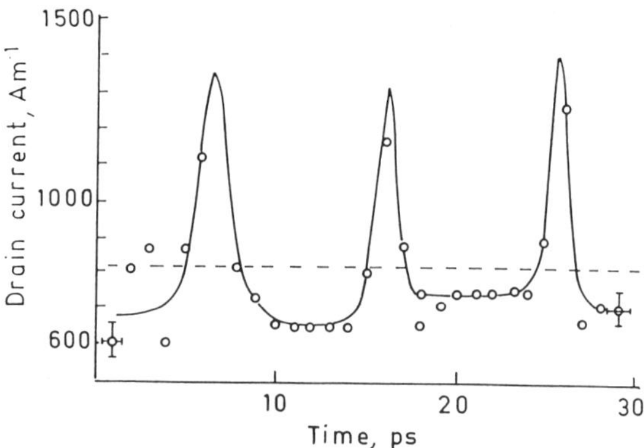

Figure 8.20 *The temporal dependence of the particle current at the drain. The peaks are caused by the absorption of Gunn domains. The dashed line represents the average drain current*

further details of Gunn domain formation. Recently Gribnikov and Zheleznyak (1991) have studies the formation and propagation of Gunn domains in transistors in two dimensions using the drift-diffusion approach.

8.9 LUMINESCENCE

It has been observed that transistors can emit light when the drain has been biased sufficiently high. Figure 8.21 shows that the light is emitted from the area immediately adjacent to the drain. This area coincides with that of the largest electric field. Spectral analysis reveals that the energy of the light is less than that of the band gap. The light cannot therefore be caused by electron–hole recombination, nor can it be due to the trapping of electrons because then it would manifest itself everywhere. The light shines for drain biases below the onset of avalanche multiplication. The energy of the light has been analysed experimentally by Zappe and Moglestue (1990) and Herzog *et al.* (1989) and been attributed to *bremsstrahlung* (German for retardation radiation). Usually this phenomenon is attributed to electrons emitting electromagnetic radiation when they are decelerated in the electrostatic field of the nucleus. It could also happen in semiconductors: scattering can cause the particle to move against the local electric field so that it loses energy and therefore emits photons. Since the energy of this photon reflects the energy of the particle, there should therefore be a close correlation between the spectrum of the bremsstrahlung and the energy distribution of the electron population.

Figure 8.22 shows the cross-section of a MESFET on which this light has been observed. The number of particles against energy has been recorded in two locations between the gate and the drain for a time span of 0.5 ps. The figure also shows the energy distribution of the particles divided by the density of states in these two places. In the logarithmic scale of the figure these distributions should be a straight line if the electrons had a Maxwellian energy distribution; this is indeed true for its tail. Figure 8.22 shows that the electron temperature rises towards the drain. The tail extends progressively towards higher energies as the carriers move into the high-field region and slopes less, indicating a higher electron

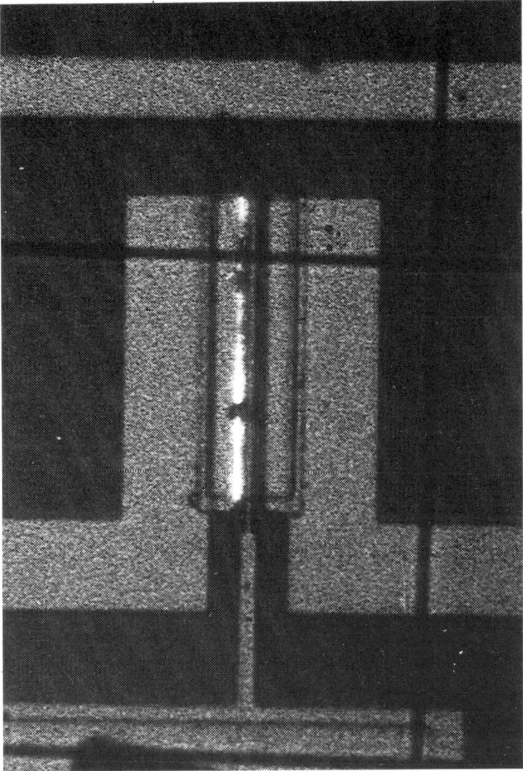

Figure 8.21 *Gate luminescence from a field-effect transistor as observed by Herzog* et al. *(1989).* $V_O = 7.6\ V$. $V_G = 0\ V$

temperature. Figure 8.23 shows that the measured energy spectrum of the bremsstrahlung correlates well with the calculated energy distribution of the particles. Assuming this correlation is true, we should, on the basis of the calculated results, expect the bremsstrahlung to reach maximum intensity nearest the drain, in agreement with observations.

8.10 HEATING

Accounting for the amount of energy exchanged between the lattice and the carriers in the form of phonons, it is possible to get a picture of the heat development rate in the simulated device. The heat development rate is calculated by mapping the accumulated absorbed and generated phonon energy throughout the simulation area for a certain time. Figure 8.24 shows the heating rate in our example transistor shown in Fig. 8.4, calculated over 8 ps. The blue colours represent cooling; black, magenta, red and yellow various degrees of heating. In the greater part of the transistor the local cooling cancels the heating approximately; the most intense refrigerating effect seen under the source and the drain is a result of the injected particles being too cold.

A comparison with a calculation of the Joule heating rate reveals that the

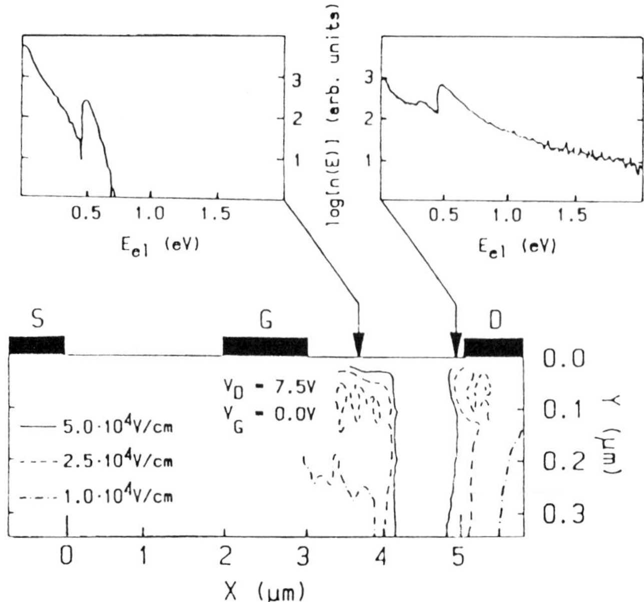

Figure 8.22 *Cross-section through the transistor for which gate luminescence has been observed (lower part) with the magnitude of the electric field calculated by the particle simulator. The two upper panels show the logarithmic energy distribution calculated from the Monte Carlo model divided by the density of states at the two indicated places between the gate and the drain. Drain bias: 7.5 V. Applied gate bias: 0 V. Gate length: 1 μm. Width: 50 μm. Source to gate and gate to drain distance: 2 μm. The epilayer is doped uniformly at 7 × 10^{23} m^{-3}*

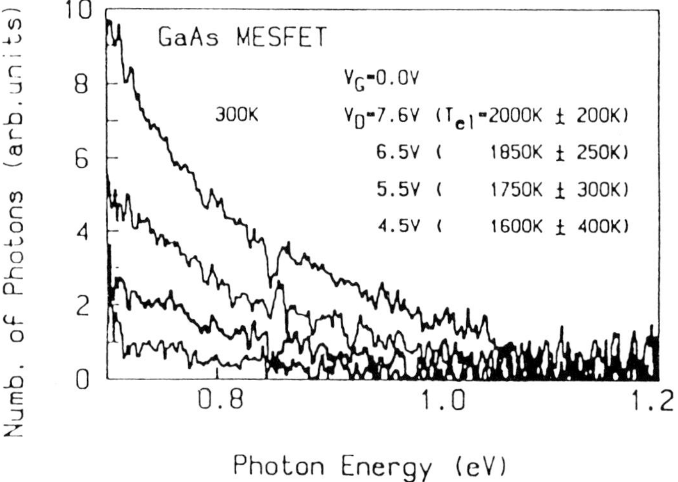

Figure 8.23 *Measured continuous background emission from the transistor of Fig. 8.22 below the band gap for several drain biases. T_{el} represents the decay constant for a fit to an exponential decay*

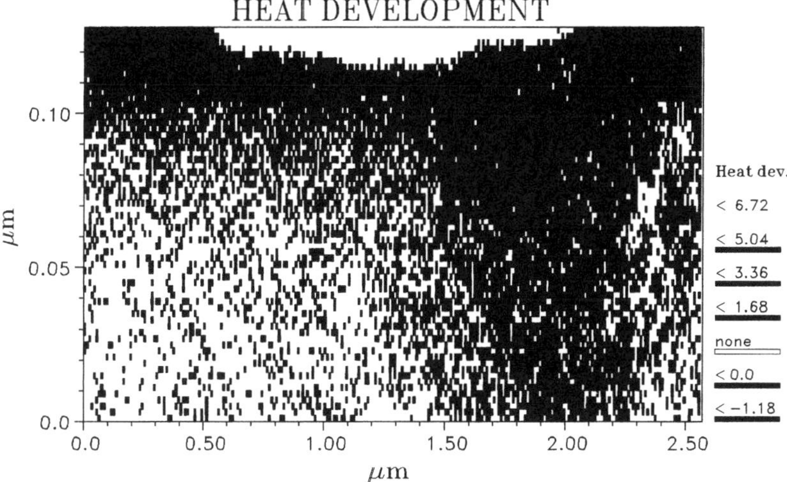

Figure 8.24 *The heating rate in the transistor of Fig. 8.4. The colour code gives ranges of heating rates in units of 10^{15} W m^{-3}, the blue hues represent cooling, the others heating. The strong cooling adjacent to the source and the drain is a result of injecting too-cold electrons. Drain bias: 2 V. Gate potential: −1 V, including the Schottky contact potential of −0.75 V. Note, this figure is reproduced in colour as Plate 8*

maximum heating obtained from the particle model takes place nearer the surface than expected, compared to an estimate based on the product of the local current density and electric field (Moglestue, 1981). The main cause of this is that the electrons have to wait for a suitable phonon to interact with after having been excited into an L minimum in the high-field region. The electron then travels some distance from where the Joule heating rate is expected to culminate, before it surrenders its energy to the lattice.

We have to emphasise that Fig. 8.24 shows the *heating rate*, *not* the actual temperature distribution. The latter depends on the efficiency of the surroundings (or mount) in removing the generated heat. Whatever way the heat can be led away we should expect the yellow and red parts of the transistor to be the hottest during operation, and this area to be the most likely where burn-out may take place.

We can also calculate the actual temperature by means of the Monte Carlo particle model. We consider the phonons as particles which scatter from each other and can split or combine in the process. We calculate the scattering rate from the higher order, or anharmonic, terms of the crystal Hamiltonian, Equation (2.29), which we have neglected. With a correct description of how the phonons can be exchanged with the surroundings we can map the local phonon distribution everywhere, just as we did for the electrons, and hence get the temperature distribution. As phonons are one to two orders of magnitude slower than electrons, a simulation involving both electrons and phonons will take correspondingly longer before the steady state for both species has been reached. The alternative approach would be to use an ordinary heat diffusion theory coupled with a heating rate calculated from our Monte Carlo model.

9

Alternating Current, Microwaves

9.1 INTRODUCTION

When the steady state has been reached in the simulation, the carrier distribution that has been obtained is correct; it satisfies the transport equation. If the bias condition of the simulated device now changes, the entire transient response will always satisfy the transport equation. Once the solution of the transport cquation has been found it remains a solution however the bias of the contacts changes. This means that transient as well as steady state characteristics can be calculated.

We may find the response of the device to a high frequency signal by applying time-variable input to one of the contacts; this will prove impractical except for high gigahertz frequencies, and then, most likely, the simulation will have to be carried out for a range of different frequencies. A more efficient method is to step the bias of one electrode or contact and Fourier analyse the transient response current. This technique will be the subject of the following section. The y and s parameters will be introduced, together with other figures of merit like gain, which the electrical engineer uses in characterisation of the device. In Section 9.4 the transistor will be described by means of an equivalent circuit consisting of a current generator, capacitances and resistors. The sizes of the components of this circuit will be calculated by relating its y parameters to the simulated ones. In this way a bridge has been established between the physicist and the electrical engineer. The engineer's and physicist's approaches to transistors are now compatible. A short discussion of how to compare experimental and calculated data will be started in this chapter and continued in Chapter 10. It is of the utmost importance to make the concepts of intrinsic data clear. The influence of stray fields on the characteristics of the device will be considered in the final section.

9.2 THE FOURIER TRANSFORM, ALTERNATING CURRENT CHARACTERISTICS

Consider a device with $N + 1$ contacts, numbered $0,1,2, \ldots, N$, $N = 1$ for diodes, $N = 2$ for transistors and triodes, $N = 3$ for tetrodes and dual-gate transistors, etc. A device with $N + 1$ contacts is often referred to as an *N-port device*. Contact 0 is biased at the reference potential which is often taken to be ground; this contact serves as the *reference contact*. The source contact serves this purpose in transistors by convention.

Assuming the device operates at steady state, the potential at contact j is V_j, $j = 1, 2, \ldots, N$. When V_j changes instantaneously by an amount, ΔV_j, at time

0, the device will respond by establishing a new steady state. This takes time due to transients: the change in the current passing through terminal i, $\Delta I_i(t)$, changes with time. $\Delta I_i(t)$ contains sinusoidal currents of all frequencies, ω.

The Fourier transform of $\Delta I_i(t)$ is

$$\Delta I_i(\omega) = \int_0^\infty \Delta I_i(t) e^{i\omega t} dt \tag{9.1}$$

The *y parameter* y_{ij} is defined as

$$y_{ij} = \partial I_i(\omega)/\partial V_j(\omega) \tag{9.2}$$

It represents the frequency response of the current of port i when the bias of port j changes by an amount ΔV_j. (Figure 9.1.) Here $\Delta V_j(\omega)$ is the Fourier transform of ΔV_j:

$$\Delta V_j(\omega) = \int_0^\infty \Delta V_j e^{i\omega t} dt = i\Delta V_j/\omega. \tag{9.3}$$

The latter step has been arrived at because ΔV_j is a constant for $t \geqslant 0$ and zero for $t < 0$.

Equation (9.1) suggests that the simulation has to be carried out for an unlimited time. This is of course not practicable, but the calculation should be pursued until the new steady state current has been ascertained. The duration of this computation is

$$T = N_t \delta T, \tag{9.4}$$

which is N_t field-adjusting time steps, each of duration δT. From Nyquist's theorem the simulated frequency response is restricted to the interval

$$\frac{\pi}{N_t \delta T} < \omega < \frac{\pi}{\delta T}. \tag{9.5}$$

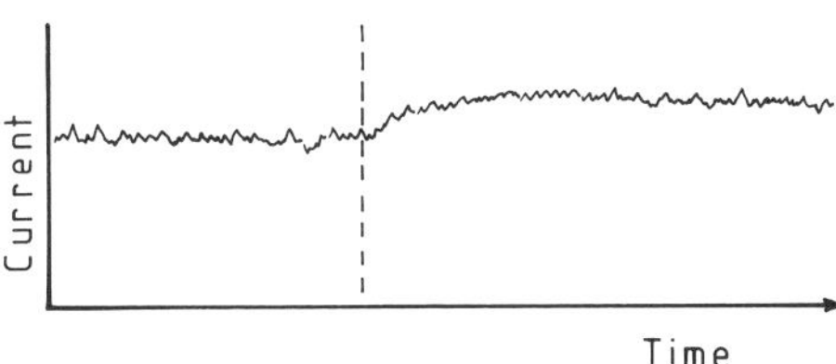

Figure 9.1 *Response current at terminal k to a change ΔV_j in the bias of port j of an N-port. The sudden change in bias happens at the point in time indicated by the vertical dashed line. The fluctuations in the current are due to noise generated by thermal fluctuations in the distribution of fields and particles inside the N-port. The units of the current and the time scale are arbitrary*

The upper limit, which usually lies in the infra-red range, is defined by the choice of the field-adjusting time step.

This would be meaningful for practical frequencies if T lay in the nanosecond range or longer, but requires a very long simulation run. As the steady state has most likely been reached before that time (neglecting the trapping and release of carriers or other rare events which do not influence the device characteristics significantly) further simulation does not add any essential new information. The current step can instead by extrapolated from T to infinity by substituting the value of the steady state current step, $\Delta I_{i'}$, for $t > T$. Nothing will be lost by this when all the noisy fluctuations are of a frequency greater then $1/T$.

The Fourier transform of the extrapolated $\Delta I_i(t)$ becomes

$$\Delta I_i(\omega) = \int_0^T \Delta I_i(t) e^{i\omega t} dt + \Delta I_i' \int_T^\infty e^{i\omega t} dt$$

$$= \int_0^T \Delta I_i(t) e^{i\omega t} dt + i \frac{\Delta I_i'}{\omega} e^{i\omega T} \tag{9.6}$$

Here $\Delta I_i'$ represents the difference in the steady state current flowing through terminal i before and after the applications of the step in the bias of port j. $\Delta I_i(t) = 0$ for $t < 0$. This extrapolation makes it possible to obtain the alternating current characteristics in a time comparable to that required to obtain the steady state current after the bias change.

The total current flowing through contact i is

$$I_i(t) = \frac{dQ_i(t)}{dt} + A\varepsilon\varepsilon_0 \mathbf{S} \bullet \frac{\partial \mathbf{F}}{\partial t} \tag{9.7}$$

where Q_i represents the charge flowing through it in the form of individual particles, and the second term the displacement current caused by temporal changes in the electric field, $\mathbf{F}$, at the contact which has area A. The latter will, apart from noisy fluctuations, usually die down when the new steady state has been reached. The unit vector $\mathbf{S}$ stands perpendicular to the contact area and is oriented into the contact.

As seen in Section 8.8 even the steady state may show great fluctuations; there it was due to wandering Gunn domains or a turbulent current flow. What is meant by steady state in this case is that, except for noisy fluctuations, one period is essentially the same as that preceding. In this case the analysis above has to be amended as follows. Let the simulation include n periods of the steady state (n should be at least 1). The step in the response current is now extrapolated to infinity by defining the step as

$$\Delta I_i(t) = \begin{cases} \Delta I_i(t) & \text{for } t \leqslant T \\ \Delta I_i'(t) & \text{for } t > T \end{cases} \tag{9.8}$$

Where $\Delta I_i(t)$ denotes the simulated step in the response current as previously and $\Delta I_i'(t)$, the extrapolated current which is periodic with period τ_π, obeys the relation:

$$\Delta I_i'(T+t) = \Delta I_i(T - t_\pi + t) \tag{9.9}$$

for $0 < t < t_\pi$.

The Fourier transform is given by

$$\begin{aligned}
\Delta I_i &= \int_0^T \Delta I_i(t)e^{i\omega t}dt + \int_T^\infty \Delta I_i'(t)e^{i\omega t}dt \\
&= \int_0^T \Delta I_i(t)e^{i\omega t}dt + \sum_{n=1}^{\infty} \exp(-i\omega n\tau_\pi) \int_{T-\tau_\pi}^{T} \Delta I_i'(t)e^{i\omega t}dt \\
&= \int_0^T \Delta I_i(t)e^{i\omega t}dt + \frac{1}{1-\exp(i\omega\tau_\pi)} \int_{t-\tau_\pi}^{T} \Delta I_i'(t)e^{i\omega t}dt
\end{aligned} \tag{9.10}$$

Here we have made use of the periodicity in $\Delta I_i'$. To make the series of the right hand term of the middle line of the equation converge, the exponential has been given an imaginary off-set, $-i\varepsilon z$, in which the real variable z goes to zero after the final summation.

9.3 SMALL SIGNAL ANALYSIS

Usually the y parameters are associated with the response of an N-port to a small increment around a selected working point of the bias of port j. By a small variation we mean one which is sufficiently small that the d.c. current characteristics can be considered linear within the range of the increment.

The y parameters are obtained by calculating the response current through all the N-ports to a voltage change in each of the contacts in turn (the reference contact is excluded). The step ΔV_j should be chosen as large as possible within the linearity of the d.c. characteristics to attain maximum accuracy.

Choosing a working point at 2 V drain bias and -1 V gate potential (including the Schottky contact potential) the four y parameters y_{11}, y_{12}, y_{21}, and y_{22} have been calculated for our example transistor. Here the index '1' stands for the gate, '2' for the drain. The parameters y_{11} and y_{21} have been obtained with a gate potential of -0.75 V, corresponding to $\Delta V_1 = 0.25$ V. The results are depicted in Fig. 9.2. The quantity y_{21} represents the frequency dependent *transconductance*, the low frequency value of which agrees with the one obtained from the slope of the transfer characteristics. (Figure 8.15.) The circular shape of the curve for the transconductance suggests that it can be parameterised as:

$$y_{21} = g_{m0}\exp(-i\omega\tau_t) \tag{9.11}$$

where g_{m0} represents the steady state transconductance and τ_t the *transit time of the transistor*, that is, the time it takes a gate signal input to reach the drain. $1/\tau_t$ gives us an idea about the upper limit of the frequency at which the transistor

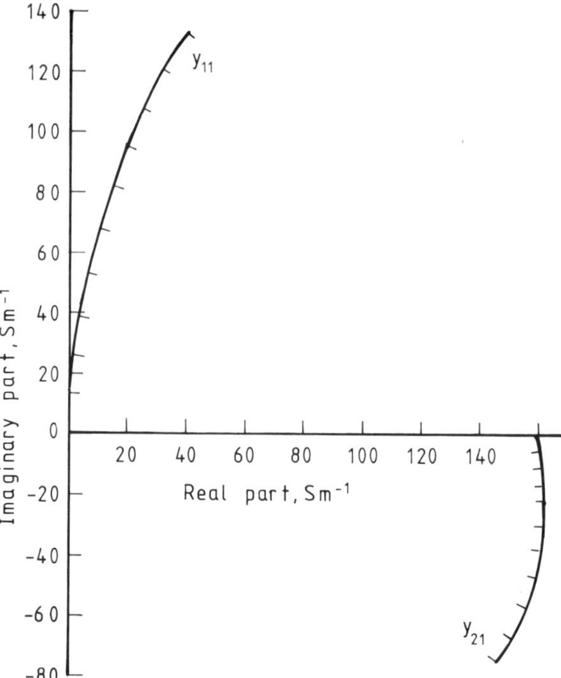

Figure 9.2 *Complex parameters y_{11} and y_{21} calculated up to 50 GHz for the example transistor of Fig. 8.1. Tick marks every 5 GHz. The calculations have been normalised to a width of one metre. The working point for this set of y parameters has been chosen to be around 2 V drain bias and −1 V gate potential (including the Schottky potential of −0.75 V)*

can operate. Equation (9.11) represents a circle with centre in the origin of the Gauss plane; y_{21}, however, deviates increasingly from this ideal law as the frequency increases. (Figure 9.2.)

The parameter y_{11} is often referred to as the *input admittance*. It follows the imaginary axis for the lower range of frequencies. The *transadmittance*, y_{22}, and the *feedback admittance*, y_{12}, have been obtained with $\Delta V_2 = 1$ V. (Drain bias 3 V, Fig. 9.3.) As the change in the drain current is relatively small, the determination of y_{12} and y_{22} is the least accurate one. The accuracy can only be improved by extending the duration of the simulation. Again the low frequency result for the transadmittance is consistent with the value obtained from the slope of the forward characteristics. (Figure 8.16.) Both y_{12} and y_{22} are often smaller in magnitude than the other two y parameters.

Many microwave engineers prefer the scattering parameters or s parameters to y parameters. These are measured by observing the wave scattered from the gate and the drain when the microwave signal enters either the gate or the drain which are connected to the wave generator through a reference resistor, Z_0, matching the impedance in the coaxial cables or coplanar lines connected to the N-port – usually this resistance is 50 Ω, a value which, in principle, is arbitrary, although commercially available microwave measuring instruments are often calibrated for this reference.

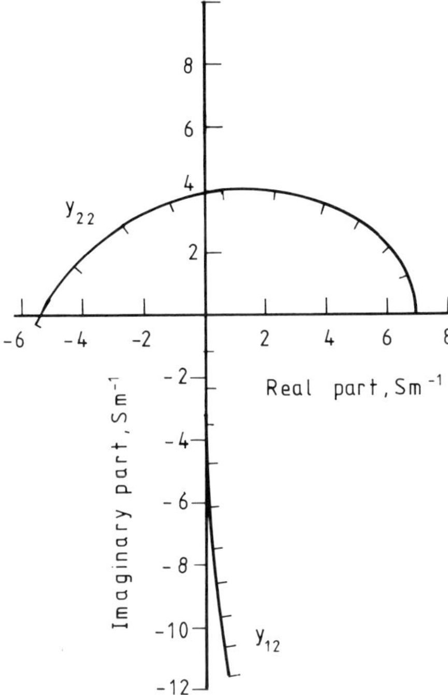

Figure 9.3 *Complex y parameters y_{12} and y_{22} calculated up to 50 GHz for the example transistor of Fig. 8.1. Tick marks every 5 GHz. The calculated results have been normalised to a width of one metre. The working point is the same as in the previous figure*

The relationship between the s and the y parameters is

$$\begin{aligned} s_{11} &= [(1 - y'_{11})(1 + y'_{22}) + y'_{12}y'_{21}]/D \\ s_{12} &= -2y'_{12}/D \\ s_{21} &= -2y'_{21}/D \\ s_{22} &= [(1 + y'_{11})(1 - y'_{22}) + y'_{12}y'_{21}]/D \end{aligned} \tag{9.12}$$

with

$$D = (1 + y'_{11})(1 + y'_{22}) - y'_{12}y'_{21} \tag{9.13}$$

and

$$y'_{kj} = y_{kj}WZ_0 \tag{9.14}$$

where W is the width of the transistor, y'_{ij} is dimensionless, while y_{ij} is normalised to the unit width.

As seen from these expressions, the y' and s parameters apply to a transistor of a definite width, while the model generally yields values normalised to a unit width. Different widths give different s parameters for transistors of otherwise identical cross-section. The y parameters are normalisable, the s parameters, however, are not, because of the required fixed reference resistance load Z_0.

9.4 THE EQUIVALENT CIRCUIT

The y parameters, or the s parameters, or other sets of parameters derived from them, describe the device formally, e.g. Anderson (1967). Let the contacts of an N-port be connected to a d.c. power source. A sinusoidal voltage signal, ΔV_j, of angular frequency ω is superimposed on terminal j. The response current of port i to this signal is ΔI_i. The signals should be sufficiently small that the current characteristics of the device be linear within ΔV_j. The *y parameters* define the relationship between the voltage and current signal:

$$\Delta I = \mathbf{y}\Delta\mathbf{V} \tag{9.15}$$

where ΔI and $\Delta\mathbf{V}$ represent the quantities

$$\Delta I = \begin{pmatrix} \Delta I_1 \\ \Delta I_2 \\ \cdot \\ \cdot \\ \cdot \\ \Delta I_N \end{pmatrix} \quad \text{and} \quad \Delta\mathbf{V} = \begin{pmatrix} \Delta V_1 \\ \Delta V_2 \\ \cdot \\ \cdot \\ \cdot \\ \Delta V_N \end{pmatrix} \tag{9.16a}$$

and $\mathbf{y}$ the $N \times N$ *admittance matrix*

$$\mathbf{y} = \begin{pmatrix} y_{11} & y_{12} & \cdots & y_{1N} \\ y_{21} & y_{22} & \cdots & y_{2N} \\ \cdot & \cdot & \cdots & \cdot \\ \cdot & \cdot & \cdots & \cdot \\ y_{N1} & y_{N2} & \cdots & y_{NN} \end{pmatrix}. \tag{9.16b}$$

For the transistor

$$\begin{pmatrix} \Delta I_g \\ \Delta I_d \end{pmatrix} = \begin{pmatrix} y_{11} & y_{12} \\ y_{21} & y_{22} \end{pmatrix} \begin{pmatrix} \Delta V_g \\ \Delta V_d \end{pmatrix}. \tag{9.17}$$

The elements of the admittance matrix depend on the selected *working point*, i.e. on the d.c. bias of the ports and generally on the frequency ω as there are capacitors and inductors in the circuit.

By means of a description starting from Equation (9.15) or (9.17) the device can be considered as a black box with N inputs and N output channels. The current flows in on one side; an electric potential is created on the other side. This function can also be reversed; the input potential creates a current according to Equation (9.15). Figure 9.4 represents this view of the transistor.

The elements of the admittance matrix can be considered as parameters without

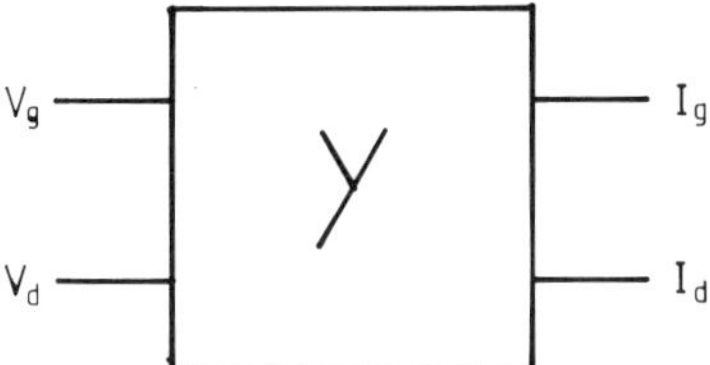

Figure 9.4 *'Black box' view of a transistor. It has two power sources marked V_g and V_d, the current flows through the current leads marked I_g and I_d*

any other physical significance than yielding a relationship between the input and the output. The Monte Carlo particle simulation shows the physical background of them in terms of particle dynamics and electromagnetic fields. This is indispensable for the electrical engineer who wants to improve on the performance of the device.

The electrical engineer prefers to describe the device in terms of an equivalent circuit consisting of current generators, resistors, capacitors and inductors. The reason for this is that the equivalent circuit elements can be used to describe a more complex circuit where the device is one part of a network. The frequency response of this network is calculated by means of commercial software. There is no fixed standard equivalent circuit to describe a transistor; authors may differ in details, namely whether to include or omit certain elements. Recently Berroth and Bosch (1990) have suggested the intrinsic circuit shown in Fig. 9.5. Note that this circuit lacks inductors. Writing down Kirchhoff's law (which says that no current can accumulate at any node) for each of the nodes, a set of relations between the gate current I_g, the drain current I_d, the gate bias V_g and the drain bias V_d can be obtained:

$$y_{11} = R_i C_{gs}^2 \omega^2 / D + i\omega (C_{gs}/D + C_{gd}) \tag{9.18a}$$

$$y_{12} = -iC_{gd}\omega \tag{9.18b}$$

$$y_{21} = g_{m0} \exp(-i\omega\tau_T)/(1 + i\omega R_i C_{gs}) - i\omega C_{gd} \tag{9.18c}$$

$$y_{22} = g_d + i\omega (C_{ds} + C_{gd}) \tag{9.18d}$$

where

$$D = 1 + \omega^2 C_{gs}^2 R_i^2 \tag{9.18e}$$

The meaning of the new symbols is clear from Fig. 9.5; ω denotes the angular frequency.

This set of equations is exact for the equivalent circuit of Fig. 9.5. Other authors have used a slightly different equivalent circuit deriving equations similar to Equations (9.18) with the restriction that

$$\omega C_{gs} R_i \ll 1. \tag{9.19}$$

Below we restrict further discussion to the example transistor.

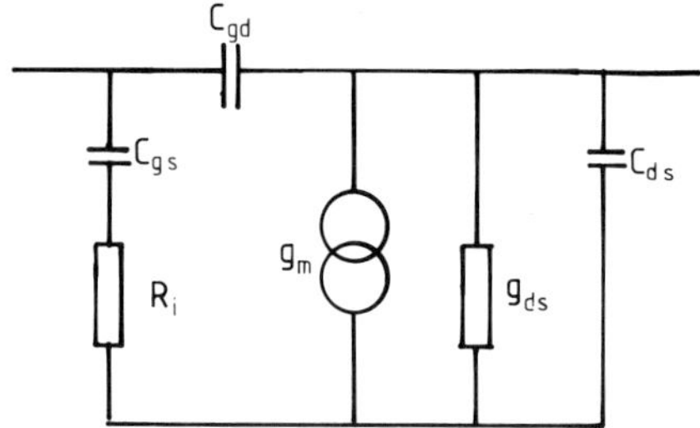

Figure 9.5 *Possible equivalent circuit for the example transistor of Fig. 8.1. The values for the various resistors and capacitors calculated from the y parameters have been quoted in Table 9.1*

Figure 9.6 shows the relationship between the intrinsic equivalent circuit and the cross-section of the actual transistor. This figure has been presented to give the reader an idea of the physical meaning of the various equivalent circuit elements.

Table 9.1 shows the values of the various elements of the equivalent circuit calculated for our example transistor at low gigahertz frequencies by means of Equations (9.18).

R_s and R_d, Fig. 9.7, represent the Ohmic resistance in the source and drain regions which do not belong to the intrinsic transistor. Then there is an Ohmic channel, or internal resistance, R_i. When no particles flow through the gate, which obviously is the case when the gate region has been depleted, Fig. 8.11, the Ohmic gate resistance is without meaning. We should bear in mind that the resistances in the intrinsic equivalent circuit of our transistor do not include those of the metal contacts because the transport inside the metal has not been considered. The resistance of the metallisation depends on the shape of the terminals.

C_{gs} and C_{gd} represent the capacitance between the source and the gate, and between the gate and the drain respectively. The drain to gate capacitance, C_{gd}, depends essentially on $V_d - V_g$. There are two opposing contributions to it. As V_g

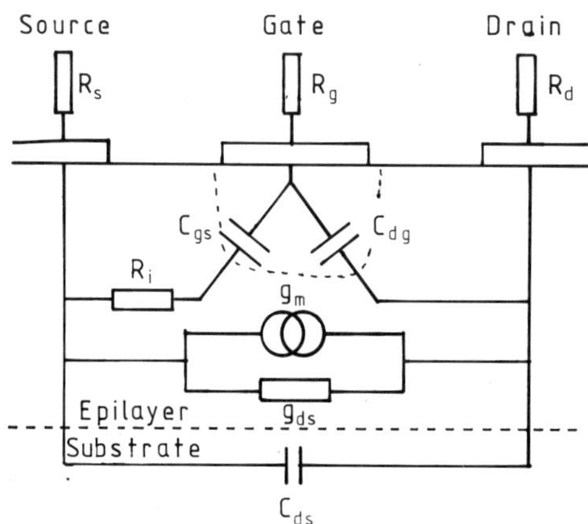

Figure 9.6 *The relationship between the equivalent circuit and the actual cross-section of the transistor*

Table 9.1 Equivalent circuit parameters for the example transistor

C_{gd}	3.86×10^{-11} F m^{-1}
C_{gs}	3.92×10^{-10} F m^{-1}
C_{ds}	-2.55×10^{-12} F m^{-1}
g_{m0}	161 S m^{-1}
g_{ds}	6.89 S m^{-1}
R_i	2.20×10^{-3} Ω m
τ	1.57×10^{-12} s

becomes more negative, the gate depletion area increases in extent, giving a positive contribution to the space charge in the vicinity of the gate. When V_d increases, the high-field region between the gate and the drain extends and more particles enter the L minima, which means an accumulation of negative charge. The former effect is the strongest, but there is a possibility that the latter could be the strongest rendering C_{gd} negative; this has, however, never been observed.

There is also a capacitance between the source and the drain, C_{ds}. The two contributions to it come from that part of the gate depletion area which is just below the gate, and from the dipole that forms between the epilayer and the substrate. The latter is weakly dependent on V_d through the formation of the substrate current. A larger substrate current means negative charge in the substrate.

The drain to source capacitance, C_{ds}, is usually small compared to C_{gd} and C_{gs}; in our case by two orders of magnitude. We have a negative value for this, which may reflect possible turbulence in the current.

As the reader will gather there is no sharp demarcation line between the three capacitances discussed here. This also applies to the internal, source and drain resistances, R_i, R_s and R_d respectively. The latter two originate partly from the metallisation of the Ohmic contacts, partly from the scattering of the carriers in the contact regions. The intrinsic equivalent circuit of Fig. 9.5 should therefore not be taken as the unique and absolute equivalent circuit of the transistor. It has been introduced only as a convenience for the simulation of larger systems containing active elements. If the transistor shows Gunn oscillations, it is necessary to introduce an additional current generator between the gate and the drain in the equivalent circuit.

9.5 GAIN

Since there are several types of gain, it is necessary to specify which type is required.

In connection with transistors we usually deal with voltage, current, power and unilateral gain. The *voltage gain* is defined as

$$G_v = \partial V_d/\partial V_g \simeq \Delta V_d/\Delta V_g. \tag{9.20}$$

Multiplying numerator and denominator by ΔI_d this can be recast into the form

$$G_v = \left| \frac{\partial I_d}{\partial V_g} \Big/ \frac{\partial I_d}{\partial V_d} \right| = |y_{21}/y_{22}|; \tag{9.21}$$

the latter step stems from the definition of the y parameters, Equation (9.2).

The *current gain* is define as

$$G_c = |\partial I_d/\partial I_g| \simeq |\Delta I_d/\Delta I_g| \tag{9.22}$$

which can be expressed in terms of y parameters in a similar way as

$$G_c = \left|\frac{\partial I_d}{\partial V_g} \bigg/ \frac{\partial I_g}{\partial V_d}\right| = |y_{22}/y_{12}|; \tag{9.23}$$

The *power gain* is measured by observing the s parameters of the device. Expressed in terms of these the power gain reads:

$$G_p = \frac{1 + |s_{11}s_{22} - s_{12}s_{21}|^2 - |s_{11}|^2 - |s_{22}|^2}{2|s_{12}|\,|s_{21}|}. \tag{9.24}$$

The maximum *unilateral gain* is defined as

$$G_u = \frac{|s_{21}|^2}{(1 - |s_{11}|^2)(1 - |s_{22}|^2)}. \tag{9.25}$$

Traditionally, gains are given in decibels:

$$G_i^d = 10 \log_{10} G_i \tag{9.26}$$

where i represents one of the subscripts v, c, p or u.

Figure 9.7 shows the voltage and unilateral gain for the example transistor introduced in the previous chapter. The gain $G_u^d > 0$ for $\omega < 2\pi f_T$ where f_T represents the maximum frequency yielding power gain greater than one (zero decibels). The maximum frequency is determined by extrapolating the straight line

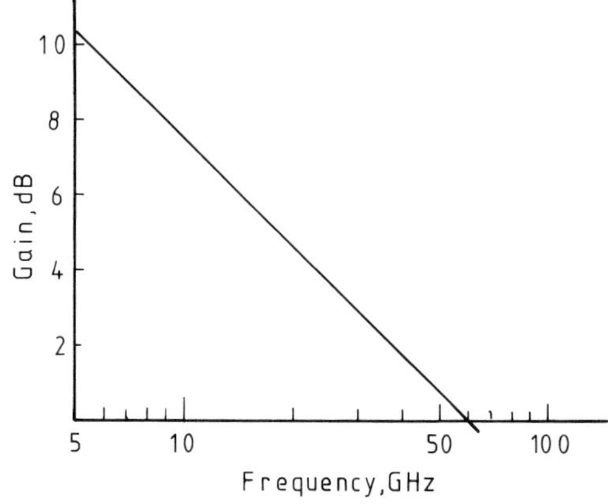

Figure 9.7 *Unilateral gain of the example transistor against frequency up to 50 GHz*

in the double logarithmic plot shown in the figure which amounts to 60 GHz for our example transistor.

9.6 THE INFLUENCE OF STRAY FIELDS

So far, the only examples of transistors that have been presented are planar transistors, namely transistors with all electrodes on top of it. The contacts have been treated as fixed potentials along the top of the simulation area, without any regard to their actual shape. This restriction to the model is not necessary.

The effect of the shape of the metal electrodes on the transistor characteristics has not received the attention by modellers it deserves. Moglestue (1983c) has studied the influence of the shape of the gate metal on the transistor characteristics by means of the Monte Carlo particle model. Figure 9.8a shows a planar transistor that has been calculated treating the contacts mathematically as fixed potential lines on top. The distance between the source and the gate is 1 μm and between the gate and the drain 2 μm. The gate length is 1 μm and the epilayer is doped uniformly n-type down to 0.3 μm with a donor concentration of $1.2 \times 10^{23}\,\mathrm{m}^{-3}$; then the doping density falls linearly to zero over 40 nm. The epilayer thickness refers to that under the gate. The substrate is intrinsic; the interface between the epilayer and the substrate is planar throughout the transistor. The transistor of Fig. 9.8b is identical to that depicted in Fig. 9.8a, but the gate of height 0.3 μm has now been considered in the simulation. The potential at the vertical sides of

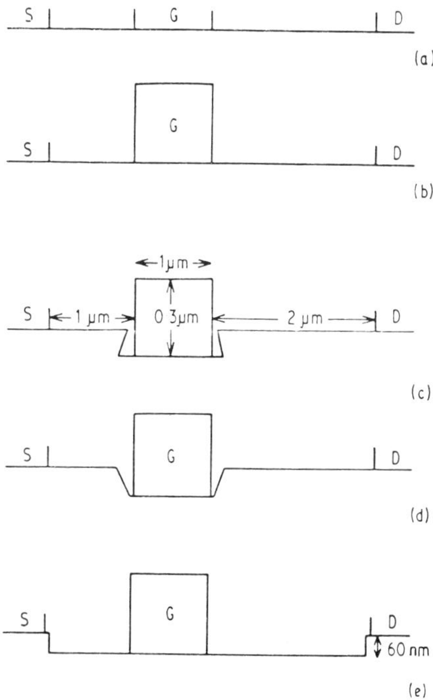

Figure 9.8 *Profiles of five transistors. a and b are identical but the shape of the gate metallisation has been included in the model in cases b–e. Transistors c, d, and e have a gate recess of different design*

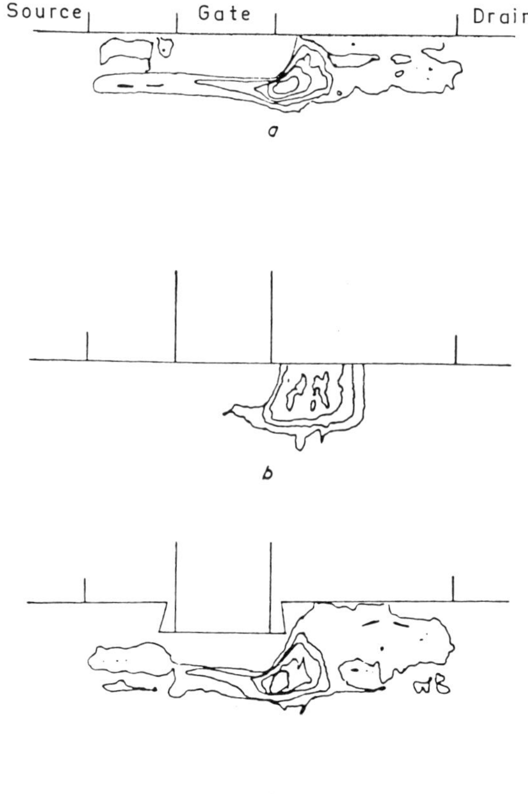

Figure 9.9 *Average steady-state internal distribution of L minimum carriers. The contour lines represent 10, 25, 50, 75 and 100% of the carrier density required for charge neutrality in the uniformly doped part of the epilayer. Drain bias: 3 V. Gate potential: −1.3 V, including the Schottky contact potential of −0.8 V. The potential of the vertical walls of the gate metal is −0.5 V. a, b and c show transistor design a, b and c respectively of Fig. 9.10*

the gate metallisation gives rise to an additional electric field penetrating the interior of the semiconductor. Figure 9.9a and b show that the distribution of *L* electrons becomes radically different when the effect of the metal shape is considered. The difference in the electron distribution and the field distribution is only due to the stray fields originating from the vertical walls of the gate electrode. The minimum noise figure reduces and the unilateral gain increases, Fig. 9.10; the stray fields thus improve the a.c. performance of the transistor. Usually stray fields are thought to impair the transistor characteristics, but this example shows that the opposite may just as easily be true.

This study is far from complete, but shows that the shape of the drain electrode should also be considered in a realistic simulation. This study suggests that the stray fields generated by the contacts having a finite shape contribute significantly to the microwave characteristics, and that transistors can be designed to take a maximum benefit of these fields. The future efforts to optimise device design should include the shape of the contacts and the environment.

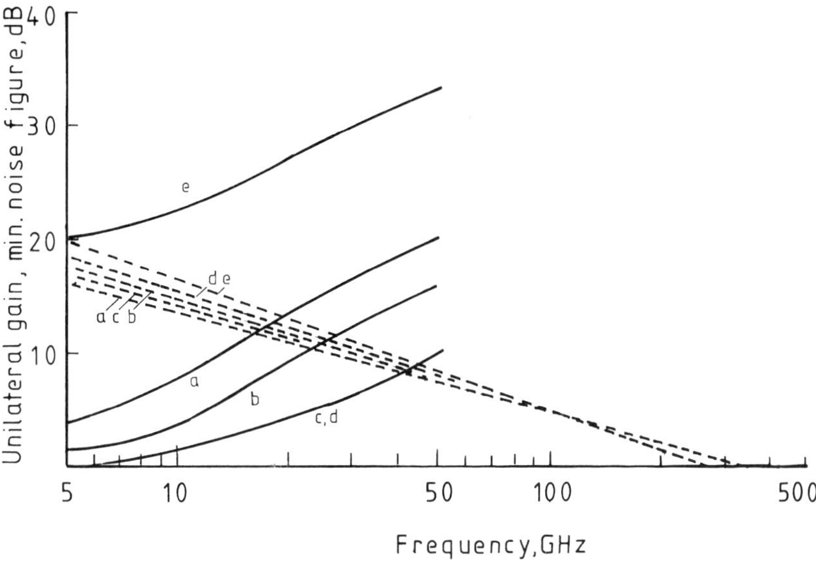

Figure 9.10 *Minimum noise figure (——) and unilateral gain (---) against the frequency for the five transistors of Fig. 9.10. The data has been normalised to the unit width, one metre*

Figures 9.8c–e show the same transistor with the gate recessed 60 nm. The uniformly doped epilayer is still 0.3 μm thick measured under the gate, and 0.36 μm under the source and the drain. Three different gate recesses have been studied; the gate metal is 0.3 μm thick for all three transistors.

The calculated electron and field distribution does not differ significantly between these three cases. Figure 9.9a shows that the heavy electrons behave very much in the same way as in case *a*. The recess seems to shield off the stray fields from the walls of the gate metal.

Transistors *c* and *d* have about the same gain and noise performance; *e* exhibits a better gain than *b* for low frequencies, but with the minimum noise figure impaired by more than 10 dB. (Figure 9.10.)

The noise figure calculated here, Fig. 9.10, is the instrinsic one; an exact explanation of what we mean by this will be given in Chapter 12.

10

Composite Material Devices

10.1 INTRODUCTION

The development of electronics moves towards faster devices, the ultimate limit to their speed set only by the laws of physics. Mobility in the channel is one restriction to the speed of the transistor, though it can be improved by joining layers of material with different band gaps. A junction where the band gaps of the two materials form a discontinuity is known as a *heterojunction*. (Figure 10.1.) The interface between GaAs and $In_yGa_{1-y}As$ or $Al_xGa_{1-x}As$ (x and y represent stoichiometric composition) or silicon and germanium are examples of such heterojunctions. The materials may be doped or semi-insulating. The idea behind such junctions is to form a quantum well at the joint where an electric field is applied perpendicular to it. The carriers in the well suffer less scattering so that their drift velocity along the interface increases relative to their bulk.

The main aim of this chapter is to describe the simulation of devices composed of different materials or alloys, e.g. heterojunction transistors. Each material should be considered as distinct because the scattering rates differ quantitatively between them. Differently doped materials, e.g. *n*- or *p*-doped silicon, however, are not different in this context when the phonon scattering rates are, with sufficient accuracy, considered independent of the doping. We have discussed how to treat ionised and intercarrier scattering for different doping in Chapter 4.

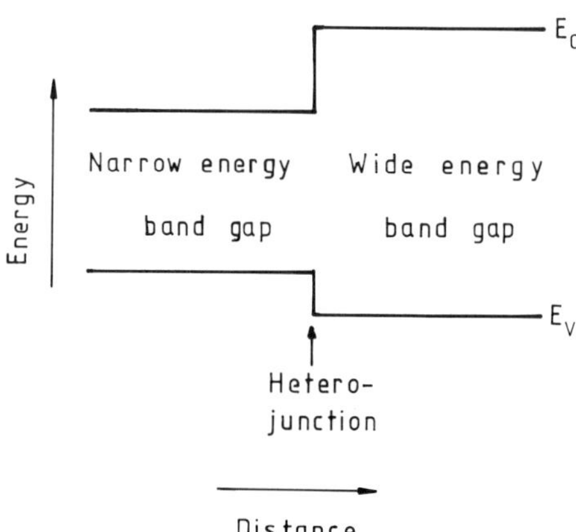

Figure 10.1 *Band structure of a heterojunction. E_c: conduction band edge. E_v: valence band edge*

The heterojunction transistor was first described by Dingle *et al.* (1978). A mobility exceeding 1170 $m^2\ s^{-1}\ V^{-1}$ at liquid helium temperatures has been reported by Pfeiffer *et al.* (1989), a figure which is the world record. The improvement in the mobility is not that spectacular at room temperature, however, because of phonon scattering.

A *heterojunction field-effect transistor* or *high electron mobility transistor* (HEMT) consists of a narrow band gap material, forming the channel, with a layer of wide band gap material on each side. The source and drain have been alloyed through the top layer to form contact with the channel. Often the interspersed layer is so thin that a proper quantum well forms through which the carriers flow. Placing the wide-gap material between layers of narrow-gap material makes an *inverted* HEMT, as shown in Fig. 8.17. The nomenclature for heterojunction transistors varies throughout the literature; sometimes it is referred to as a MODFET (*modulation doped field-effect transistor*).

The description of the confinement of the carriers to a quantum well requires an approach based on quantum mechanics. The necessary physics, in addition to that discussed in the previous chapters, will be introduced here.

Describing such devices by the conventional drift-diffusion or hydrodynamic model is problematic because of the discontinuity in the conduction or valence band at the junction due to the fact that the carriers gain or lose kinetic energy by crossing it. Here the Monte Carlo particle model is at its most useful as it employs a first principles physical description of the transport. This will be discussed in the next section. A treatment will be given based on Schrödinger's equation and the Newtonian mechanics of Section 5.8. Criteria for access into the wide-gap material will be given as an heuristic example. The steady state characteristics of the inverted HEMT depicted in Fig. 8.17 will be discussed in view of the field and carrier distribution and the influence of the wide-gap material barrier. This exposition is intended as an illustrative example to explain the principal simulation techniques.

Time-dependent tunnelling and criteria for resonant tunnelling will be discussed in Section 10.4. Criteria for admission of individual carriers through the above narrow barriers and a simple presentation of the dynamics inside tunnel barriers will also be given. Quantum transport in narrow channels will be discussed in the subsequent section, where we shall formulate a criterion for entry into a sub-band from the quasicontinuum. The treatment of confined carriers will be extended to wires and quantum dots in Section 10.5. In Section 10.7 we shall discuss a complex example transistor with quantum transport. This transistor has been manufactured and characterised experimentally. We shall see that this gives us an opportunity to compare our simulated results with experimental data; the calculated characteristics agree well with the measured ones. The chapter closes by discussing points to be aware of when comparing theoretical predictions with experimental data.

10.2 THE HETEROJUNCTION

Two materials with energy gap of different width joined together form a *heterojunction*. Neither the conduction band nor the valence band edge align so both the electrons and the holes experience an abrupt *potential barrier*. The discontinuity of the valence and conduction bands at the joint is independent of any electric or magnetic field. In this section we shall disregard any possibility of tunnelling through barriers; the treatment of tunnelling will be postponed until Section 10.4.

When the alloy $Al_xGa_{1-x}As$ or $In_yGa_{1-y}As$ (x and y representing the stoichiometric content of gallium and indium respectively) is joined to GaAs, the electrons and the holes will experience a step in potential between the two materials. For electrons in the Γ minimum this step is (Shur, 1990) given by

$$\Delta E_G = \begin{cases} \Delta E_{gg} & \text{for } x < .45 \\ 0.476 + 0.125x + 0.143x^2 + 1.5y - 0.4y^2 & \text{for } x \geq .45 \end{cases} \tag{10.1}$$

where

$$\Delta E_{gg} = 1.247x + 1.5y - 0.4y^2 \tag{10.2}$$

represents the difference in energy between the Γ minima for the GaAs and the alloy. The valence band edge has energy

$$E_V = 0.4\Delta \mathrm{E}_{gg} \tag{10.3}$$

and the conduction band edge energy

$$E_C = \Delta E_G - E_V. \tag{10.4}$$

The off-set energy of the other minima of the conduction band and the maxima of the valence bands can be found from the literature.

Schrödinger's equation describing the carriers near a heterojunction has been introduced in Section 6.6. It was then shown that the motion parallel to the interface is like that of free electrons with a distinct effective mass. If there is a lattice mismatch between the two materials, e.g. between GaAs and $In_yGa_{1-y}As$, the crystalline uniaxial strain or stress causes the effective mass (Foulon *et al.*, 1990) and the acoustic deformation potential to shift.

Perpendicular to the interface, the conduction or valence bands divide into sub-bands. Whether or not this has to be considered depends on the difference between the energy levels of the sub-bands. This point will be discussed later.

Let us introduce a local coordinate system with its z axis perpendicular to the heterojunction. The energy associated with the perpendicular motion is given by

$$E_t(1 + \alpha E_t) = \tfrac{1}{2}\hbar^2[m_{33}^{*\,-1}k_z^2 + 2(m_{13}^{*\,-1}k_x + m_{23}^{*\,-1}k_y)k_z]. \tag{10.5}$$

For an isotropic non-parabolic conduction band structure around the centre of the Brillouin zone

$$E_t(1 + \alpha E_t) = \frac{\hbar^2}{2m^*}k_z^2 \tag{10.5'}$$

where α represents the non-parabolicity factor, m_{ij}^* the effective mass tensor components of the carrier and E_t the energy associated with the motion perpendicular to the barrier.

For simplicity we shall assume a spherical band structure which contains all the essential points we want to explain. The only difference between this approach and that based on the general band structure lies in the amount of algebra involved. When the junction represents a positive step, V_B (e.g. moving from GaAs into $Al_xGa_{1-x}As$) the particle loses kinetic energy by an amount corresponding to the step in potential energy

$$E_t' = E_t - V_B \tag{10.6a}$$

and can only be admitted if $E_t' > 0$, otherwise it will be reflected. When the step is negative (e.g. from $Al_xGa_{1-x}As$ to GaAs) the particle gains in kinetic energy:

$$E_t' = E_t + V_B \tag{10.6b}$$

The component of the wave vector perpendicular to the junction after admission becomes

$$k_t' = \{2m_t^*(E_t \pm V_B)\,[1 + \alpha(E_t \pm V_b)]\}^{\frac{1}{2}}/\hbar \tag{10.7}$$

where the + and − signs refer to gain and loss of kinetic energy respectively. This description is, strictly speaking, only valid if the Schrödinger equation describing the particle is separable into a parallel and perpendicular part respectively.

At the time of the last scatter event the distance, z_0, of the particle from the junction is, in the absence of a magnetic field, given by

$$r_z = \frac{eF_t t_0^2}{2m_t^*} + \frac{\hbar k_{t0}}{m_t^*} t_0 + z_0 \tag{10.8}$$

where t_0 represents the time it takes to arrive at the potential step, which is shorter than the free flight allocated to that particle for this particular flight. The component of the wave vector perpendicular to the joint was k_{t0} at the time of the last scatter; on arrival it is

$$k_t = k_{t0} + eF_t t_0/\hbar. \tag{10.9}$$

The final position and wave vector at the end of the flight are calculated from Equations (5.42) and (5.43) respectively, with the position and wave vector at the barrier as initial coordinates. The duration of the remaining flight most likely has to be amended because of the change in kinetic energy on crossing or reflection from the junction.

The wave vector parallel to the interface does not change on admission or reflection. The velocity associated with it, however, changes on admission due the difference in the effective mass between the two materials. A particle cannot transfer from one conduction band minimum to another on crossing because such a change would involve the introduction of an additional wave vector representing the transfer between different conduction band minima, which means violation of the law of conservation of momentum.

In *graded junctions* the stoichiometric composition of the semiconductor varies with position $\mathbf{r}$. The conduction or the valence band edge can formally be described as $E_B(\mathbf{r})$. The particle feels the gradient $\nabla E_B(\mathbf{r})$ as a force acting on it in addition to the Lorentz force. Now the scattering rates become position dependent due to the local variation in the stoichiometric composition. However, we tabulate them for that part of the graded junction where they are the largest. To estimate the time of free flight and select scattering we use the same procedure as explained in Chapter 5, but with the largest possible value for the scattering rates applying to the graded junction. When the scattering mechanism has been chosen, an additional random number representing the ratio of the correct scattering rate to the maximum one is generated to give the executability of the chosen scattering. Most often this ratio is easy to obtain.

10.3 THE HETEROJUNCTION TRANSISTOR

Figure 8.17 shows an example of an inverted HEMT. In this section we shall discuss the aspects and the results of simulating it by means of the Monte Carlo particle model.

The width of the transistor is much larger than the length, so that a model which is two-dimensional in geometrical space is sufficiently accurate. The cross-section of the simulated area is divided into areas with boundaries following those of the simulated area or the heterojunction interfaces. The areas are numbered to distinguish them. Each area has a material index attributed to it indicating which material the particle moves in for the sake of selecting the scattering mechanism and the time of flight. The heterointerfaces are assumed to be sharp, i.e. defined withint one atomic layer. Ikarashi *et al.* (1990) have shown by high resolution electron microscopy that this assumption is reasonable. In the case of junctions with a graded stoichiometric alloy composition we can describe the transport by assuming an additional force acting on the particles. This force is given by the product

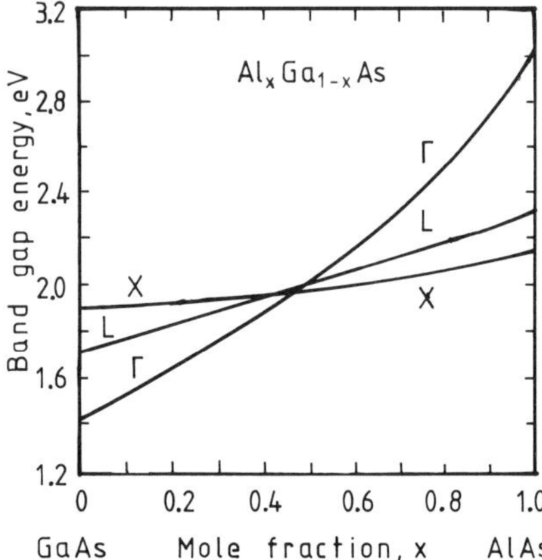

Figure 10.2 *Band gap energy of the various conduction band minima in $Al_xGa_{1-x}As$ against the stoichiometric composition x reckoned from the valence band. From Casey (1978)*

of the particle's charge and the gradient of the conduction or valence band edge when no field is present. At the end of each individual flight it is investigated whether the particle has moved into a different area by checking the area and material index. If these have altered, it is examined whether access across the border is possible and the wave vector and the position at the end of the flight are calculated accordingly as explained in the previous section. Particles straying outside the simulation region, except through the contacts, are returned by reflection from its external boundary.

The device shown in Fig. 8.17 has already been described in Section 8.8 in connection with the wandering Gunn domains. The Poisson equation has been solved every 15 fs over a rectangular 20 by 10 nm grid with the short edge perpendicular to the heterofaces. This is sufficiently fine to consider the solution of the field equation as quansicontinuous. Figure 10.3 shows the potential distribution for 3 V drain bias and −0.75 V gate potential (inclusive of the Schottky contact potential). The equipotential lines tend to follow the interface between the epilayer and the $Al_{0.3}Ga_{0.7}As$ layer; the corresponding magnitude of the electric field is shown in Fig. 10.4. The direction of the field can be obtained from the gradient of the potential in Fig. 10.3. The maximum of the electric field follows the Gunn domain as it moves towards the drain. Here the electrons are most likely to be excited into the L minima of the conduction band. As L electrons, they see a much lower barrier to the AlGaAs layer, Fig. 10.2, and therefore find it easier to enter. Once they have passed into the GaAs substrate as Γ electrons, Fig. 8.18, they find that the local field is nowhere strong enough to give them sufficient energy to overcome the barrier again, either by transfer into a higher energy conduction band minimum or directly. These particles are therefore trapped in the substrate, but prevent further electrons from entering it by electrostatic repulsion. Particles still succeed in entering the wide-gap material from the epilayer, but will be returned to it.

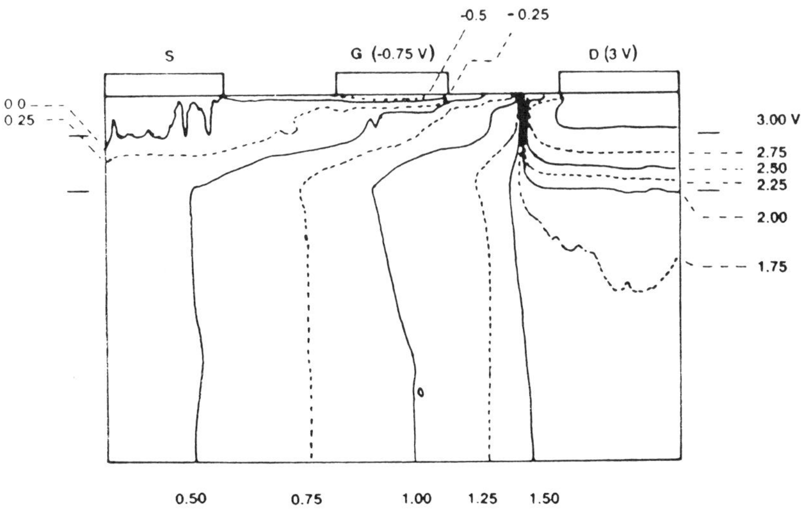

Figure 10.3 *The potential distribution in the inverted HEMT of Fig. 8.17 with a drain bias of 3 V and a gate potential of −0.75 V. The parameter represents the potential in volts. The horizontal bars indicate the positions of the heterojunctions*

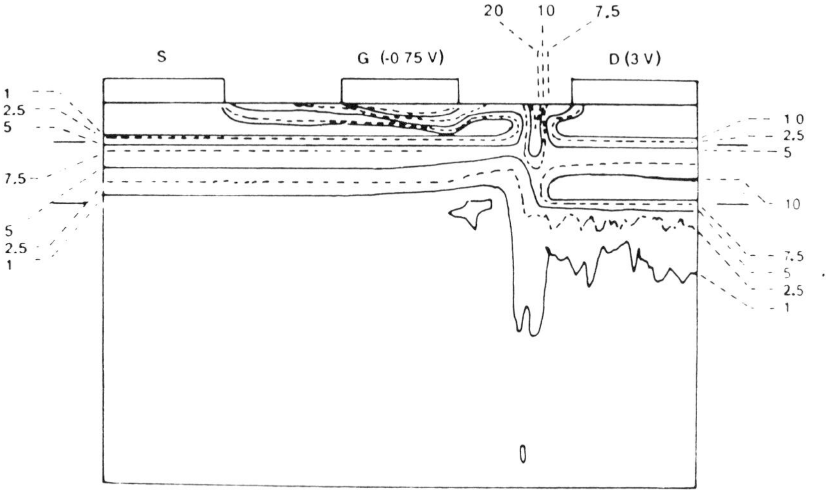

Figure 10.4 *The magnitude of the electric field throughout the transistor at the same operating conditions as in Fig. 10.3 The parameter represents the magnitude of the electric field in MV m^{-1}. The horizontal bars indicate the position of the heterojunctions*

Figure 10.5 shows the transfer characteristic for 3 V drain bias. The current represents the average over several Gunn oscillations and shows weak pinch-off. There is no longer any formation of substrate currents – apart from the trapped electrons that move back and forth with the periodic creation of the Gunn domains – to explain this. The soft transition to zero is caused by band bending in the gate area of the heterojunction; a stronger gate potential is needed to close the channel than that needed for the ordinary MESFET of the same geometrical dimensions.

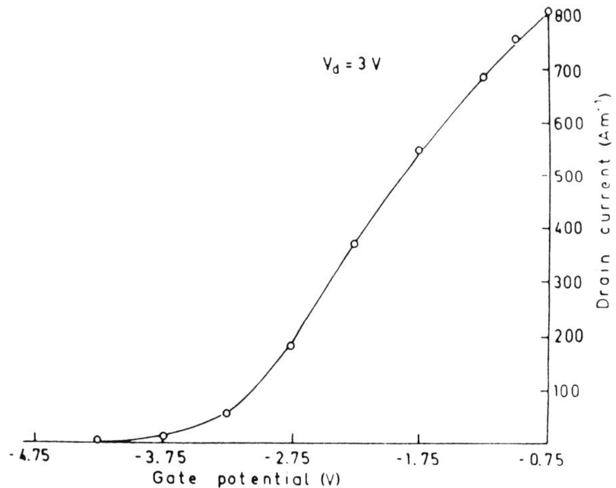

Figure 10.5 *The steady state transfer characteristics of the inverted HEMT of Fig. 8.17. The gate potential includes the Schottky contact potential of −0.75 V*

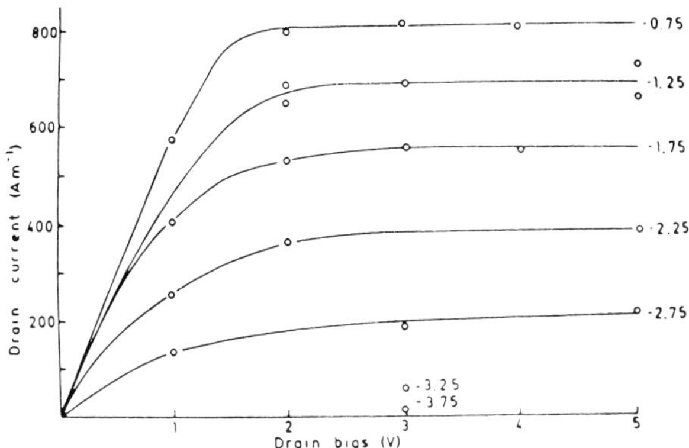

Figure 10.6 *The forward characteristics for the HEMT of Fig. 8.17 for different gate biases. The parameter represents the gate potential in volts, including the Schottky contact potential of −0.75 V*

The forward characteristics are shown in Fig. 10.6. Again the current represents the average over several Gunn oscillations. The current saturates well; the trans-admittance is almost zero which is a most desirable property for amplification purposes (had it not been for the Gunn oscillations in the drain current).

Except at the source and the drain, a quantum well forms along the top heterojunction. Here the conduction band divides into sub-bands which get closer the higher the levels rise. Figure 10.7 shows the level of the three lowest sub-bands against position along the interface. Everywhere the separation between the sub-bands is so small that they overlap due to their broadening. We shall explain how this originates in Section 10.5. This means that quantised transport does not need to be considered. This is quite common for this type of HEMTs. Only in the high-field region need the level of the lowest sub-band be considered where it represents a barrier to the transport along the channel.

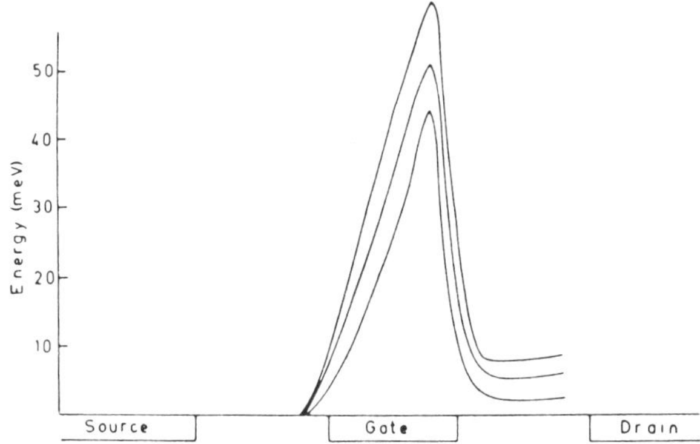

Figure 10.7 *The energy levels of the first three sub-bands of the quantum well forming at the upper heterojunction of the transistor described in Fig. 8.17. The energy distribution refers to the moment when the Gunn domain is about to leave the gate*

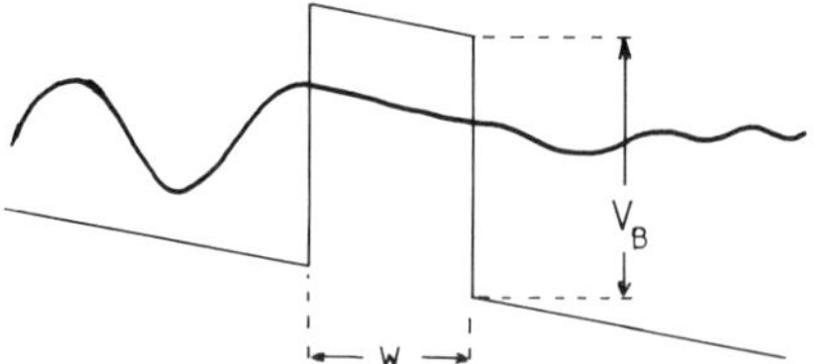

Figure 10.8 *Cross-section through a potential tunnel barrier. The arrow indicates the direction the particle travels in; the wavy line the wave function*

10.4 TUNNELLING

An electron moving along the z axis encounters a barrier of height V_B and width w. (Figure 10.8.) The wave equation of this system reads

$$-\frac{\hbar^2}{2m_t^*}\frac{d^2\Psi_t}{dz^2} + U(z)\Psi_t = E_t\Psi_t \tag{10.10}$$

with

$$u(z) = \begin{cases} U_0(z) & \text{for } z < -w/2 \\ & \text{and } z > w/2 \\ U_0(z) + V_B & \text{for } -z/2 \leqslant z \leqslant z/2 \end{cases} \tag{10.11}$$

with $U_0(z)$ representing the position-dependent potential.

Equation (10.10) is devided into three regions; it is solved separately for each and the solutions are joined at $z = \pm w/2$ requiring that the logarithmic derivative with respect to z is continuous:

$$\lim_{z\to \pm w/2^-}\frac{\Psi_t(z)}{d\Psi_t(z)dz} = \lim_{z\to \pm w/2^+}\frac{\Psi_t(z)}{d\Psi_t(z)dz} \tag{10.12}$$

The superscripts $-$ and $+$ indicate that z approaches the limit $\pm w/2$ from within and without the barrier respectively. During flat band conditions, i.e. with $U_0 = 0$, the wave equation can be solved analytically. The solution is

$$\Psi_t = \begin{cases} A_1e^{ikz} + B_1e^{-ikz} & \text{for } z < -w/2 \\ A_2e^{i\kappa z} + B_2e^{-i\kappa z} & \text{for } -w/2 \leqslant z \leqslant w/2 \\ A_3e^{ikz} & \text{for } z > w/2 \end{cases} \tag{10.13}$$

with

$$k \equiv \sqrt{2m_t^*E_t}/\hbar \tag{10.14a}$$

and

$$\kappa = \sqrt{2m_t^*(E_t - V_B)}/\hbar \tag{10.14b}$$

neglecting the non-parabolicity of the band structure.

The factor $e^{i\kappa z}$ represents a wave travelling towards positive z. Assuming the incident particle moves this way, a term $B_3\, e^{-ikz}$ for the transmitted wave ($z > w/2$) does not exist. When $E_t < V_B$ the *transmission probability* is ($\kappa = i\sqrt{2m_t^*(V_B - E_t)}/\hbar$)

$$\mathfrak{J} = \left|\frac{A_3}{A_1}\right|^2 = \frac{4(\kappa/k)^2}{[(\kappa/k)^2 - 1]^2 \sinh^2(kW) + 4(\kappa/k)^2 \cosh^2(kW)}. \tag{10.15}$$

E_t represents the energy associated with the motion perpendicular to the barrier. Even when $E_t > V_B$ the transmission probability

$$\mathfrak{J} = \frac{4(\kappa/k)^2}{[(\kappa/k)^2 + 1]^2 \sin^2(kW) + 4(\kappa/k)^2 \cos^2(kw)} \tag{10.16}$$

is not 1 for all energies. (Figure 10.9.) When $U_0(z) = F_t z$, i.e. when the electric field perpendicular to the barrier is constant, an 'analytical' solution of the wave equation can still be found in terms of the Airy functions:

$$\Psi_t = \begin{cases} A_1 Ai(\zeta) + B_1 Bi(\zeta) & \text{for } < -w/2 \\ A_2 Ai(\zeta) + B_2 Bi(\zeta) & \text{for } -w/2 \leqslant z \leqslant w/2 \\ A_3 Ai(\zeta) & \text{for } z > w/2 \end{cases} \tag{10.17}$$

with

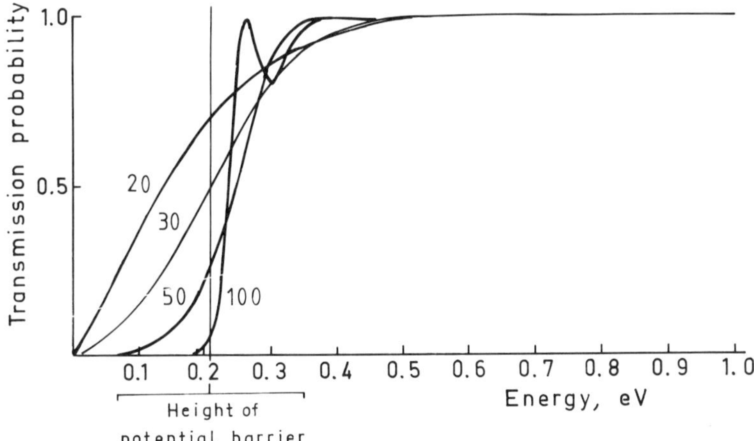

Figure 10.9 *Transmission probability of a tunnel barrier of $Al_{0.3}Ga_{0.7}As$ during flat-band condition against energy of the incident particle. The barrier height is 220 meV, which equals that of the heterojunction between GaAs and $Al_{0.3}Ga_{0.7}As$. The parameter represents the barrier width in ångstroms*

$$\zeta = \left(\frac{2m^*}{(\hbar eF_s)^2}\right)^{\frac{1}{3}}(eV_b - E - eF_s x) \tag{10.18}$$

The transmission probability becomes

$$\mathfrak{T} = \left|\frac{A_3 Ai(w/2)}{A_1 Ai(-w/2)}\right|^2. \tag{10.19}$$

To distinguish tunnel barriers for bulk material during the simulation, an additional qualifier code has to be attributed to each simulation area. This qualifier takes the value 0 for areas that are treated as bulk, 1 for areas where tunnelling is possible and 2 where quantum transport has to be considered. Kizilyalli *et al.* (1987) have studied the transfer of electrons between layers in a superlattice by means of Monte Carlo particle modelling.

When a particle has entered or traversed a tunnel barrier in the course of an allocated time of free flight, the flight is re-examined to investigate whether the particle is permitted to traverse. A flat random number $r_{\mathfrak{T}}$ is chosen, if $r_{\mathfrak{T}} \leqslant \mathfrak{T}$ with $\mathfrak{T}$ calculated from one of the Equation (10.15), (10.16), or (10.19) the particle is permitted to pass the barrier, otherwise it is reflected.

Tunnelling takes time. Studying dynamic properties like a.c. characteristics this may show a significant effect. Theorists like Anwar *et al.* (1989) have calculated the duration of transport through the barrier by considering the particles as a wave packet. Buot and Jensen (1990) and Jensen and Buot (1991) have solved the problem by means of *Wigner functions*. A calculation of the particle dynamics based on this formalism requires more algebraic operations than one based on classical Newtonian mechanics, and is therefore slow. A very simple model will be suggested here.

When $E_t > V_B$, the wave vector component associated with the transversal motion across the barrier is given by Equation (10.7):

$$k_z' = \{2m_t^*(E_t - V_B)[1 + \alpha(E_t - V_B)]\}^{\frac{1}{2}}/\hbar. \tag{10.7$'$}$$

When $E_t < V_B k_z'$ becomes imaginary, we have to consider the particle current instead (Kane, 1969). The velocity of the particle becomes

$$v_z = \frac{-i\hbar}{2m_t^*}\left(\Psi_t^* \frac{d\Psi_t}{dz} - \Psi_t \frac{d\Psi_t^*}{dz}\right) \tag{10.20}$$

which for the rectangular, field-free barrier, Equation (10.11), becomes

$$v = \frac{4\hbar\kappa^2 k^3 A_1}{m_t^*[(k^2 - \kappa^2)^2 \sinh^2\kappa w - 4k^2\kappa^2 \cosh^2\kappa w} \tag{10.21}$$

which shows that the particle passes through the barrier with a finite speed. The corresponding wave vector is obtained from

$$k_z' = m^* v/\hbar. \tag{10.21$'$}$$

The motion can be calculated from Equations (5.42) and (5.43) without any modification to the Lorentz force.

Since it is necessary to distinguish tunnelling particles from non-tunnelling ones, an additional particle coordinate is needed. This means that N_s memory positions in the computer are needed where N_s represents the number of superparticles. This extension of the required memory can be avoided if the time coordinate is stored with a negative sign as flag when the field-adjusting time step has been used up for the particle. In this way the simulator recognises that the particle is tunnelling when it resumes following it.

As in the case of a thick barrier, Section 10.3, the time of entry and exit from a barrier has to be known. A possible refinement to the method will be to split the transmission into two parts: the probability of entry into the barrier is $1 - |B_2/A_1|^2$; the chance of reflection from the exit is $1 - \mathfrak{T}$. Random numbers are used to decide on reflection or transmission as explained. A quantum mechanical treatment shows that the wave function splits into a transmitted and a reflected part after hitting the barrier. However, since an electron cannot split, we have to choose which part really represents it, the transmitted or the reflected part - the quantity $\mathfrak{T}$ serves this purpose. $\mathfrak{T}$ represents the fraction of the impinging electrons which are successful in passing the barrier.

The author has simulated the transport through an n-type doped GaAs resistor with an $Al_{0.3}Ga_{0.7}As$ barrier across it. It was found the carriers accumulate in front of the barrier and create a void behind it. The space charge thus created forms a strong dipole field making the valence and conduction band edges run rather steeply in the barrier region. The field there is so strong that many carriers gain enough energy by acceleration to overcome it.

Resonant tunnelling can take place through a pair of barriers so close together that the conduction (valence) band quantises into sub-bands. (Figure 10.10.) The particles which have the greatest chance of penetrating the double barrier are those entering the quantum well between them at a sublevel. The particle does not need to enter the first barrier at the correct energy as long as it attains the right one from the electric field during transmission.

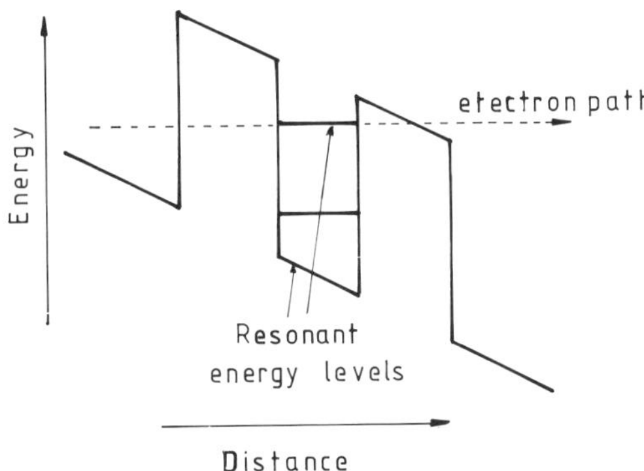

Figure 10.10 *Schematic band structure of a pair of tunnel barriers sufficiently close that the conduction (valence) band of the material between them divides into sub-bands*

10.5 CONFINED STATES. TRANSPORT IN QUANTUM WELLS

The conduction or valence bands divide into sub-bands between a pair of barriers. Figure 10.11 shows the energy levels against the thickness of a well generated by a layer of GaAs between two thick sheets of $Al_{0.3}Ga_{0.7}As$. Regardless of how narrow the well is, one quantum level always exists; when its width approaches zero the quantum level rises to lie near the continuum. When the well widens, more energy levels are permitted – these approach one another in energy. In an infinitely wide well, the levels come so close together that they can be considered as forming a quasicontinuous band, as in bulk material. Bulk material can, on the contrary, be considered as an extreme case of a wide quantum well.

To each energy level, E_t, we may associate a wave vector k_z:

$$E_t(1 + \alpha E_t) = \frac{\hbar^2 k_z^2}{2m_t^*} \tag{10.22}$$

where α represents the non-parabolicity factor and m_t^* the effective mass of the carrier. This equation says that k_z take discrete values, one for each energy level. The values are discrete and sharp. It is only possible to enter from another level or from the quasicontinuum above by means of a phonon that renders the particle with the exact fitting k_z; but the chance of this happening is nil. The quantum well can therefore not be filled from above; on the other hand once a particle stays in a sub-band it remains there forever. This is an unphysical situation.

Quantum physics, however, allows a *broadening* of the *energy* levels by an amount, ΔE_t depending on the time Δt, the particle remains in the level. The relationship between Δt and ΔE_t is given by Heisenberg's uncertainty principle:

$$\Delta t \Delta E_t > \hbar \tag{10.23}$$

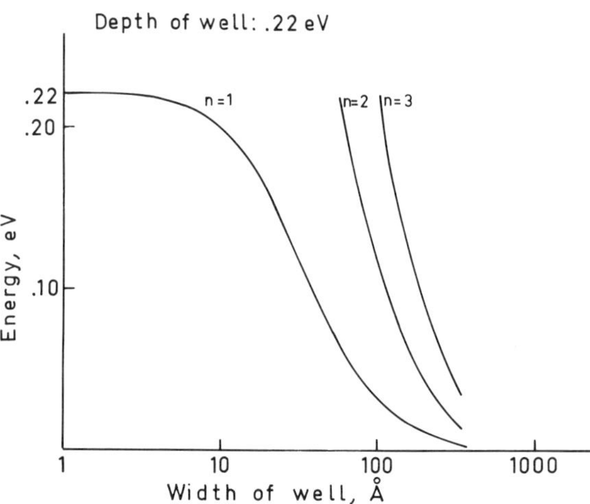

Figure 10.11 *The energy levels of the first three sub-bands against the width of a quantum well of depth 220 meV*

Interpreting Δt as the inverse scattering rate into the energy level, this relation reads

$$\Delta E_t/\lambda_t > \hbar \tag{10.24}$$

where λ_t represents the scattering rate of transfer.

Assume that after an inter-sub-band scattering the energy associated with the momentum perpendicular to the well is E_t'. Then the event can only take place if

$$E_t - \Delta E_t \leqslant E_t' \leqslant E_t + \Delta E_t \tag{10.25}$$

otherwise the choice of this event has to be considered as self-scattering according to Section 5.5.

The reader may argue that Δt really should be the actual time the particle has spent in a certain energy level prior to selection of the event bringing it out of this level, which is not necessarily the time since the previous scatter event. Such a scheme will make the allowed range of E_t' narrower. While other criteria for inter-sub-band transfer may be proposed, the one discussed here has been used successfully by the author. We shall return to this point in the next section.

Wires of a few square nanometres' cross-section are known as *quantum wires*. These can be manufactured by depositing a fine grate of metal strips on a thin layer of doped narrow band gap material on top of a wide-gap material acting as a barrier. (Figure 10.12.) Biasing the grate, the semiconductor below the metal strips becomes depleted of carriers so that the mobile carriers can only move below the spaces between them.

This system has to be described by solving Schrödinger's, Boltzmann's transport and the field equations self-consistently. In the effective mass approximation the Schrödinger equation is of the form

$$-\frac{\hbar^2}{2}\left(\frac{1}{m_1^*}\frac{d^2}{dx_2} + \frac{\partial}{\partial \mathbf{r}}\,\mathbf{m}^{*-1}\frac{\partial}{\partial \mathbf{r}}\right)\Psi + [U_1(x) + U_t(r)]\Psi = (E_1 + E_t)\Psi \tag{10.26}$$

which is separable into a perpendicular part in the coordinates $\mathbf{r}$,

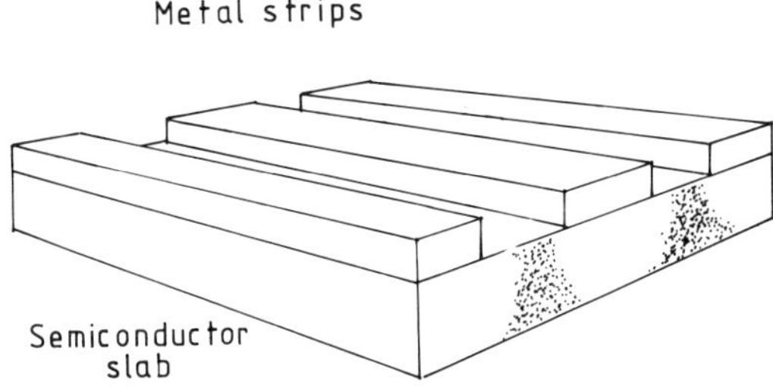

Figure 10.12 *Possible geometrical shape of a quantum wire device. The biased metal strip causes depletion of the narrow-gap semiconductor so that the carriers get confined to the spaces between them. Dots represent charged carriers*

$$\left\{-\frac{\hbar^2}{2m_1^*}\frac{\partial}{\partial \mathbf{r}}\mathbf{m}^{*-1}\frac{\partial}{\partial \mathbf{r}}+U_t(\mathbf{r})\right\}\Psi_t(\mathbf{r})=E_t\Psi_t(\mathbf{r}), \tag{10.27a}$$

and a part along the strips, the x direction

$$\left\{-\frac{\hbar^2}{2m_1^*}\frac{d^2}{dx^2}m_1^{*-1}+U_t(x)\right\}\Psi_1(x)=E_1\Psi_1(x) \tag{10.27b}$$

where $\Psi = \Psi_t\Psi_1$; m_1^* represents the effective mass along and $\mathbf{m}^*$ the two-dimensional mass tensor perpendicular to the wire.

The latter describes the motion along the wires which is similar to that of free electrons with mass m_1^*; $U_1(x)$ represents the potential along the wire which gives rise to the drift electric field. The energy is quasicontinuous in k_x.

Equation (10.27a) describes the energy states perpendicular to the wire; the energy takes discrete values. Here $U_t(r)$ represents the potential defined by the barriers confining the carriers to the wire, which depends on the bias of the strips.

If the wires are sufficiently close to one another, carriers can cross into those neighbouring by tunnelling.

Depositing an additional layer of metal strips at right angles to the first ones at the underside of the geometry shown in Fig. 10.12 it is also possible to confine the electrons in the x direction as well. The Schrödinger equation (10.26) no longer becomes separable because of the effective mass tensor, but the energy levels of the carriers are discretised in all directions. We now have a *quantum dot*. Donors and acceptors in semiconductors with energy levels in the band gap are also examples of quantum dots. A negative charged impurity in n-type doped material representing a potential hill, is also a special case of a *quantum antidot*.

10.6 PROBLEMS ASSOCIATED WITH COMPARISON WITH EXPERIMENTAL DATA

A good Monte Carlo particle model should be able to reproduce measured data. When it does, it can with confidence be used as a computer-aided tool to design other devices.

Modern devices are manufactured on wafers. We start with a substrate consisting of a slice from a semiconductor ingot, a single crystal which is as perfect as possible. The devices or integrated circuits are fabricated by growing successive layers of doped and undoped semiconductor material and metal on it; the structurisation of the wafer is carried out by a combination of lithography and selective etching through masks. Each wafer contains many devices of identical nominal design, yet their electrical characteristics differ between individual devices. These differences are due to deviations in device geometry, crystal defects or inhomogeneity in doping. Modern manufacturing technology is challenged by the endeavour to reduce this spread in quality over the wafer and between wafers.

To some degree the electrical properties of devices reflect the way they have been fabricated: the thickness of doped layers can vary by a few atomic layers, while the edges of the metal contacts and the distance between them can show variations of a few nanometres. These variations between individual components are not important for large components, but when the geometrical detail amounts to tenths of a micrometre or less the relative influence of these variations becomes significant. These facts have to be kept in mind when comparing measured and simulated data.

The devices are manufactured by depositing layers of differently doped material on top of one another. This is done either by evaporation, metal-organic chemical vapour deposition (MOCVD), molecular beam epitaxy (MBE) or metal organic molecular beam epitaxy (MOMBE). The latter two techniques allow control of the thickness of the layer to one atomic layer, but growth is slow compared to that of the other two techniques. Doping can vary over the wafer and may be caused by inhomogeneity in the temperature during deposition or an inhomogeneous gas flow.

Doping can also be introduced by ion-beam irradiation. The dopants are shot towards the surface then are diffused into the wafer by their own kinetic energy. The exact doping profile depends on the energy of the ion beam and the temperature of the wafer and the crystalline orientation of the substrate. Metal oxide field-effect transistors (MOSFETs) are manufactured by exposing selective areas to ion implantation.

Both the host material and the dopant source should be as chemically pure as possible. Absolutely pure material cannot as yet be manufactured, but the supplier can guarantee the stated degree of purity. This, however, is not enough: the ingot from which wafers are cut contains crystal faults, and the process of sawing it into slices may introduce additional imperfections. Crystal faults generated in the substrate may propagate into the epitaxial layers.

The metallisation, i.e. the deposition of the electrical contacts, is carried out by means of optical or X-ray lithography or by electron beam writing. The latter is very slow compared to the former two. The processes are akin but the electron beam produces sharper metal edges and therefore better uniformity of nominally equal devices over the wafer. By this technique feature sizes of about a hundred nanometres or less can be obtained. Care has to be taken to eliminate the effect of scattered electrons during writing. The resolution of the optical or X-ray lithographic process is defined by the lens system and the wavelength of the radiation. The positioning of the metal structures relies on using alignment marks; the accuracy of placing the photolithographic masks or adjusting the electron beam is defined by the sharpness of these masks. The geometry of the electrodes can be checked by scanning electron microscopy (SEM).

Often the gate is recessed into the semiconductor. The recess is excavated by selective etching or ion milling after the various layers have been deposited.

The Ohmic contacts are manufactured by heating the wafer for a short while after depositing the metal in order to make it fuse into the semiconductor so that the Schottky barrier vanishes. It is not easy to establish how far the metal penetrates into the semiconductor and how the metal atoms are eventually distributed. This problem has already been discussed in Section 8.3. It is possible to map the doping distribution by means of selective ion mass spectroscopy (SIMS) but this method is destructive and yields somewhat distorted results because atoms become displaced during the process.

Often it is found that the electrical properties of devices change as they age: atoms from the contacts migrate to change the contact resistance; impurities diffuse to alter doping profiles and the barrier at the heterojunctions.

The manufacture of modern devices is thus a very complex issue. In order to achieve a reliable comparison, it is important that the modeller really understands the various processing steps. He may have his own practical experience of the problems and the uncertainties concerning the layered structure and doping profile of the wafers; or he may obtain it from close discussion with the manufacturer.

A proper understanding of how the measurements have been carried out and what they mean is just as indispensable. Again, an understanding of the functioning of the measuring equipment is needed; the modeller and the experimentalist should agree on the significance of what is being observed. The definition of terms like intrinsic, biases, etc. should be made clear. For example, are the quoted biases really measured at the electrodes of the device, or is there a potential drop between the point the bias has been applied and the electrodes due to metallic resistance? Do we carry out the corrections in the right way? Do we understand intrinsic as referring to the semiconductor parts of the device, as it has been used in this book, or have the metal contacts been included as well? The differences are subtle, but may be significant. The resistance of the metal contacts adds to the internal source, gate and drain resistances of the transistor. The shape of the metal may influence the characteristics of the device through stray fields, as has been indicated in simulation carried out by Moglestue (1984a) and discussed in Section 9.6.

10.7 EXPERIMENTAL VERIFICATION OF THE PARTICLE MODEL

In the last section the uncertainties of experimental data have been discussed. However, the theorist also encounters uncertainties in the model which have to be analysed. In general, the model describes an idealised device. If this bears little relation with the one it is meant to describe, the comparison with experimental data is doubtful. In this section we shall see examples of facts we should consider.

Some authors are so confident of Monte Carlo modelling that the success of simpler approaches has been measured by comparison with Monte Carlo simulations. This is not a goal we should endeavour to achieve; it is better to compare results with actual measurements if we can. Ultimately, nature herself is the highest judge in assessing theories.

The first verification of the Monte Carlo particle model was reported in 1979 by Moglestue and Beard who calculated the direct current characteristics of the planar dual gate field-effect transistor shown in Fig. 10.13. Both gates had a length of 1 μm; the distance between the electrodes can be read from the figure. The contacts were resting on a 0.18 μm epilayer uniformly doped n-type at 10^{23} m^{-3} and

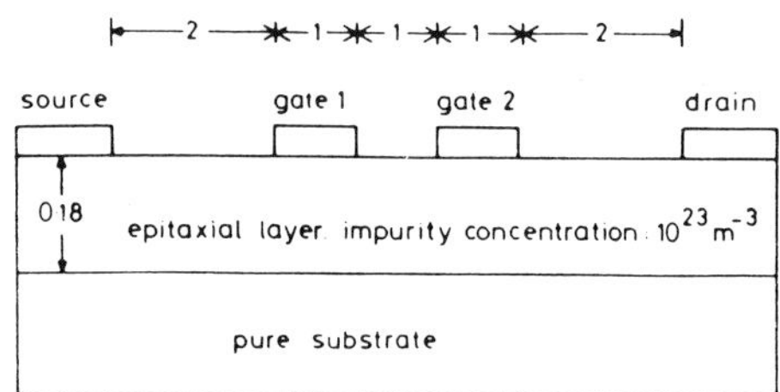

Figure 10.13 *Geometry of the dual-gate transistor for which a comparison between Monte Carlo particle modelling and experimental data has been made for the first time. The geometrical dimensions can be read from the figure.*

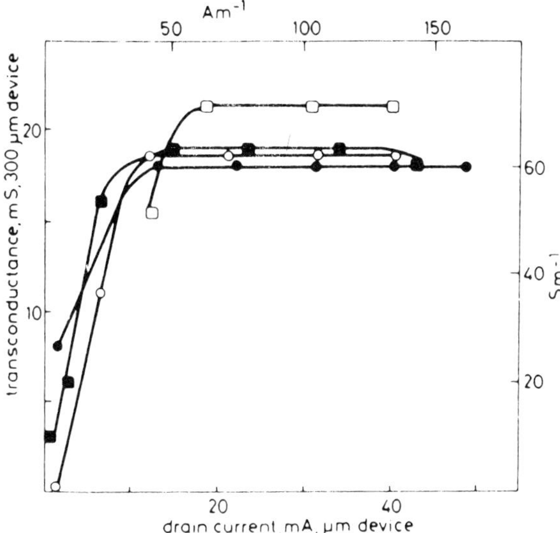

Figure 10.14 *Comparison of the calculated and measured transconductances of the two gates against the drain current. Circles refer to the gate closest to the source, squares to the other gate. Open symbols indicate simulated points, full symbols measured data*

the substrate was intrinsic. A device of the same nominal geometry had been built and characterised electrically. Any attempt to compare the calculated drain current failed because of the large spread in the measured data, in spite of all devices being of the same nominal geometry; this indicated a spread in the thickness of the epilayer. However, the measured transconductances against the gate bias for both gates fell within a sufficiently narrow range for a comparison to make sense. (Figure 10.14.) The calculated transconductance showed a satisfactory agreement with those measured.

Figure 10.15 shows the cross-section of a carefully manufactured heterojunction transistor. The reader may wonder why we should use such a complicated structure to verify the model and argue that a simpler one should be used. The answer to this is that we know how the transistor has been made and are confident that it really has been made to specification. The simpler structures have usually been made by a simpler manufacturing process which cannot guarantee the accuracy of the specified geometries that well. The epitaxial layers of the heterojunction transistor have been grown by molecular beam epitaxy; the accuracy of the thickness of the various layers can be controlled within 10 Å. The doping level was checked by Hall mobility measurements during manufacture and the geometrical arrangement of the electrode measured by means of scanning electron microscopy.

The etch-stop layer, which has been introduced to ensure uniformity in the threshold gate bias over the entire wafer, represents a resistance to the electrical current. It is 30 Å thick, and represents a potential barrier of 220 meV to the electrons. Electrons below this energy have, depending on energy, up to a 50% chance of being let through during flat-band condition. (Figure 10.9.) The possibility also exists that, even if the electron has sufficient energy to overcome the barrier, it will be reflected.

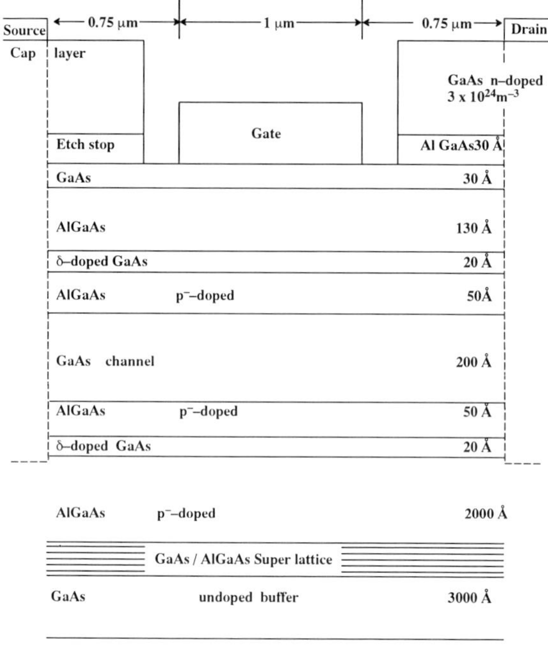

Figure 10.15 *Geometry of a transistor with one etch-stop layer. The p-type layers are doped at* $5 \times 10^{22}\ m^{-3}$ *acceptors; the thickness of the top epilayer is 400 Å. The distance between the gate metal and the recess wall is 125 nm. The dotted lines indicate the extent of the alloyed Ohmic contacts assumed for the simulation*

The other $Al_{0.3}Ga_{0.7}As$ layers are too thick for tunnelling. The only way through for the particle is to obtain enough energy to overcome it. This energy is different whether the electron is in a Γ, L or X minimum of the conduction band. (Figure 10.2.) On entry into the wide band gap material it loses kinetic energy equal to the difference in potential at the heterojunction. Leaving it, it gains correspondingly in kinetic energy. Particles with insufficient energy are reflected.

The transistor has been simulated with the same model as that used to simulate those discussed in Chapter 8 with the additions presented in Sections 10.3 and 10.4. The transport is considered as bulk-like everywhere; the free surfaces are assumed to have trapped charges of sheet density corresponding to a surface potential of -0.6 V in the absence of other biases.

Figure 10.16 shows the measured transfer characteristics of the transistor of Figure 10.15 (full curve). The measurements have been performed by means of a HP 4145 B Semiconductor Parameter Analyser Station. The measured gate threshold voltage was (-1.10 ± 0.05) V – the deviation represents the spread in the threshold voltage over the entire two-inch diameter wafer which contained many transistors of identical nominal design. This small deviation in the threshold voltage proves good uniformity and therefore makes a comparison between the theory and the measured data meaningful. We see from the figure that the agree-

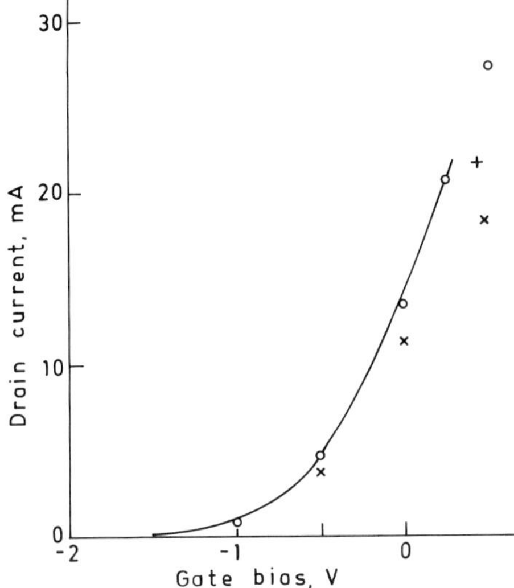

Figure 10.16 *Measured (full line) and calculated (discrete points) transfer characteristic of the transistor depicted in Fig. 10.13. Dashed line: measured data extrapolated to 0.5 V gate bias. x: model with surface charges and no alloyed Ohmic contacts. +: model with surface charges in the gate recess between the gate and the drain, and alloyed Ohmic contacts. ○: model with alloyed Ohmic contacts and surface charges* except *in the bottom of the recess. Drain bias: 2 V*

ment is nothing to boast about (× in Fig. 10.16) for only the calculated threshold is right.

In the absence of any etch-stop barrier the local electric field at its place is so small that the majority of the electrons would have insufficient energy to overcome it. From the expectation that the carriers would be in thermal equilibrium in the vicinity of the barrier, the current would be considerably lower than that we actually achieved from the simulation. However, the presence of the AlGaAs layer causes carriers to aggregate at one side and a void to build up at the lee side. This charge distribution causes the local electric field to rise, which in turn accelerates the particles making it easier for many of them to tunnel or jump across the barrier. This explains why we saw such high drain currents.

To achieve a better agreement between theory and experiment, it was postulated that metal from the source and the drain, which are Ohmic contacts, had been alloyed into the epilayers. Previously we have treated the contacts as an interface, allowing the free passage of particles (as we did in Chapter 8). The source and drain regions, those indicated by the dotted line in Fig. 10.15, are uniformly doped with the same donor density as that of the cap layer and the particles inside them experience no barriers. The justification for this assumption is that metal alloys into the semiconductor when depositing the Ohmic contacts. We do not know exactly how this alloying takes place, but the structure of the contact areas can be rather complex, as discussed in Section 8.3, and is different for each individual transistor. With this modification to the boundary conditions, a perfect agreement within the experimental error was obtained for the transfer characteristics except for the most forward-biased gate (the four lowest ○ and the + in Fig. 10.16).

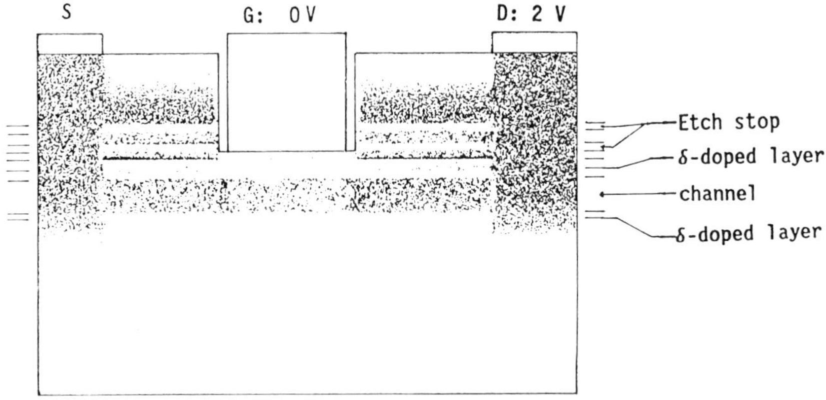

Figure 10.17 *Instantaneous carrier distribution in the transistor shown in Fig. 10.15. Drain bias: 2 V. Gate bias: 0 V (corresponds to a gate potential of −0.75 V, the Schottky contact potential)*

The free surfaces of the transistor contain unsatisfied chemical bonds and therefore reconstruct. As a result they trap electrons. From the outset the trapped surface charges have been included in the model to make it more realistic; they never get released during the simulation but give rise to a surface potential of −0.6 V, Chandra *et al.* (1979). The surface charges at the bottom of the gate ditch create their own depletion as a supplement to the one caused by the gate itself.

Figure 10.17 shows the carrier distribution throughout the transistor with alloyed Ohmic contacts and surface charges. The depletion at the drain side of the gate recess wall is more pronounced than at the source side, in spite of the same surface charge density on both sides – this is a 'wind effect': the particles tend to move in the direction of the drain where some of them head for the wall.

The effect of the surface charges in the bottom of the recess is not serious when the gate depletion is stronger than that caused by the surface charges. However, when the gate bias is more positive than 0.15 V, i.e. when the gate potential including the Schottky contact potential becomes less than that of the surface charges, the surface charges take over the definition of the aperture of the channel. At strong forward bias of the gate the depletion created by these charges dominates so that the channel does not open as wide as it would were they absent. Removing them from the vicinity of the gate at the drain side, the calculated drain current also agreed with the experiment for the greatest positive gate biases, as shown in Fig. 10.16 (the uppermost ○). The physical justification for this assumption lies in the fact that the electric field here tilts the band structure of the GaAs sufficiently to release the trapped charges.

The next step was to model a more complicated transistor. The design, shown in Fig. 10.18, is similar to Fig. 10.15 but with the following additions:

i) A 50 Å GaAs and 30 Å $Ga_{0.3}Al_{0.7}As$ layer acting as an etch stop have been inserted between the original etch stop and the cover epilayer. This enables the manufacture of both *enhancement* and *depletion transistors* on the same wafer in order to build fast digital circuits (Berroth *et al.*, 1990).

ii) The thicknesses of the atomic doped planar (**δ**) layers have been reduced to 17 Å.

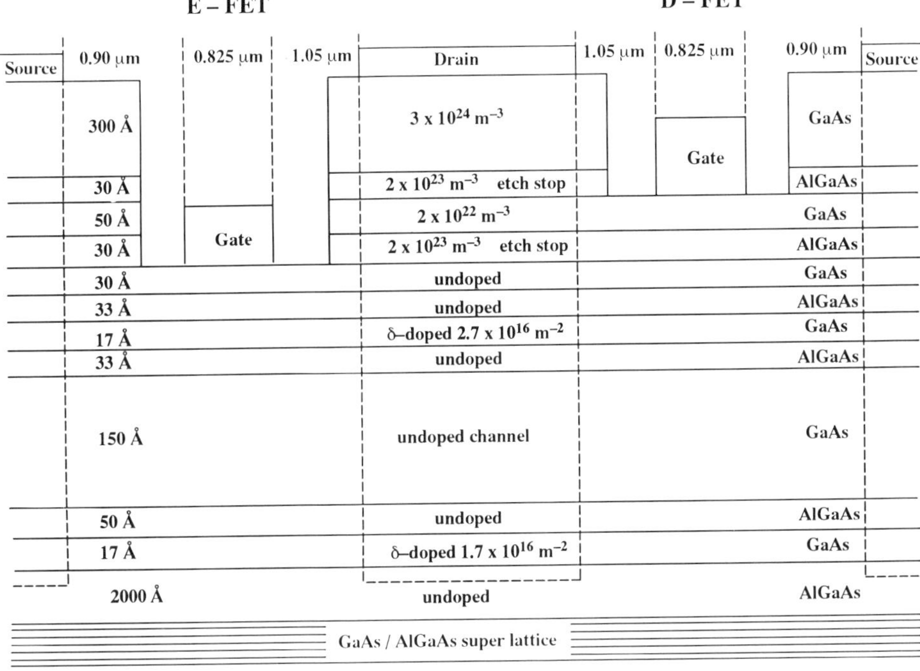

Figure 10.18 *The vertical structure of the depletion (D) and the enhancement (E) transistor. The dotted line indicates the extent of the Ohmic contacts*

iii) The thicknesses of the 130 Å and 50 Å layers above the channel have both been reduced to 33 Å.
iv) Most importantly, the channel is now 150 Å wide instead of 200 Å – it becomes a *quantum channel.*

Figure 10.19 shows the measured transfer characteristics of both the enhancement and depletion transistors (full curves). The corresponding simulated characteristics (not shown) from this model were too low; as the only essential difference between these two transistors and the previous one lies in the thickness of the active channel, it was decided to include the quantisation of the conduction band in the channel into sub-bands as described in Section 10.5.

With this addition to the model, the simulated transfer characteristic for the depletion transistor agreed well with observations. (Figure 10.19.) The calculated current against gate bias curve for the enhancement transistor has the correct form, but has been translated towards a more negative voltage. Agreement will be achieved when a Schottky contact potential of −0.95 V is assumed rather than the −0.75 V used for the simulations. (The Schottky contact potential has been introduced to the model as a boundary condition for the gate potential in the solution of Poisson's equation.) A Schottky contact potential of −0.95 V for the enhancement transistor is reasonable because the lowest quantum state in the narrow GaAs layer on which the gate rests is situated near the conduction band edge of the $Al_{0.3}Ga_{0.7}As$. Translating the curve 0.2 V, the agreement is

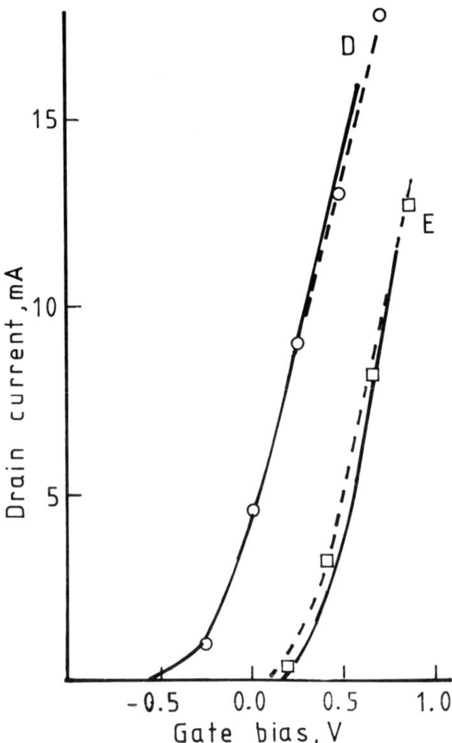

Figure 10.19 *Measured (full curve) and calculated (circles and squares) transfer characteristics for the depletion (D) and the enhancement (E) transistor. The calculations have been performed assuming a Schottky contact potential of −0.75 V for the D-FET and −0.95 V for the E-FET. Drain bias: 2 V. The calculation has been carried out accounting for quantised transport in the channel*

perfect within the experimental errors both for the forward and the transfer characteristics.

From Fig. 10.18 we see that the GaAs layer adjacent to the gate contact metal is so thin that it forms a quantum well; the Schottky contact potential is expected to increase by the offset from the bottom of the well to its lowest energy level, which is less for the depletion transistor than for the enhancement, shown in Fig. 10.11. Considering that the accuracy in the thickness of the various layers is within 10 Å, the lowest energy level for the layer of the depletion transistor is near enough the bottom of the well that we can consider the Schottky contact potential to be −0.75 V. For the enhancement transistor, however, the lowest energy level lies near the top of the well; the contact potential is therefore expected to be near −0.95 V. Assuming this is correct we see that the simulated transfer characteristics agree well with the measured ones.

The amendments we have made to the model do not change the Monte Carlo particle modelling in principle. We have only introduced additional physics, *namely* quantisation of the conduction band in the wells, tunnelling and alloyed contacts. The significance of these additions to the model is to improve it to reproduce the measured results.

Note that no transport mechanism or geometrical data has been altered. We may argue that we do not know exactly how the metal from the Ohmic contacts enters into the semiconductor, what doping profile it generates and how much the heterojunction barriers in the Ohmic contact regions have been reduced. This question remains unanswered, but we have obtained a model of the contacts which yields good agreement between the particle model and the observed current characteristics.

We have given examples of how to amend the boundary conditions to attain agreement with experimental data. As we have seen, the overall agreement is good, which indicates that we are on the right path to a realistic description. The reader may argue that this suggests parameter fitting. It does not. Parameter fitting means that intrinsically meaningless parameters have been chosen to make an analytical expression fit experimental results. This has certainly not been done here. All the amendments introduced can be justified physically, in full accordance with the philosophy of particle modelling.

The discussion presented here and in the previous section has attempted to illustrate the importance of knowing the structure of the device under study and how it has been measured, how to interpret the experiments and how to be aware of the physical processes taking place inside the device.

10.8 FIELD EMISSION

Field emission represents another interesting aspect of particle modelling. Figure 10.20 shows a typical *field emitter*. A tip made from metal or from a semiconducting material, e.g. silicon (Gray *et al.*, 1986), protrudes through a thin metal film with a hole in it; this film acts as a grid. An anode has been placed some distance, typically 1 μm above it. The height of the tip and the radius of the hole are about 0.5 μm; the curvature of the top of the emitter is 30 nm. When there is sufficient electrical potential difference between the emitter and the anode a current will flow in the evacuated space between them (Gray, 1989a); this current can be controlled by biasing the grid. The system functions in the same way as the vacuum valve (Gray, 1989b; Dalacu and Kitai, 1991) and is relatively easy to manufacture.

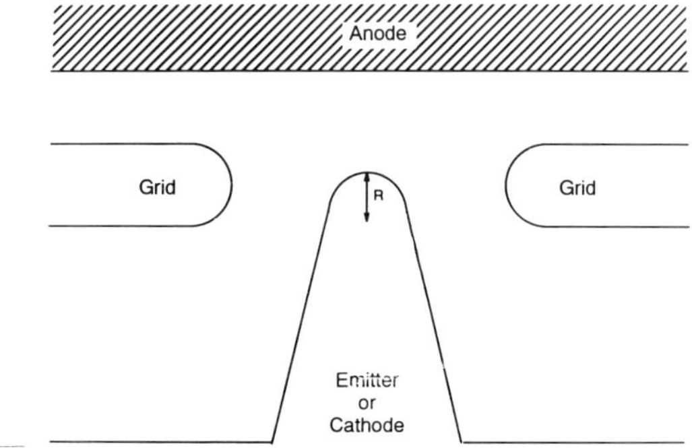

Figure 10.20 *Geometry of a field emitter. The anode and the grid are metallic; the emitter or cathode can be made from either metal or a semiconductor. The large bias of the anode causes electrons in the emitter to tunnel into the vacuum causing a current to flow, which can be controlled by the grid. The device is analogous to the electron valve*

It can be applied in flat display panels (Kaneko *et al.*, 1991), or flat panel television. It is fairly robust and promises to function in the high giga- and lower terahertz range (Gray, 1989c; Neidert *et al.*, 1991) and may form the basis for future high speed electronics. It can also serve as a safety device protecting delicate solid state circuits from the effects of sudden bursts caused by electrostatic discharges or lightning.

The emitter itself has a spherical top with radius R, typically of the order of tens of nanometres. The electric field inside it is almost zero - exactly zero if it were made from an ideal metal - because the electrons inside it tend to shield off the field. The field outside will have maximum strength at the top because the distance to the anode is shortest there. The field creates a triangular barrier against the electrons entering into vacuum, the height of which is given by the work function, ϕ_m, and with a slope given by the strength of the field, F. (Figure 10.21.) The magnitude of the work function, though difficult to ascertain since it depends on the way the cathode has been manufactured, is of the order of a few electron volts.

The current flowing towards the anode (and the grid) is caused by electrons tunnelling through this barrier. The field is strong enough for a significant part of the tail of their wave function to penetrate the barrier giving the electron a finite chance to travel through it. The Schrödinger equation for the tunnelling particle reads

$$-\frac{\hbar^2}{2m_0}\nabla^2\Psi + e\phi(r)\Psi = E\Psi \qquad (10.28)$$

with

$$\phi(r) = \begin{cases} 0 & \text{for } r < R \\ F(R-r) + \phi_m & \text{for } r \geqslant R. \end{cases} \qquad (10.29)$$

The distance r is measured from the centre of the spherical summit of the emitter. We assume $\phi(r)$ vanishes inside it; in reality it decays rapidly towards the interior. The actual shape of $\phi(r)$ inside the semiconductor or metal is immaterial to the

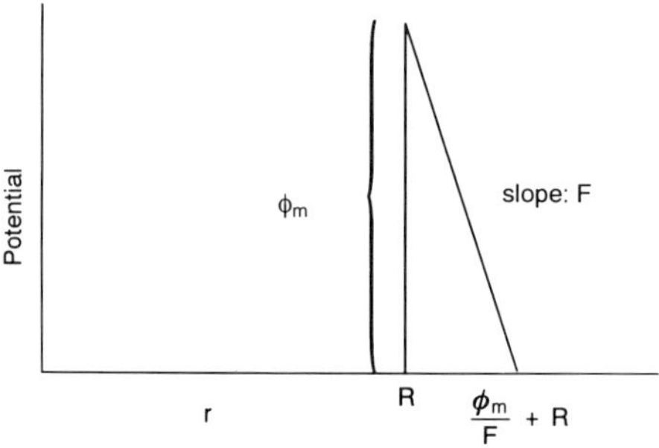

Figure 10.21 *Shape of the potential barrier separating the electrons in the material from vacuum under high electric field. The electrons can enter the vacuum by tunnelling through the barrier. ϕ_m represents the work function of the emitter material and F the electrostatic field*

problem; the only thing that matters is the energy of the particle on reaching the surface.

In polar coordinate form Schrödinger's equation reads:

$$\left\{-\frac{\hbar^2}{2m_0}\left[\frac{1}{r^2}\frac{\partial}{\partial r}\left(r^2\frac{\partial}{\partial r}\right)+\frac{1}{r^2\sin\theta}\frac{\partial}{\partial\theta}\left(\sin\theta\frac{\partial}{\partial\theta}\right)+\frac{1}{r^2\sin^2\theta}\frac{\partial^2}{\partial\phi^2}\right]+e\phi(\mathrm{r})\right\}\Psi=E\Psi \tag{10.30}$$

which is separable into a radial and a spherical part:

$$\Psi = r\Psi_r(r)\,Y_l^{|m|}(\theta,\phi) \tag{10.31}$$

where $\Psi_r(r)$ only depends on r and $Y_l^{|m|}$ represents the spherical harmonic given by Equation (3.28). Here l and m represent the angular and magnetic quantum numbers respectively. Inserting this into Equation (10.30) we get the two wave equations

$$\left\{\frac{d^2}{dr^2}+\frac{2m_0}{\hbar^2}\left[E-e\phi(r)-\frac{1(1+1)\hbar^2}{2m_0r^2}\right]\right\}\Psi_r(r)=0 \tag{10.32}$$

and

$$\left\{\frac{1}{\sin\theta}\frac{\partial}{\partial\theta}\left(\sin r\frac{\partial}{\partial\theta}\right)+\frac{1}{\sin\theta}\frac{\partial^2}{\partial\phi^2}\right\}Y_l^{|m|}(\theta,\phi)=l(l+1)Y_l^{|m|}(\theta,\phi). \tag{10.33}$$

Here m_0 represents the mass of the free electron which is appropriate outside the emitter. The tip has spherical symmetry and if the hole is round with its centre coinciding with the axis of the emitter we should expect that the wave function has no angular structure; this implies that $Y_l^{|m|}$ is constant which is only possible when $l = m = 0$, according to Table 3.2. The condition $l = 0$ reduces Equation (10.32) to

$$\left\{\frac{d^2}{dr^2}+\frac{2m_0}{\hbar^2}\left[E-e\phi(r)\right]\right\}\Psi_r(r)=0, \tag{10.34}$$

the solution of which is the Airy function

$$\Psi_r(r) = Ai(\xi) \tag{10.35}$$

with

$$\xi(r) = (2m_0e\phi\hbar^{-2})^{1/3}[r - E/(e\phi)]. \tag{10.36}$$

The associated Airy function $Bi(\xi)$ is not an acceptable solution because it diverges for growing r. (Figure 3.15.) The one-dimensional Equation (10.34) is the cornerstone of the Fowler–Nordheim (1928) theory.

The field-emitter device can be modelled treating the vacuum formally as another material where no scattering takes place. The probability that a particle tunnels through the barrier is

$$\mathfrak{J} = |Ai[\xi(R)]/Ai[\xi(R + \phi_m/F)]|^2. \tag{10.37}$$

A flat random number between 0 and 1 is generated to choose whether the particle is allowed to tunnel. If this number is less than $\mathfrak{J}$ the particle moves into the vacuum. Experience shows, however, that often the chance of the particle escaping from the tip is very small indeed, and if the superparticle represents a large number of real ones, the vacuum current can be erratic so that a very large simulation run is required to get a sufficiently accurate estimate of it. The reason why is that the superparticle is too coarse; we have to resort to simulating all the particles instead, for which we most likely do not have the computer capacity. Instead we let the superparticle divide on entering the surface: a particle of charge $\mathfrak{J}e_s$, where e_s represents the superparticle charge, enters into the vacuum. Of course it is unphysical to let $\mathfrak{J}e_s$ be less than the elementary charge of the electron. The part left behind still carries its original charge; the perturbation in the charge distribution inside the emitter made by this error is negligible. Of course the particles flying towards the anode or the grid have different individual charges which should be accounted for when solving the field equation. Their presence, however, represents a very small correction to the electric field in vacuum.

Under certain circumstances it is possible to get field emission from the edges in transistors or even through a planar surface in the vicinity of an edge. This could represent a possible breakdown mechanism which the electrical engineer should bear in mind when developing high-field devices.

11

Ambipolar Devices

11.1 INTRODUCTION

In many devices the concentration of holes and electrons are more or less of the same order of magnitude. In this case both types of carriers have to be considered in the Monte Carlo model. The scattering rates and the band structure of the two species of charged particles differ considerably, as we have seen in Chapters 3 and 4. We now have to define superholes in a similar way as superelectrons; each superhole represents a number of real ones. Two kinds of mobile particles enter the charge distribution term of the field equation; for each field-adjusting time step, transport histories of all the electrons and the holes have to be simulated. During simulation it is, for each time step, recommended to follow all the electrons first, then the holes. The reason for this will become clear when studying recombination of electron–hole pairs.

In principle, ambipolar transport does not differ from that studied so far. In this chapter we shall start discussing the current through a forward biased p–n junction. Then in Section 11.3 the semiconductor photodetector will be described. The results of our simulation agree qualitatively well with experimental data. The particles have been induced into photodetectors by absorption of photons, in contrast to the p–n junctions and bipolar transistors where the particles enter via the contacts. Section 11.4 discusses the response of a field-effect transistor to a high-energy α-particle passing through it. This is a natural extension of the previous section because here the electrons and the holes have been generated by the passing particle. We then proceed to a field of growing interest, the simulation of a bipolar transistor when the acceptor density of the base is of the same order of magnitude as that of the donors in the emitter and collector regions. In this connection we shall also remark on what to do if these concentrations differ by orders of magnitude. The chapter then closes with a discussion of the recombination of electrons and holes in a laser. The simulated results of these two sections have not been published previously and are included only to demonstrate other abilities of the particle model.

11.2 THE P–N JUNCTION

When *p* and *n* doped semiconductors are joined together, we get a *p–n junction.* It acts like a rectifier as it allows the electric current to flow only one way. The Fermi level of the two semiconductors join so that a potential barrier is generated. (Figure 11.1a.) No current can flow until the p–n diode is sufficiently forward biased that either the conduction or valence bands become horizontal. (Figure 11.1b.) The current increases when the junction is forward biased further,

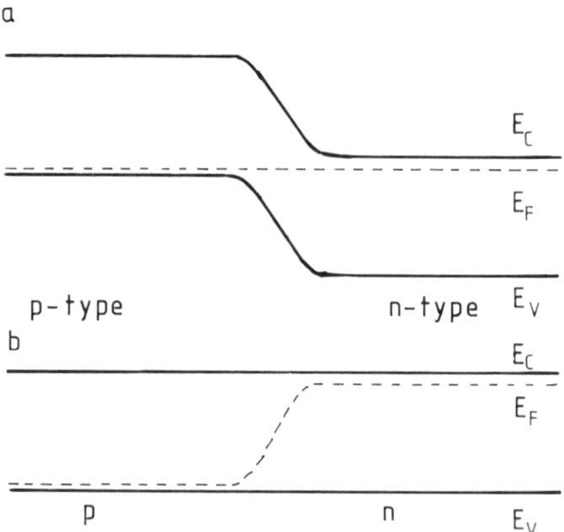

Figure 11.1 *The band structure of a p–n junction. a) unbiased, b) biased such that the current flows through it. The donor density in the n-side equals the acceptor density at the p-side*

but the band structure remains flat-band due to the electric field being screened off dynamically. The fall in the potential takes place elsewhere. Figure 11.2 outlines the typical current against bias for such a junction; points a corresponds to situation b of Fig. 11.1.

According to simple textbook theory, e.g. Sze (1969), the electrons will flow from the cathode to the anode, the holes in the opposite direction. Such an approach does not account for the formation of an electron–hole plasma. The plasma that forms at both sides of the junction as the holes and electrons pass each other has received interest from Azimov *et al.* (1980). Perhaps the easiest way of indicating the presence of such a plasma is to measure the luminescence from it. This has been done for *p–n* junctions in silicon by Dunstant (1982), Ong *et al.* (1983) and by Tanaka *et al.* (1983) in gallium arsenide. Radiation from the plasma in heterojunctions has been observed by Alferov *et al.* (1974) and Guckel *et al.* (1982). Beneking (1982a, b) has presented a theoretical description of the transport in the abrupt *p–n* junction based on the drift-diffusion approach. This, however, does not yield much information on the establishment and the dynamic screening of the plasma building up at the junction. This can only be obtained by solving Boltzmann's transport and Poisson's field equation self-consistently, in other words by Monte Carlo particle modelling. This was first done by Moglestue (1986c). Consider a diode as shown in Fig. 11.3. The two sides of the GaAs *p–n* junction are both doped at 10^{23} m^{-3}, *p*-type on the left side, *n*-type on the other side. Ohmic contacts have been placed at both ends, each of them 320 nm from the junction. An Ohmic contact potential of 0.5 V has been applied in the forward direction. The applied bias is obtained by adding the difference in quasi-fermi energy between the anode and cathode. The two contacts are allowed to absorb both species of carriers, but the cathode was only allowed to inject electrons, the anode only holes.

The Monte Carlo particle simulation started from charge neutrality everywhere with one electron (hole) for each donor (acceptor). As the simulation progressed, electrons and holes accumulated to form a *plasma* on both sides of the junctions,

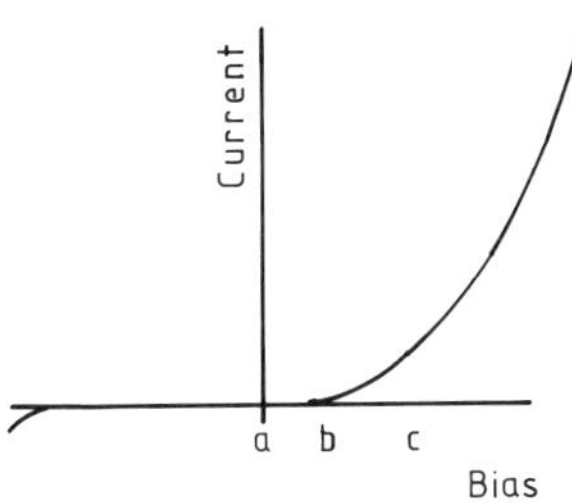

Figure 11.2 *Typical current against bias characteristic of a p–n junction. Point a corresponds to a and b in Fig. 11.1. During reverse bias the current starts to flow when the dielectric breaks down*

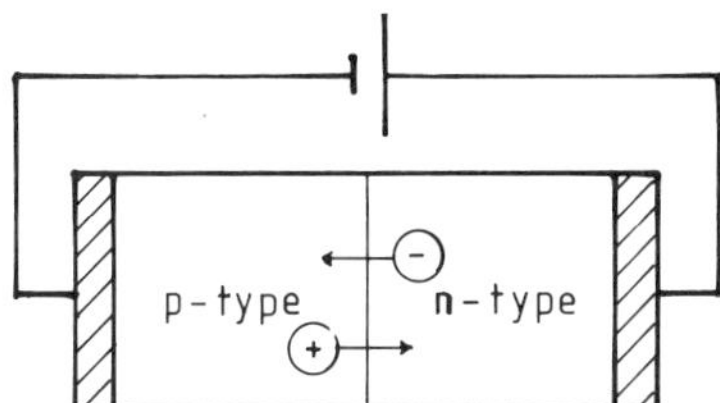

Figure 11.3 *Geometry of the simulated p–n junction diode. The diode is forward biased at an Ohmic contact potential of 0.5 V. ⊕ and ⊖ represent electrons and holes respectively; the arrows indicate their drift direction. The hatched areas represent the metal electrodes*

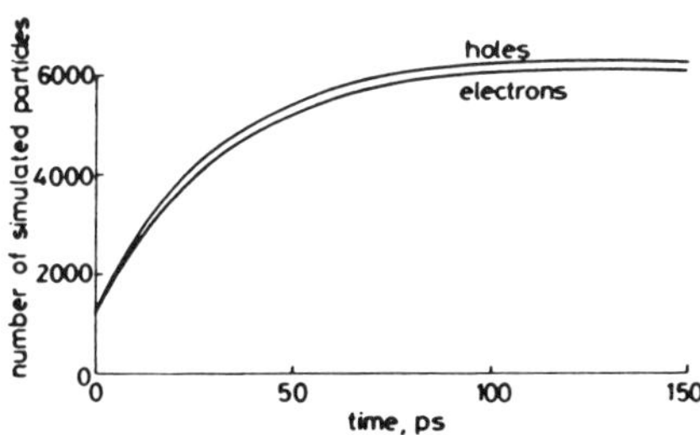

Figure 11.4 *Number of simulated electrons and holes against time since the start of simulation. The requirement of the minimum particle concentration to obtain charge neutrality everywhere was used as our start condition*

Fig. 11.4. The plasma population saturated after 100 ps, then every injected hole and electron could pass unhindered through the junction; the current attained its steady state value.

Figure 11.5 shows that the plasma extends to both sides of the junction during steady state. The difference in the electron and the hole population reflects the

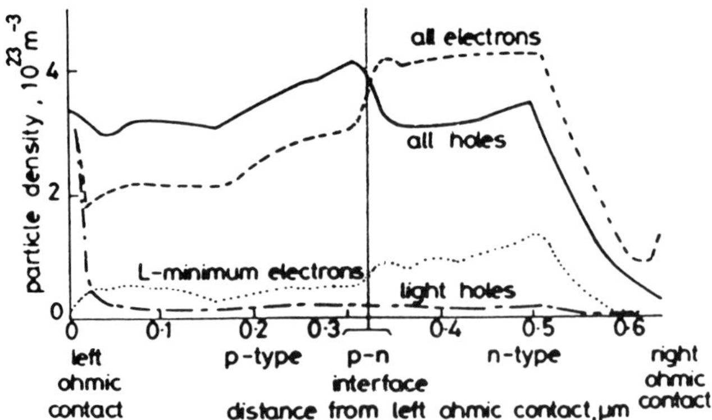

Figure 11.5 *Calculated density of holes, L electrons, light holes and all electrons throughout the junction diode of Fig. 11.3 during steady state (i.e. at 150 ps). The quantity of split-off holes and X electrons is negligibly small*

doping. The maximum electron or hole density reaches about four times that required to neutralise the charge generated by the ionised donors or the acceptor atoms. The plasma density falls gently at the p-side, while it rises to a maximum at the n-side before falling sharply towards the anode. Repeating the simulation for a diode of double length, i.e. with the contacts 640 nm from the junction, the same gentle slope at the p-side shows and the plasma density still peaks at the same place at the n-side. Our result is thus not an artefact of the choice of distance of the junction from the contact. As the condition for Gunn domain formation, Equation (8.10), is fulfilled the plasma peak can be attributed to such a Gunn domain. The electrons tend to move it one way, the holes the other way; the slower holes cause the plasma population to extend furthest into the n-side but the Gunn domain remains stationary. A simulation extending over a time of the order of one nanosecond confirms this.

Figure 11.6 shows that the potential mostly falls near the cathode. There is hardly any potential difference over the plasma peak at the n-side. This means that the electric field has vanished; there is dynamic screening of the plasma. The population reaches such a density that the screening length equals the average distance between particles. *The plasma is saturated.* Then any additional particles can pass through it freely.

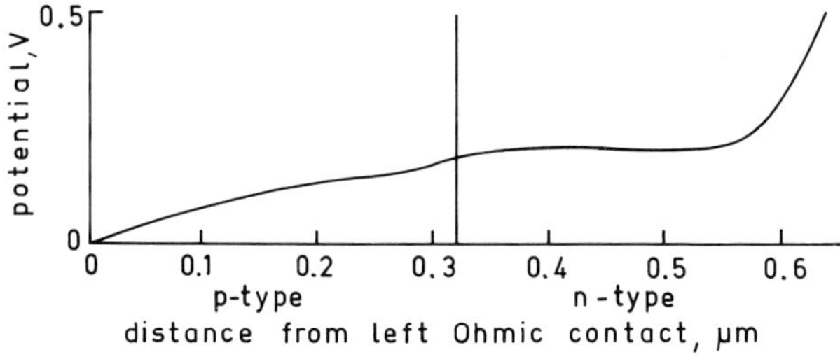

Figure 11.6 *Calculated electrostatic potential against distance throughout the p–n junction of Fig. 11.3 during steady state*

This effect had been predicted by neither Guckel *et al.* (1982), nor Bannov *et al.* (1984) nor Brennan and Hess (1984) – the last-named authors considered a quasiballistic approach to transport. The formation of plasma may not be of importance to steady state transport but can be of great importance for millimetre wave application as it represents an additional impedance to the current through the junction. An oscillating potential contributes towards reducing or increasing the extent of the plasma, thus absorbing or releasing carriers to the current during the cycles.

The transit time for the carriers is of the order of picoseconds, which is too small for any significant recombination over the half a micrometre which has been simulated. Recombination has therefore been excluded from the modelling.

11.3 PHOTODIODES AND DETECTORS

The very first Monte Carlo particle simulation of a photodiode was reported by Moglestue (1984b, c). The diode consisted of a 2 μm thick *n*-type doped film of GaAs doped at $10^{22}\,m^{-3}$, on top of which two Ohmic contact strips had been deposited 2 μm apart. The response of the diode to a 25 fs light pulse was simulated. It was found that when the *response current* died down, a significant hole–electron plasma was left behind in the interior of the material. This was not anticipated from the theoretical analysis of Beneking (1982a, b) which stated that the holes and electrons move towards the anode and the cathode respectively and that all the photogenerated particles are gone when the response current pulse is over. Raman-spectroscopic analysis by Tanaka *et al.* (1983), however, indicated the presence of such a plasma after the response current had vanished.

Later, more systematic theoretical and experimental studies of *metal–semiconductor–metal* (MSM) *photodiodes* with Schottky contacts have been performed. Monte Carlo particle simulations of such structures have been carried out in two

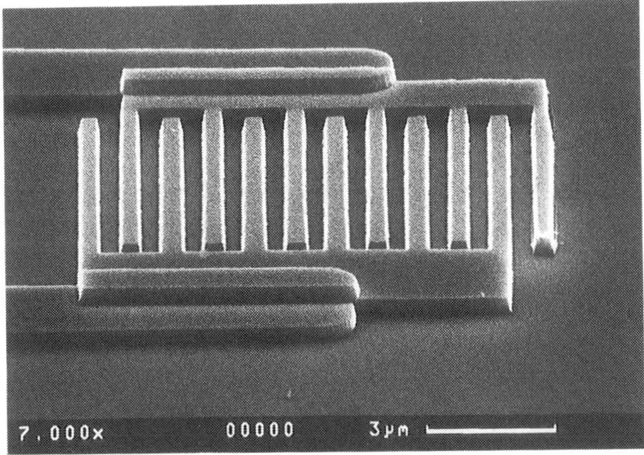

Figure 11.7 *Photograph showing a typical metal–semiconductor–metal Schottky contact photodetector. The distance between the fingers and their width are both 1.5 μm, their length 10 μm. Detectors of the same design, but with different numbers of fingers and finger width, length and separation have also been built and measured*

dimensions for the electrons only, treating the holes as a fluid. Koscielniak *et al.* (1989, 1990) have treated both holes on electrons in their one-dimensional Monte Carlo simulation. We have carried out a more systematic two-dimensional Monte Carlo particle study of MSM diodes that were built at the Fraunhofer Institute of Applied Solid State Physics at Freiburg im Breisgau, Germany, and measured at the Max Planck Institute of Research in Solid State Matter at Stuttgart and the University of Stuttgart, Germany (Moglestue *et al.*, 1990, 1991a; Rosenzweigh *et al.*, 1991; Lambsdorff *et al.*, 1991; Lambsdorff, 1990; Klingenstein *et al.*, 1991).

The photodiode consists of a thick layer of GaAs, unintentionally n-type doped at $5 \times 10^{19}\,\mathrm{m}^{-3}$, on to which two comb-shaped Schottky contact electrodes with their teeth gripping into each other have been deposited. (Figure 11.7.) The teeth, referred to as fingers, have width W, length L_g and have a distance d_D between them. A two-dimensional description is justified because $W \gg L_g$.

Figure 11.8 shows the cross-section through the detector between two adjacent fingers; the anode is kept at a fixed voltage, the cathode is connected to ground. The light penetrates the semiconductor through the areas not covered by the metal; we shall neglect the bending of the light due to its wave nature.

Due to absorption, the intensity of the light at depth y decays according to the law

$$I(y) = I_0 \exp(-\alpha_\lambda y) \tag{11.1}$$

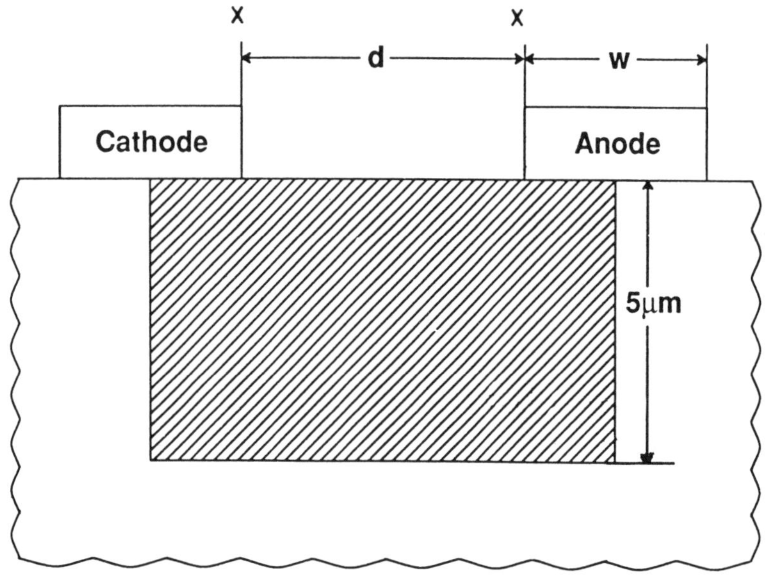

Figure 11.8 *Cross-section of a photodiode. Distance between anode and cathode: 1.5 μm. Uniform unintentional n-type doping density: 5×10^{19} m^{-3}. The part exposed to the light lies between X_A and X_C in the coordinate system with origin in the lower left hand corner of the hatched simulation area*

where I_0 represents the intensity of the light immediately beneath the surface and α_λ the absorption coefficient, which has been tabulated against wavelength by Aspnes *et al.* (1986). Light absorption consists of converting a photon into a hole–electron pair. The energy of the photon must exceed that of the band gap. Direct semiconductors, materials like GaAs where the maximum of the valence band and the minimum of the conduction band both lie at the Γ point, are ideal as photodetectors and photovoltaic generators. As indirect semiconductor like silicon absorbs light by the assistance of an intervalley phonon taking care of the conservation of momentum or by a multiphonon process and is therefore less efficient as a detector. A photodetector from silicon has therefore to be made thicker whereby the charge collection will take longer. When the photon is absorbed, a hole–electron pair is created; the electron receives an energy E_e given by

$$E_e(1 + \alpha E_e) = \frac{\hbar^2 k_e^2}{2m^*} \tag{11.2a}$$

where m^* represents its effective mass, k_e its wave vector and α the non-parabolicity of the conduction band near the Γ point; the hole receives the energy

$$E_h = \frac{\hbar^2 k^2}{2m_0}\{A' \pm [B'^2 + C'^2(k_x^2 k_y^2 + k_y^2 k_z^2 + k_z^2 k_x^2)/k^4]^{1/2}) \tag{11.2b}$$

where m_0 denotes the rest mass of a free electron and A', B' and C' are the same constants as those we encountered in Section 3.5. The upper sign in front of the term in square brackets refers to heavy holes, the lower sign to light holes. The energy of the electrons is reckoned upward from the bottom of the conduction band, that of the holes downwards from the top of the valence band. Conservation of energy requires

$$E_e + E_h + E_G = \hbar\omega_L \tag{11.3}$$

where E_g represents the band gap energy at the Γ point and ω_L the angular frequency of the light. Conservation of momentum requires

$$\mathbf{k}_e + \mathbf{k}_h + \omega_L \mathbf{j}/c = \mathbf{0}, \tag{11.4}$$

with c denoting the velocity of light, and $\mathbf{j}$ is a unit vector in the direction of the photon.

The position of the absorbed photon or the generation of an electron and hole pair, (x_p, y_p), has been chosen by means of two flat random numbers, r_x and r_y, such that

$$x_p = (x_A - x_C)r_x \tag{11.5a}$$

corresponding to a uniform intensity of the light along the surface and

$$r_y = \frac{1 - \exp(-\alpha_\lambda y)}{1 - \exp(-\alpha_\lambda t_D)} \tag{11.5b}$$

corresponding to an intensity according to Equation (11.1) at depth y from the surface of the detector. The symbols x_A and x_C represent the coordinates of the anode and cathode edge respectively of the free surface of the detector exposed to the light. The simulation has been carried out down to a depth of $t_D = 5.12\,\mu\text{m}$, where there is hardly any light, which means that the value of the denominator is practically 1. The denominator has been introduced in Equation (11.5b) to ensure that the probability density, $p_a(y)$, of absorption over the depth,

$$p_a(r_y) = \frac{\exp(-\alpha r_y)}{1 - \exp(-t_D \alpha_\lambda)} \tag{11.6}$$

normalises, in other words that

$$\int_0^{t_D} p_a(y)dy = 1 \tag{11.7}$$

in accordance with Section 5.3.

We are now going to describe our simulation of the MSM photodetector with finger separation and width $1.5\,\mu\text{m}$. The simulation started from dark under steady state, the dark current only consisting of electrons. Light was then shone on the detector for 70 fs, creating 1.8×10^9 pairs per metre finger. The wavelength of the light was 620 nm, corresponding to a photon energy of 2 eV. With this energy $\boldsymbol{\alpha}_\lambda = 4.279 \times 10^6\,\text{m}^{-1}$ in Equation (11.1) (Aspnes *et al.*, 1986), $I_0 = 5.1 \times 10^{21}\,\text{m}^{-3}$. The flash generated 10 000 pairs of superparticles, each representing $1.8 \times 10^5\,\text{m}^{-1}$ real ones. Each pair started with an energy of 0.58 eV, which is 2 eV minus the band gap energy of 1.42 eV. Poisson's equation was solved every 40 fs over a rectangular mesh measuring $20 \times 23\,\text{nm}^2$; the flash thus lasted for $1\frac{3}{4}$ time steps.

Figure 11.9 shows the simulated response particle current. It consists of electrons flowing out of the anode and holes out of the cathode. As the two species of particles flow in opposite directions in the external circuit, the currents add (electrons flow against the direction of the current by convention). The figure shows that the response current initially consists mainly of electrons; the holes are responsible for the tail. The initial dominance of electrons is due to their higher mobility or lower effective mass. It is customary to attribute a *time-decay constant*, τ, to each contribution, describing the decay as $e^{-t/\tau}$ where t represents time. As seen from the figure, this is a rather inaccurate approximation which should not be considered to be more than a crude guide in spite of its popularity with experimenters. If the decay were exponential, the graph in Figure 11.9 would consist of straight lines.

As in the case of the Ohmic contact photodetector, a considerable residue of electrons and holes remains after the decay of the response current. Figure 11.10 shows the evolution of the photogenerated electron and hole populations, the distribution of the potential and the magnitude of the electric field. Note that the

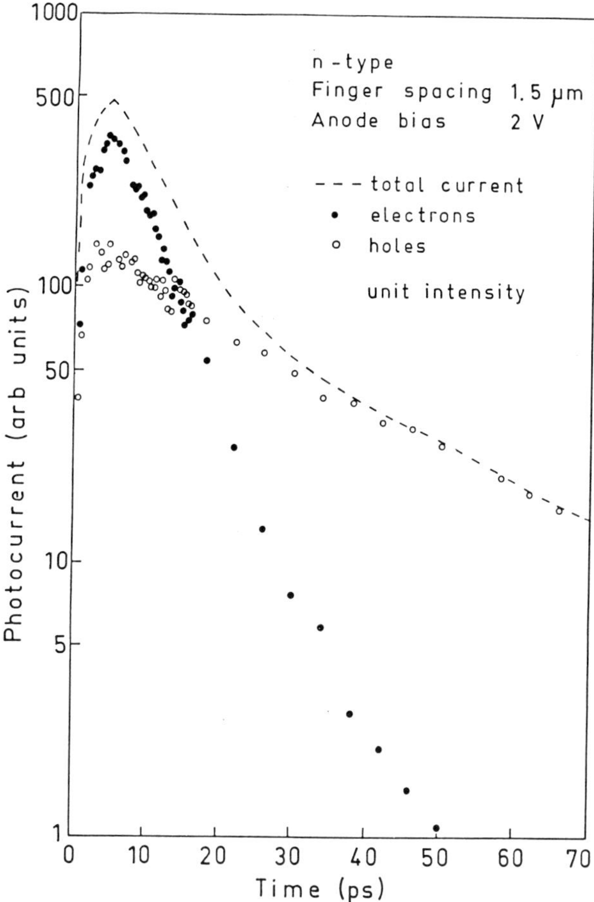

Figure 11.9 *The calculated electron, hole and total response photocurrent of the diode with 1.5 µm distance between the fingers. Anode bias: 2 V. Light intensity: 5.48 × 10^9 W m^{-2}*

linear geometrical scale differs for the two orthogonal directions. The nominal anode bias is 2 V, the cathode grounded. The simulation has been carried out with a Schottky contact potential of −0.75 V, which means that the cathode and the anode potential were held at respectively −0.75 and 1.25 V. Towards the deep interior the potential approaches 0 V. In addition, a surface charge density corresponding to a potential of −0.6 V was assumed for the exposed semiconductor surface. The reason for the existence of this charge has been explained in Section 8.3.3. The effect of these surface charges and the negative potential of the cathode is that the material becomes depleted of carriers prior to illumination at least down to the bottom of the simulated part of the detector and that the photogenerated electrons tend to sink into the interior; the holes, however, are attracted electrostatically to the surface.

Both the potential and the electric field do not change appreciably during the flow of the response current. The electric field peaks at the corners of the electrodes.

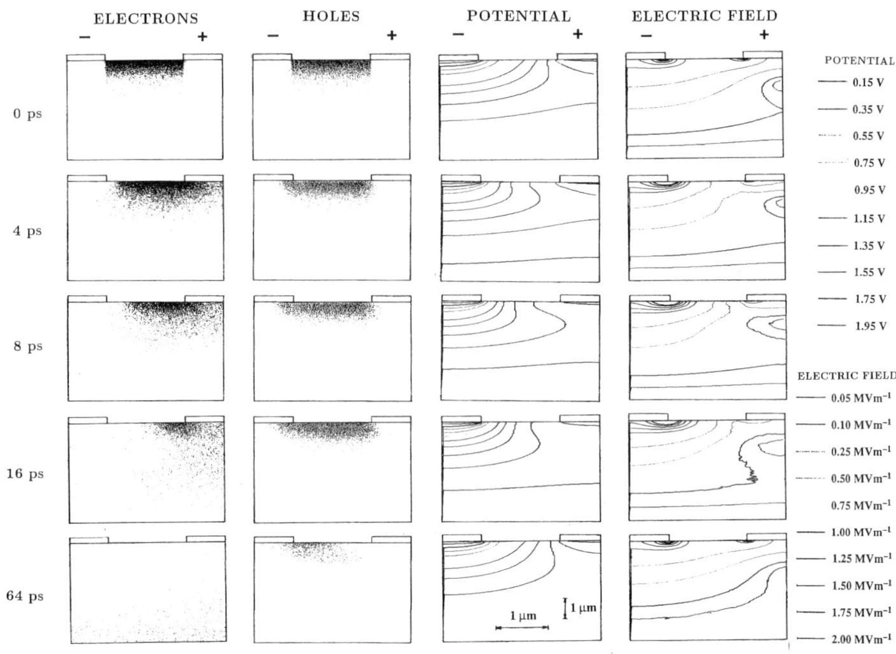

Figure 11.10 *Calculated distributions of electrons, holes, electric potential and magnitude of electric field at 0, 4, 8, 16 and 64 ps after the 70 fs light flash. The corresponding electron–hole pair density at the surface is* $I_0 = 5.1 \times 10^{21}\ m^{-3}$ *at time 0 (at light intensity of* $5.48 \times 10^9\ W\,m^{-2}$*). The distance between the anode and the cathode is 1.5 μm. Anode bias: 2 V. The linear scale in the two orthogonal directions is indicated in the bottom panel. Note, this figure is reproduced in colour as Plate 9*

Increasing the intensity of the light ten-fold changes the picture drastically. Still, the same number of superparticles is created by the flash, but now each superparticle represents ten times as many real ones. Figure 11.11 shows the particle response current in the same arbitrary units as in Fig. 11.9. The response current is not ten times as strong as we might have expected. The electron and hole tails now coincide. Figure 11.12 shows that the electrons and the holes coexist in the area below the free surface and the anode. The electric field considerably reduces here. A plasma forms that is held together by mutual Coulombic attraction between the two species of particles. The population is sufficiently concentrated to screen itself from the externally applied field; *dynamic screening*, as in the *p–n* junction discussed in the last section, has been established. This is accomplished within a few picoseconds. Such a plasma may be considered as a super-exciton. Dynamic screening is a phenomenon which requires a rather complicated theory based on Boltzmann's transport and Poisson's field equation. The Monte Carlo particle model solves these equations, however, and is therefore capable of describing it. If one particle attempts to escape, it will be pulled back by the net electrostatic attraction by the rest of the plasma. However, electrons may stray into the anode, where it is absorbed, contributing to the current flowing through the external circuit and ending in the power source. There will be one positive excess charge in the plasma; to regain charge neutrality, the plasma expels one

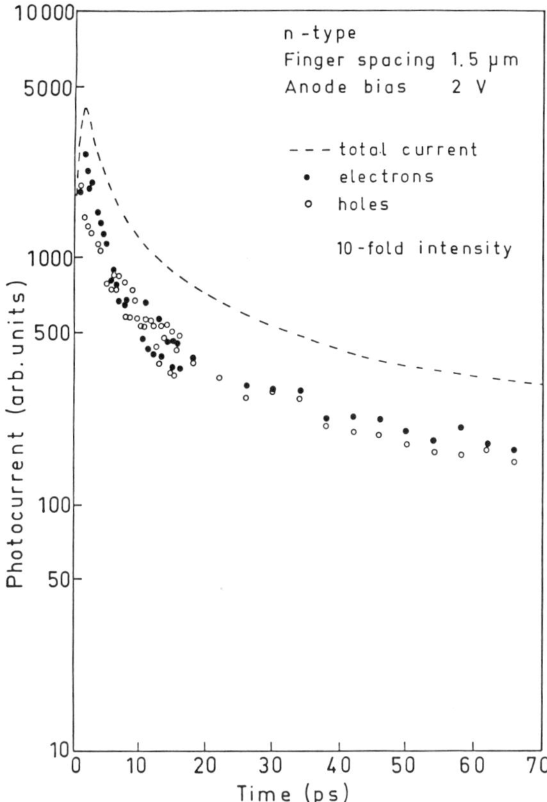

Figure 11.11 *The calculated eletron, hole and total response photocurrent of the diode with 1.5 μm distance between the fingers. Anode bias: 2 V. Light intensity: 5.48 × 10^{10} W m^{-2}, ten times that of Fig. 11.9*

hole that moves towards the cathode to be absorbed there. From the figure it appears that the centre of mass of the holes moves towards the anode, thus in the direction opposite to that in which they are expected to move. This is caused by the fact that the expelled holes are removed from the left.

Increasing the light intensity by another factor of ten so that each superparticle represents $1.8 \times 10^7\,m^{-1}$ real particles, the dynamic screening is established within the first picosecond. The plasma now extends all the way towards the cathode. Figure 11.13 shows that holes now even escape through the anode.

Returning to the case of unit intensity, decreasing the anode bias has very much the same effect on the dynamic screening. Now the electric field becomes small enough to allow it to form for smaller carrier concentrations.

All our simulated results for MSM diodes represent the first 72 ps after the end of the light pulse. In this span of time no significant recombination is expected to take place. In fact, recombination has not been considered. The residual plasma will eventually vanish by recombination but this will take nanoseconds or even longer.

The response current has been measured by Klingenstein *et al.* (1991) by means

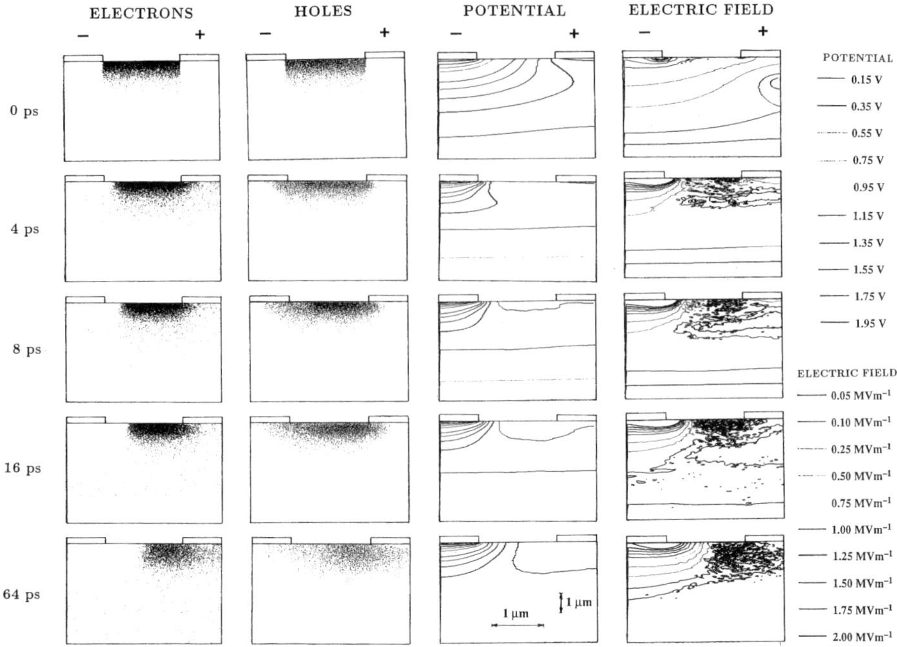

Figure 11.12 *Distribution of electrons, holes, potential and magnitude of electric field for the same diode as in Fig. 11.10. Anode bias: 2 V. Light intensity: 5.48 × 10^{10} W m^{-2}, ten times that of Fig. 11.10 or 11.9. Note, this figure is reproduced in colour as Plate 10*

of electro-optic sampling, a technique described by Valdmanis *et al.* (1987). The photodetector was excited by a 70 fs pulse from a mode-locked dye laser and sampled by a time-delayed pulse of the same duration derived from the same source. The response curve was found to be in good qualitative agreement with our theoretical prediction. A quantitative comparison meets with two problems:

i) The intensity over the spot is not uniform, covering many fingers; however the intensity differs over the illuminated area. We may assume it has a Gaussian intensity distribution over it, but it is uncertain whether this is true.
ii) The tail of the measured response current is stronger than the calculated one. There is still uncertainty as to the reason for this. Perhaps there are crystal faults allowing currents to leak from the contacts along the fault planes. The proximity of the plasma to the contacts can assist carrier injection by tunnelling. A suggested explanation that carriers are trapped and released cannot hold because the time scales for the trapping and release of carriers are too large for the duration of the response current.

An improvement in the agreement may be obtained by including such physical effects in the model. This brings us back to the discussion we had in Section 10.7 that we must know the system to get an accurate description of it. We must know what we are measuring and how to interpret the results. We can attempt to include various physical effects in the model; an improved agreement indicates a better

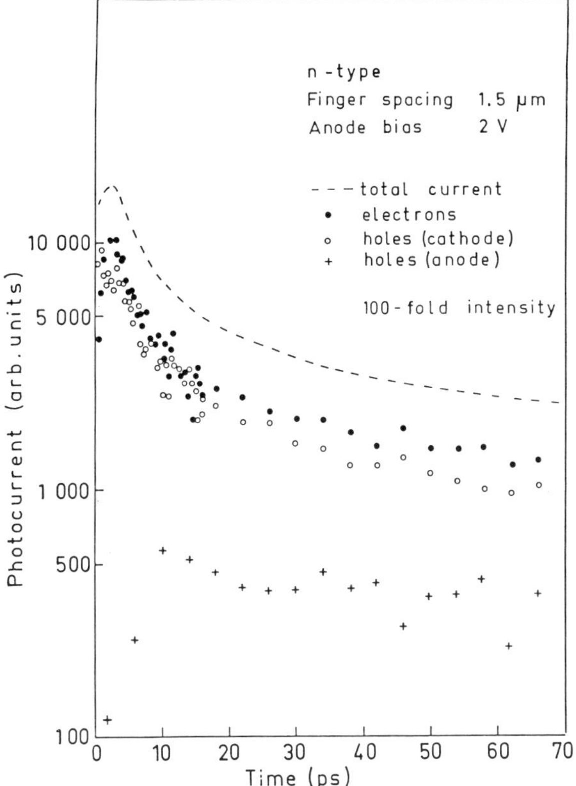

Figure 11.13 *The calculated electron, hole and total response photocurrent of the diode with 1.5 μm distance between the fingers. Anode bias: 2 V. Light intensity: 5.48 × 10^{11} W m^{-2}, 100 times that of Fig. 11.9*

description. This attempt to improve should stop when the quantitative agreement falls within the uncertainty of the experimental data.

11.4 EFFECTS OF α-RADIATION ON A TRANSISTOR

Electronic circuits applied in earth satellites and space travel are exposed to high energy cosmic radiation. The effect of such radiation is to create false pulses in the circuit which may corrupt the information it is processing. In computers such an event can have fatal results on a calculation. Although the effect of single cosmic radiation hits can be eliminated through the parity check, a combination of events may generate an unwanted error passing the check. The chance of such events increases as the individual components of the logic circuits get smaller. Cosmic radiation may even effect computers operating at the terrestrial surface. The circuit is most likely to recover if the cosmic radiation is not too intense, but the recovery is not instantaneous. Of course we endeavour to construct resilient circuits but to do so we must understand what happens when a cosmic particle strikes. The Monte Carlo particle model will prove useful in investigating such effects on individual components.

The effect of exposing transistors to α-radiation has been studied experimentally by Buot *et al.* (1987). It was found that a transistor might even burn out under unfavourable circumstances. Moglestue *et al.* (1991b) have modelled the effect of an α-particle hitting a GaAs field-effect transistor through the gate. The results of this work will be discussed here to illustrate yet another aspect of ambipolar transport.

The cross-section of the simulated transistor is similar to the one shown in Fig. 8.4, but with 1 μm gate length and 1 μm separation between the source and the gate, and the gate and the drain. The epilayer is 0.18 μm thick uniformly doped n-type at $1.5 \times 10^{23}\,\mathrm{m}^{-3}$. The source and drain have been grounded and the gate biased at -1.8 V, which is beyond the threshold bias (when the drain is biased). The corresponding gate potential is -2.6 V, assuming a Schottky contact potential of -0.8 V. Surface depletion due to surface charges has also been included in the model.

What exactly happens when an α-particle of megaelectron volts of energy hits the transistor perpendicular to the surface through the centre of the gate has never been properly modelled. We have assumed that the α-particle leaves a straight track of 0.1 μm diameter of hole–electron pairs of concentration $10^{24}\,\mathrm{m}^{-3}$ at time 0. The penetration depth depends on the energy of the α-particle; in the simulation the track stops at the bottom of the simulated area. Although the problem is a three-dimensional one, we have used a two-dimensional approach. The plasma column becomes a sheet; this simplification is not too serious as the two-dimensional approach yields the same general features as a three-dimensional model with less computer effort.

Figure 11.14 shows the evolution of the distribution of the electrons and the

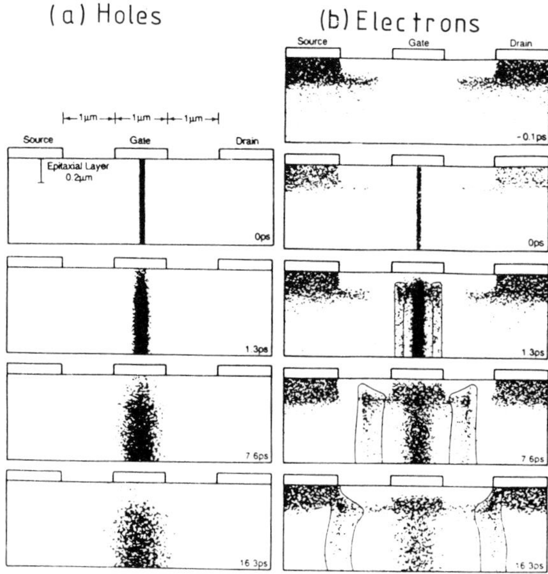

Figure 11.14 *The simulated evolution of (a) the hole and (b) the electron distribution before, during and after a 2 MeV α-particle penetrates through the centre of the gate of the field-effect transistor of Fig. 8.4 under normal incidence. The circle areas are dominated by L electrons; there are hardly any outside them*

holes. The left panels show the evolution of the distribution of the holes; the right panels that of the electrons. Inside the circled areas there is an overwhelming majority of L electrons; outside hardly any. Initially all the electrons reside in the Γ minimum of the conduction band. A shroud of heavy, or L, electrons soon develops by emitting phonons to the lattice; it moves near saturation velocity towards the source and the drain which are equivalent in this symmetric configuration of the transistor. The centre of the core, which consists almost entirely of light electrons, expands very slowly. Dynamic screening causes the electric field inside it to vanish; the plasma is held together by mutual electrostatic attraction between the two species of carriers. Thermal diffusion is responsible for its slow expansion.

Figure 11.14 also shows that the plasma extends right up to the gate; its repulsive effect on the electrons is compensated for by the attraction to the holes. The gate current, Fig. 11.15, shows an initial sharp peak consisting of both electrons and holes. This current then passes through a minimum in order to culminate at a secondary, broad peak which eventually decays. This secondary peak consists of holes only.

Also the source and the drain current, which are equal, start to flow immediately; they peak at about the same time as the gate current. The first part of these two currents is caused by the hot electron front pushing the indigenous electrons in the epilayer in front of them by Coulombic repulsion. The peak is reached when the hot electron shroud reaches the contacts after about 30 ps. (Figure 11.15.) The final decay then sets in; towards the end of the life of the response the source and the drain currents amount to half the gate current. The transistor has recovered about 60 ps after the event.

Figure 11.16 shows that the heat development rate is largest in the epilayer and reaches its maximum near the source and the drain contacts just before the hot electrons constituting the shroud are absorbed. The heating rate does not reflect

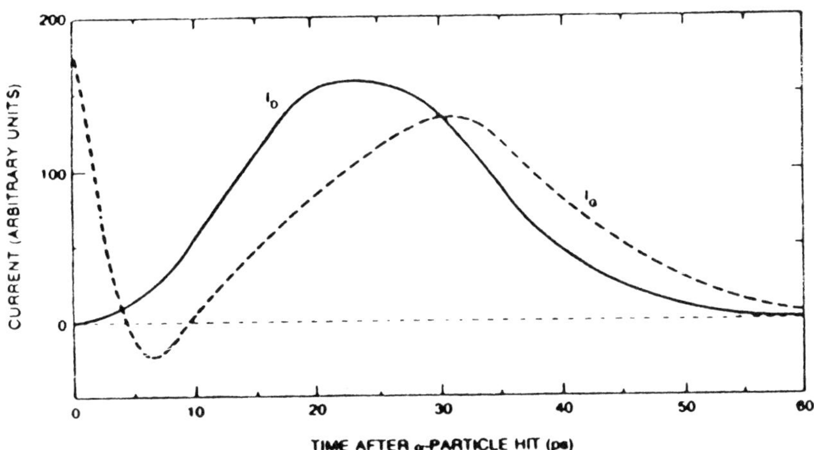

Figure 11.15 *Calculated source, drain and gate current against time after the passage of the α-particle through the centre of the gate. The source and the drain are equal by geometrical symmetry and the biasing of the transistor. Both the source and the drain are connected to earth; the applied gate bias is −1.8 V which corresponds to a gate potential of −2.6 V (Schottky contact potential: −0.8 V)*

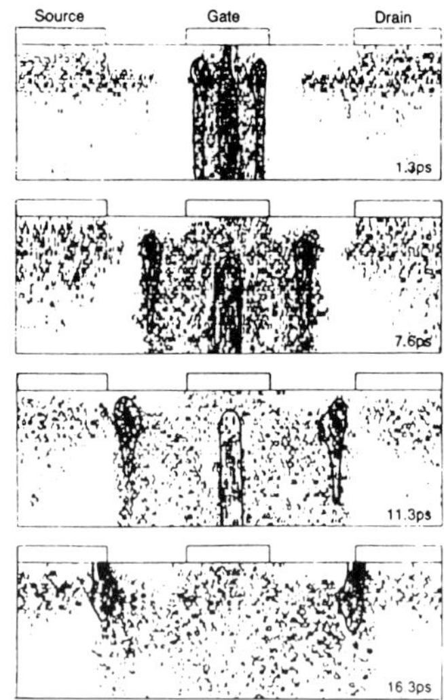

Figure 11.16 *Heating rate at different times after the passage of the α-particle. The rate outside the fully drawn enclosure is 1.3 W m^{-1}, those inside it above. The black solid areas correspond to a rate between 2.6 and 6.5 W m^{-1}*

the actual temperature which depends on the efficiency with which the heat can be removed from the transistor. The temperature rise from a single event is only a few degrees. Heat diffusion is a slow process: if the α-particle radiation is sufficiently intense, even if the transistor recovers electrically from each event, the heat can accumulate so that the material of the transistor begins to melt. Buot *et al.* (1985, 1987) have indeed observed burn-out in the same place where the Monte Carlo particle model predicts the maximum heat generation rate.

11.5 THE BIPOLAR TRANSISTOR

Figure 11.17 shows a cross-section through a bipolar transistor. The emitter and the collector regions of the gallium arsenide are doped n-type; the base is p-type, as indicated in the figure. The base has been separated from the emitter by a thin layer of intrinsic $In_{0.5}Ga_{0.5}As$. All contacts are Ohmic. The current flowing from the emitter to the collector is very sensitive to the base bias. The transistor acts like a current amplifier.

Modelling of bipolar transistors involves following both holes and electrons. Park *et al.* (1984) simulated a bipolar transistor using a one-particle model. Some authors, e.g. Katoh and Kurata (1989), Katoh *et al.* (1989) and Hu *et al.* (1989), used a combination of a Monte Carlo particle model for the majority carriers and a drift diffusion approach for the minority carriers, the holes. Rockett (1988) used a fixed-field model and Pelouard *et al.* (1988) developed an analytical model based on Monte Carlo results. In this section we shall present an example to demonstrate how both types of carriers can be modelled by means of the Monte Carlo approach. All contacts are allowed to absorb both types of carriers: the emitter and the collector can only inject electrons; the base only holes.

Figure 11.18 shows the distribution of the electrons and the holes in the tran-

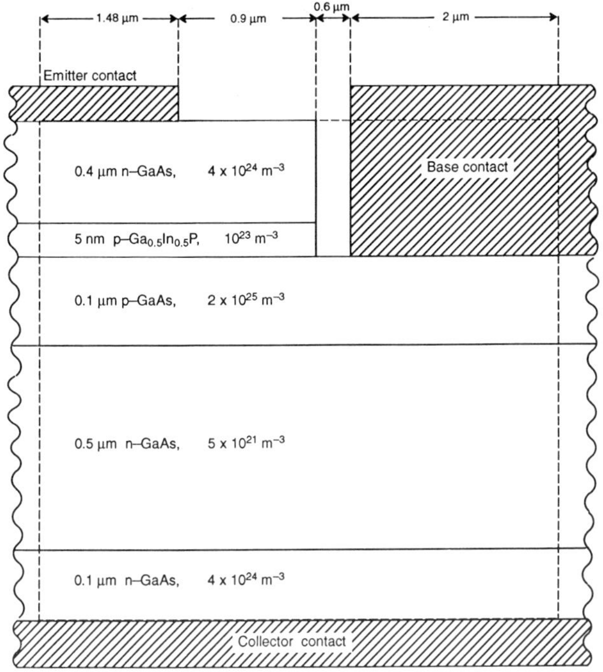

Figure 11.17 *Cross-section through a bipolar transistor*

sistor during steady state. Figure 11.19 demonstrates a typical calculated base current characteristic showing the expected rise with base voltage.

In the simulations behind Figs. 11.18–20 the superholes and superelectrons represent the same number of real particles, when the doping density of the p- and n-doped material differs by orders of magnitude; the two species of superparticles can also be made to represent different numbers of real particles.

Unfortunately the bipolar transistor can be awkward to simulate because of the tendency of an electron–hole plasma to form in large parts of the transistor, as shown in Fig. 11.18. As we have seen in Section 11.2 such a plasma formed in the p–n junction built up to a certain density before further particles could pass through without being absorbed. This will also happen in bipolar transistors; inside the plasma the external field will be screened off, and thus the field distribution throughout the transistor is likely to differ from the one expected from drift-diffusion or hydrodynamic calculations where such plasma formation is neglected. We have also discussed the effect of such dynamic screening in connection with photodetectors in Section 11.3.

Due to the formation of the electron–hole plasma it takes more computer time before steady state has been reached compared to a simulation with only one type of carrier. A possible way to accelerate the computer run is to *revalue* the superparticles, i.e. to increase the charge of each of them. This contradiction of our advice of Section 8.5 has in some cases led to a saving of computer time, but the transient arising from such an approach is unphysical and can therefore not be used as the basis for any analysis. Only the eventual steady state is correct. This approach does not always work, as we can see from Fig. 11.20: holes tend to prefer

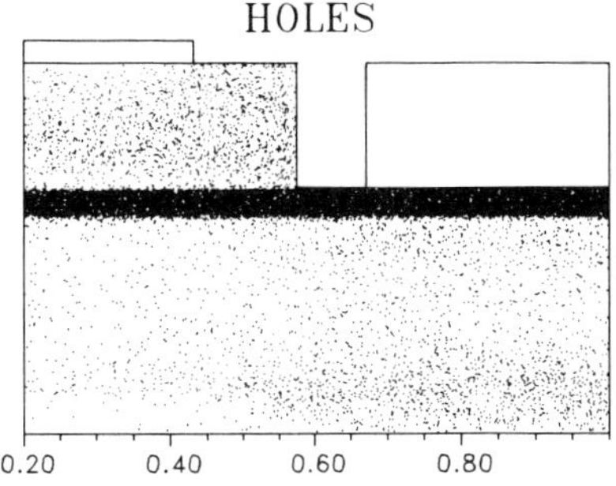

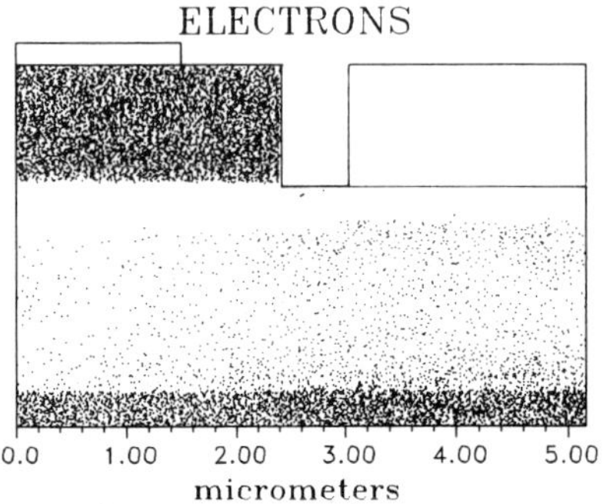

Figure 11.18 *Steady state distribution of holes (upper panel) and electrons (lower panel) throughout the bipolar transistor of Fig. 11.17. Base potential: 3 V, emitter and collector both at zero potential*

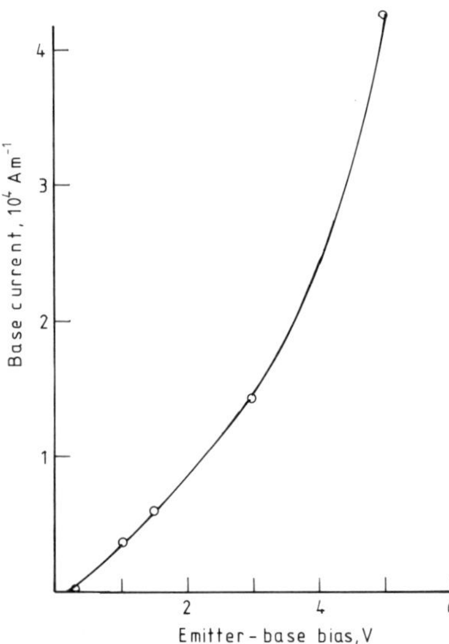

Figure 11.19 *Simulated base current against the base potential for the transistor shown in Fig. 11.17*

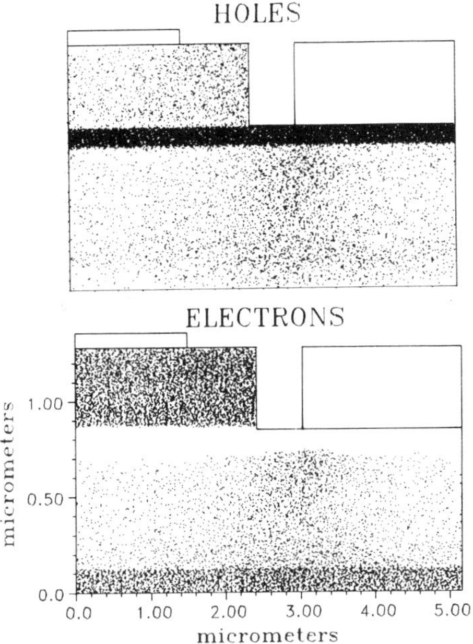

Figure 11.20 *Transient distribution of holes (upper panel) and electrons (lower panel) after a sudden increase in the base potential from 3 to 5 V. Emitter and collector both at zero potential. Plasma formation nucleated near the corner of the base contact which slowly extends throughout the base region*

to enter the base electrode through the corner nearest the edge of the recess it rests in, generating a plasma cloud which extends slowly; this revaluation scheme therefore cannot accelerate the simulation.

11.6 SPONTANEOUS RECOMBINATION OF ELECTRON–HOLE PAIRS

Recombination of electron–hole pairs is usually a slow process; it can either be direct, emitting a photon, or take place via traps or crystal faults. In the latter case either the electron or the hole must first be bound to an ion and then the recombination can take place successively by emitting phonons or photons. The physics of trap-assisted recombination is rather complex, but follows the rules outlined in Section 4.8. This kind of recombination has usually a very low rate so that it does not play an important role and can therefore be omitted from the model. However, recently it has been possible to grow semiconductor layers by means of molecular beam epitaxy at around 200°C, a lower temperature than that used previously. There are indications (Kaminska *et al.*, 1989) that the material now contains so many crystalline defects that trap-assisted recombination will be considerable.

In principle, electron–hole pair recombination can be considered as scattering from the point of view of selection among the other things that could happen to the particle and for the estimation of the time of free flight. In the case of recombination it is necessary to find the particle with which it combines which could mean ploughing through many particles to find the correct one. Fortunately there is an easy and fast algorithm for this, the so-called *P^3M algorithm* (Hockney and Eastwood, 1988): each particle has an address or identification in the computer memory. Let the device be divided into a grid which many not be the same as the one used to solve Poisson's equation. The cell size in this grid could be related to the local screening distance. Each such cell contains a number of particles. At the start of each field-adjusting time step a list of all the particles of the opposite charge within each cell is compiled afresh. This is faster than continuously updating it by keeping account of the particles leaving and entering the cell. One list for each cell is sufficient.

Direct radiative recombination has been discussed by Agrawal and Dutta (1986). The exposition given here follows their book in outline, but the expressions for the recombination rates derived here are better suited for particle modelling. The rate of spontaneous emission or absorption of photons will be calculated from time-dependent perturbation theory, in principle in the same way as we obtained the scattering rates in Chapter 4. An electron initially in state $|e\rangle$ in the conduction band is transferred to the state $|h\rangle$ in the valence band by emitting a photon. The transfer can only take place if the state $|h\rangle$ is free. From the golden rule, Equation (4.17), assuming the photon emission to be instantaneous, the emission rate is

$$\lambda_{sp} = \frac{2\pi}{\hbar} |H_{ph}|^2 D_S^a(E_h)\delta(E_G + E_e + E_h - \hbar\omega) \tag{11.8}$$

where H_{ph} represents the matrix element

$$\mathcal{H}_{ph} = -\frac{e}{m_0}\mathbf{A}\bullet\mathbf{p} \tag{11.9}$$

with m_0 representing the rest mass of the free electron, $D_S^a(E_h)$ the density of states per unit interval of energy in the valence band, E_h and E_e the electron and hole energies respectively which are reckoned from the conduction and valence band edges as defined in Section 3.6, E_G the band gap energy, and e the elementary electronic charge. Furthermore, $\mathbf{p}$ denotes the momentum of the photon and $\mathbf{A}$ its vector potential:

$$\mathbf{A} = -\hat{\mathbf{e}}\left(\frac{2\hbar}{\varepsilon_0 n_r^2 \omega V_{cr}}\right)^{1/2} \sin(\omega t - \mathbf{k}_{ph}\bullet\mathbf{r}), \tag{11.10}$$

where n_r represents the refractive index of the semiconductor material, ε_0 the permittivity of vacuum, V_{cr} its volume, $\hat{\mathbf{e}}$ a unit vector along $\mathbf{A}$, $\mathbf{k}_{ph}$ the wave vector of the photon, related to its momentum through the relation $\mathbf{p} = \hbar\mathbf{k}_{ph}$, $\mathbf{r}$ its position and ω its angular frequency.

For spontaneous emission

$$D_S^a(\mathbf{E}_h) = 2\frac{V_{cr}}{8\pi^3}\int_0^{2\pi} d\phi \int_0^{\pi} \sin\theta \, d\theta \int_0 k^2 dk/dE \tag{11.11}$$

where the factor 2 arises from the two possible spin states of the hole. With the parabolic approximation to the structure of the valence band, Equation (3.69), the density of states per unit interval of energy becomes

$$D_S^a(\mathbf{E}_h) = \frac{\sqrt{2}\, V_{cr} \mathbf{m}_h^{*3/2} \sqrt{E_h}}{\pi^2 \hbar^3}. \tag{11.12}$$

A generalisation to the more accurate band structure, Equation (3.68), or to the general one is in principle straightforward. Calculating $|\mathcal{H}_{ph}|^2$ we substitute Equation (11.10) for **A** in Equation (11.9):

$$|\mathcal{H}_{ph}|^2 = \frac{e^2 \hbar}{2 m_0^2 \varepsilon_0 n_r^2 \omega V_{cr}} \, |\langle e | \hat{\mathbf{e}} \bullet \mathbf{p} | h \rangle| \tag{11.13}$$

where the factor 2 stems from averaging the sine over time.

The wave function $|e\rangle$ is of the form $u_{\mathbf{k}e}(\mathbf{r}) \exp(i\mathbf{k}_{ph} \bullet \mathbf{r})$ where $\mathbf{k}_e$ and $\mathbf{r}$ refer to the electron and $u_{\mathbf{k}e}(\mathbf{r})$ represents the Bloch function reflecting the crystalline structure of the semiconductor, and $|h\rangle$ resembles one of the orbitals of Table 3.2. Using Kane's model (1957) for the valence and conduction bands

$$|\langle e | \hat{\mathbf{e}} \bullet \mathbf{p} | h \rangle|^2 = \frac{4\pi^3 m_0^2 E_G (E_G + E_{so})}{3 m_e^* (E_G + 2E_{so}/3)} \, \delta(\mathbf{k}_e + \mathbf{k}_h) \tag{11.14}$$

where m_e^* represents the effective mass of the electron in the conduction band, E_G the band gap energy, E_{so} the split-off band energy, $\mathbf{k}_e$ and $\mathbf{k}_h$ the wave vector of the electron and the hole, respectively, and the δ-function ensures that the electron and the hole have the same wave vector on recombination. It expresses the *k-selection rule*. The momentum of the photon can be neglected in comparison with that of the carriers.

Using Equations (11.14), (11.13) and (11.12) in Equation (11.8) yields the following expression for the *spontaneous emission rate* of photons:

$$\lambda_{sp} = \frac{\sqrt{2}\, e^2 m_h^{*3/2} \sqrt{E_h} E_G (E_G + E_{so})}{24 \pi^4 \hbar^3 \varepsilon_0 n_r^2 \omega m_e^* (E_G + 2E_{so}/3)} \, \delta(\mathbf{k}_e + \mathbf{k}_h). \tag{11.15}$$

The energy of the emitted photon, E_{ph}, is given by

$$E_{ph} = \hbar\omega = E_G + E_h(\mathbf{k}_h) + E_e(\mathbf{k}_e). \tag{11.16}$$

Furthermore, the k-selection rule requires that the wave vector of the electron and the hole are equal in magnitude but oriented in opposite directions:

$$\mathbf{k}_e = -\,\mathbf{k}_h. \tag{11.17}$$

As already stated, the simulator treats the spontaneous emission of photons formally like a scattering mechanism. Only the electrons in the conduction band can cause photon generation; the holes cannot because they represent merely electronic vacancies in the valence band. On photon emission the electron fills a vacant position by ridding itself of its excess energy in the form of a photon, i.e. a photon is generated while the hole–electron pair is being annihilated. This can only take place if the hole already exists, the probability of which is

$$f_h(E_h) = \{1 + \exp[(E_h - E_{Fh})/(k_B T_e)]\}^{-1} \qquad (11.18)$$

where E_{Fh} represents the quasi Fermi energy of the hole population, T_e its temperature and k_B Boltzmann's constant. For simplicity we have assumed that the holes are Fermi–Dirac distributed and that the disequilibrium function is unity.

Having selected spontaneous photon emission in our simulation, another flat random number representing $f_h(E_h)$ has to be generated to decide whether the electron–hole recombination really should take place.

If the value of this number is less than that of $f_h(E_h)$ we have to find a hole of wave vector given by Equation (11.17) which is within approximately one screening length of the electron. In the actual simulation we do not eliminate the pair immediately. A more computer efficient method is as follows:

> Simulate the transport histories of all the electrons in turn for the given field-adjusting time step. When recombination has been selected for a particle, treat it initially as self-scattering but note the position and the wave vector of the electron and follow it further until the time step is up. When all the electrons have been followed, turn to the holes. Check the list of the selected electrons for each hole to see whether there is one in the vicinity of the matching wave vector. When a matching hole has been found, both it and the relevant electron are eliminated from the simulation. In the contrary case, the particles have to continue their transport histories, which we have already provided for. We could also make use of the P^3M algorithm to find the hole – which method is the fastest depends on the length of the list of recombination candidates.

Adhering strictly to the k-selection rule will result in never finding a hole to recombine with. However, since this does not happen in reality so the rule has to be relaxed. From Heisenberg's uncertainty principle, Inequality (10.24), the uncertainty of the electron's energy, ΔE_e, is

$$\Delta E_e > \hbar\lambda_{sp}, \qquad (11.19)$$

ΔE_e corresponds to a wave vector

$$\Delta k = \sqrt{2m_e^*\Delta E_e}/\hbar \qquad (11.20)$$

neglecting non-parabolicity. The k-selection rule is therefore relaxed to finding a hole with a wave vector such that

$$|\mathbf{k}_e + \mathbf{k}_h| < \Delta k. \qquad (11.21)$$

11.7 STIMULATED PHOTON EMISSION, LASERS

It is also possible to simulate photon emission by means of another photon, i.e. stimulated photon emission. Stimulated emission is made use of in lasers. In this process a photon participates in creating another by annihilating a hole-electron pair. The relaxed k-selection rule has to be obeyed in this simulation, too.

The interaction between the photon and the electron-hole pair is given by Equation (11.9) with the magnetic vector potential from one photon given by Equation (11.10). The energy of the photon is

$$E_{ph} = \hbar\omega = \hbar k_{ph} c/n_r \tag{11.22}$$

where ω represents its angular frequency, k_{ph} the magnitude of its wave vector, n_r the refractive index of the semiconductor medium and c the velocity of light in vacuum (the velocity inside the semiconductor is c/n_r).

The density of states per unit interval of photon energy is obtained from Equation (11.11) without the factor 2 and with E_{ph} given by Equation (11.22):

$$D_S^a(\mathbf{E}_{ph}) = \frac{V_{cr}}{8\pi^3} 4\pi \frac{k_{ph}^2 n_r}{\hbar c}. \tag{11.23}$$

The stimulated emission rate is calculated from Equations (11.8) making use of Equations (11.13), (11.14) and (11.23):

$$\lambda_{st} = \frac{e^2 n_r \omega E_G (E_G + E_{so})}{96\pi^4 \varepsilon_0 m_e^* c^3 \hbar (E_G + 2E_{so}/3)} \tag{11.24}$$

where the symbols have the same significance as above.

The scattering rate from one single photon is not very large, but it is proportional to the intensity of light present. Of course the electron state in the conduction band must be occupied and the electron state in the valence band must be free. The respective probabilities, $f(E_e)$ and $f_h(E_h)$, are given by Equations (3.82) and (3.87). The rate for stimulated photon emission is therefore

$$\lambda_{st}^E = \lambda_{st} f(E_e) f_h(E_h). \tag{11.25}$$

The opposite process, however, can also take place; this will also be true for stimulated emission - the new photon is being reabsorbed elsewhere. The opposite process results in creating a hole and an electron which can only take place if the states the pair is entering into are free - the probability of this is $1 - f(E_e)$ for the electron and $1 - f_h(E_h)$ for the hole. The corresponding electron-hole generation rate is

$$\lambda_{st}^A = \lambda_{st}[1 - f(E_e)]\,[1 - f_h(E_h)] \tag{11.26}$$

We get a net gain of photons when $\lambda_{st}^E > \lambda_{st}^A$ or

$$\lambda_{net} = \lambda_{st}^{E} - \lambda_{st}^{A} = \lambda_{st}[f(E_e) - f_h(E_h) - 1] > 0. \tag{11.27}$$

The factor within the square bracket represents the *optical gain*; the laser emits light when it is positive. It is up to the designer to construct the laser such that the losses or absorption of generated photons is kept as low as possible. The photons should stay inside the cavity which is placed between mirrors in all directions except in the one from which the light should emerge. The mirrors consist of surfaces or interfaces to a medium with a refractive index smaller than that of the semiconductor core. The light is confined by total reflection from the mirrors. It can only escape through the mirror if the photons travel towards the mirror under an incident angle within a cone of aperture

$$\theta = \arcsin\ (n_r'/n_r) \tag{11.28}$$

where n_r and n_r' represent the refractive index of the medium of the cavity and outside respectively. For gallium arsenide the light can only escape if the angle of incidence is less than about 15° from the normal when $n_r' = 1$.

Our exposition suggests that we can simulate photons as a separate species of particles. As such they should rather be described by the magnetic vector potential, Equation (11.10), to find the strength of the field because we are interested in the interaction between them and the electrons. As we work with superparticles we have to let the superelectron and the superhole represent the same number of real particles. We also have to introduce the *superphoton* which represents the same number of real photons as the superparticles because one superelectron–hole pair generates many real photons. The stimulated emission rate is proportional to the number of photons present. We do not need to keep track of the individual (super)photons in addition to the electrons and the holes as long as we know, or have a realistic estimate, of their number and distribution. However, both the absorption and emission of photons should be treated separately, just as we treat phonon absorption and emission as separate events.

12

Noise

12.1 INTRODUCTION

The Monte Carlo particle model is the only known model that describes physical noise from first principles. Physical noise originates from the fluctuations in the distribution of the carriers and the particle velocities and hence from the induced changes of the local electric field. The particle nature of the current produces shot noise; the fluctuations in the electric field give rise to displacement currents at the terminals where the noise is observed. As the model simulates real transport histories of individual particles, the fluctuations that come out of it reflect true noise. Figure 12.3 shows, as an example, the accumulative particle count at the end of a simulated uniformly doped gallium arsenide resistor bar responding to a sudden small change in the applied voltage over it. The fluctuations from the straight line during steady state, the slope of which represents the particle current, are mainly due to shot noise and plasma oscillations.

However, the size and the speed of the computer allow simulation of only N_s out of N particles; we shall see in the penultimate section of this chapter that the calculated noise therefore gets exaggerated by a factor $\sqrt{(N/N_s)}$. In addition to this there will be another small correction due to extraneous effects arising from the finite frequency of solving the field equation and the finite size mesh over which it has been solved. This model noise stems from the fact that the electric field is constant and uniform within each mesh cell during the entire field-adjusting time-step.

A weakness of the Monte Carlo particle model is that the exaggerated noise may overwhelm the signal. The remedy to this is to increase the number of particles N_s or, which amounts to give the same effect, compute the average histories of several runs of each particle. Even with $N_s = N$ the noise may drown the signal. In this case the noise is real; the corresponding real device would also exhibit more noise than signal. The temptation to choose $N_s > N$ yields an unphysical result and should therefore be resisted.

Before embarking on the rather lengthy study of noise we shall discuss the intrinsic minimum noise figure in the next section. We shall derive an expression for this figure which differs from the one more familiar to the microwave engineer because ours is only concerned with the physical processes taking place inside the semiconductor and neglecting anything happening in the metallic contacts. We shall thus derive an expression for the truly intrinsic noise properties of the device. Our theory will be verified by computer experiments.

Turbulent flow, which is chaotic, also gives rise to fluctuations in the terminal currents of the device. These manifest themselves as noise in measurements although they are not genuine noise in the sense of the theory presented in Section 12.4. We shall come back to this in the last section of this chapter.

12.2 THE MINIMUM INTRINSIC NOISE FIGURE

The minimum noise figure is an important figure of merit for transistors. One aim of designing transistors is to reduce the noise figure as much as possible. The noise is generated by the transport of the carriers, both in the transistor itself and in the metal contacts. The latter is often an artefact of the fabrication process and may vary between otherwise nominally identical devices. The former gives rise to the intrinsic noise, which can be studied by the particle model: this will be the focus of our attention here. Previous authors such as Van der Ziel and Chenette (1978), Tajima and Shibata (1978), Pucel *et al.* (1975) and Rigand *et al.* (1973) modelled noise by means of the drift-diffusion approach assuming the existence of internal noise generators. Bächtold and Strutt (1968) calculated *the minimum noise figure* to be

$$F_{\min} = 1 + \frac{2\pi\langle I_{nc}^2\rangle \operatorname{Re}(y_{11})}{k_B T_{LT} \Delta\omega |y_{21}|^2} \tag{12.1}$$

where y_{11} and y_{21} represent the input admittance and the transconductance respectively, $\langle I_{nc}^2\rangle$ a fluctuation current generated from an assumed noise source, k_B Boltzmann's constant, T_{Lt} the lattice temperature and $\Delta\omega$ an angular frequency interval. The *Fukui* (1979) *formula* is a popular alternative expression for the minimum noise figure:

$$F_{\min} = 1 + k'\omega[g_{m0}(R_g + R_s)]^{\frac{1}{2}}. \tag{12.2}$$

Here k' represents a numerical factor, R_g and R_s the gate and source resistance respectively and g_{m0} the steady state transconductance. The gate resistance, R_g, and that of the source, R_s, include the resistance due to the metal gate and source contacts respectively and depend on their shape. The factor k' is a fitting parameter which can only be determined by measurement, and differs for different transistors. It does not shed any light on the physics of noise - Equation (12.2) is therefore useless for our present analysis.

The definition behind both Equations (12.1) and (12.2) is the usual one, namely the ratio of signal to noise power at the gate, $W_g/\Delta W_g$, divided by the corresponding ratio at the drain, $W_d/\Delta W_d$:

$$F_{\min} = \frac{W_g/\Delta W_g}{W_d/\Delta W_d}. \tag{12.3}$$

When ΔW_g and ΔW_d refer to noise power at the gate and the drain respectively, generated by processes internal to the device only we get *the intrinsic minimum noise figure*. The intrinsic minimum noise figure was first discussed by Moglestue (1985). Because we do not make use of external noise sources we have to derive an expression for $F_{\min}$ which is useful for Monte Carlo modelling.

The signals are sinusoidal in the Fourier domain. The input and output power are respectively

$$W_g = \tfrac{1}{4} V_g^2/Z_g \tag{12.4a}$$

and

$$W_d = \tfrac{1}{4} Z_d I_d^2 \tag{12.4b}$$

where Z_g and Z_d denote the impedance of the gate and drain part of the transistor respectively, V_g the gate bias and I_d the drain current. The output part of the transistor may be considered as a current generator due to internal gain

$$I_d = y_{21} V_g, \tag{12.5}$$

with y_{21} representing the transconductance so that

$$W_d = \tfrac{1}{4} |y_{21}|^2 V_g^2 Z_d. \tag{12.6}$$

The source bias is free of noise because we do not consider noise generated in the supply but there are fluctuations in the gate potential, ΔV_g, due to the motion of the electrons in the epilayer. The fluctuations in the input power are therefore

$$\Delta W_g = \tfrac{1}{4} \Delta V_g^2 / Z_g. \tag{12.7}$$

The output noise power consists of the amplified input noise power $(W_d/W_g)\Delta W_g$ and the internally generated noise power

$$\Delta W' = \tfrac{1}{4} \langle \Delta I_d^2 \rangle Z_d, \tag{12.8}$$

where $\langle \Delta I_d^2 \rangle$ represents the deviation from the mean in the drain current. The output power (at the drain), ΔW_d, is the sum of $\Delta W_d'$ and $\Delta W'$:

$$\Delta W_d = \Delta W_g W_d / W_g + \tfrac{1}{4} \langle \Delta I_d^2 \rangle Z_d. \tag{12.9}$$

Combining Equations (12.3) and (12.9) yields

$$F_{\min} = 1 + \frac{\langle \Delta I_d^2 \rangle}{|y_{21}|^2 \Delta V_g^2}. \tag{12.10}$$

This can be reformatted introducing the input admittance $y_{11} = \partial I_g / \partial V_g \simeq \Delta I_g / \Delta V_g$, where ΔI_g represents the average deviation in the gate current which, in most cases, is a pure displacement one:

$$F_{\min} = 1 + \left| \frac{y_{11}}{y_{21}} \right| \frac{\langle \Delta I_d^2 \rangle}{\langle \Delta I_g^2 \rangle} \tag{12.11}$$

All entities entering this equation can be obtained from the Monte Carlo simulator; the two y parameters as explained in Section 9.3 and the fluctuation currents $\langle \Delta I_d^2 \rangle$ and $\langle \Delta I_g^2 \rangle$ are extracted for each field-adjusting time-step.

The difference between Equations (12.11) and (12.1) is fundamental as Equation (12.1) includes external noise sources, while the former only reflects noise

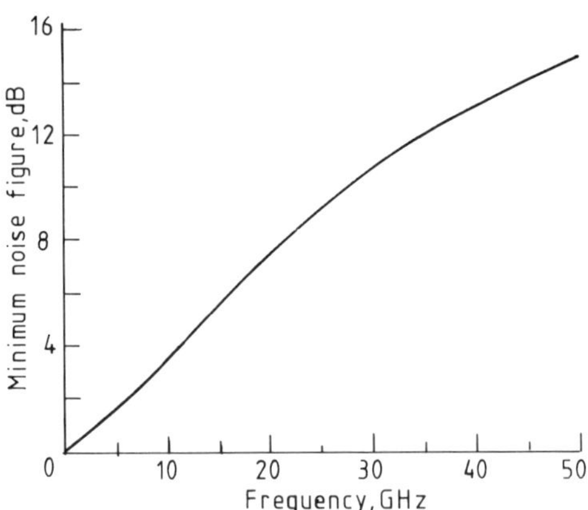

Figure 12.1 *Simulated minimum noise figure against frequency for the transistor shown in Fig. 8.4*

generated by the carriers inside the transistor they move through. A direct comparison cannot therefore be made. To enable a direct comparison the transistor should be described by means of an equivalent circuit with internal noise sources (intrinsic noise generation), sources in the terminal interfaces (requires a good contact theory) and with external sources (due to measuring equipment and induction in the external circuit). A description of the contact noise sources is no better than a guess since a satisfactory theory for contacts has not yet emerged. Some of the difficulties with such a theory were discussed in Section 8.3.2. It is, therefore, at the present stage difficult to give any estimate for the relative contributions to the noise from the contacts. When a microscopic, realistic theory of contacts is available these contributions can be included in the Monte Carlo particle model. The intrinsic noise figure represents the limit of what can be achieved in the minimum noise figure.

Due to the finite capability of the computer, the transport histories of only N_s of the N particles can be simulated; this means both $\langle \Delta I_g^2 \rangle$ and $\langle \Delta I_d^2 \rangle$ are exaggerated by the same factor. This cancels in Equation (12.11) so the model yields the true intrinsic noise figure.

The fluctuation currents can be estimated directly from the simulation. It includes the fluctuations of the particle flow and of the displacement current which is not zero even during steady state although the average displacement current vanishes. Figure 12.1 shows the minimum noise figure calculated for the example transistor of Chapter 8. Our calculated intrinsic minimum noise figure shows typical behaviour: it increases with the frequency of the signal and eventually it will exceed the gain, defining an upper frequency for which the transistor can be used as an amplifier which is generally lower than the cut-off frequency at which the gain equals one (or zero decibels).

12.3 THEORY OF NOISE

In this section a theory to describe the noise spectrum of a uniform current will be presented. The theory is based on the correlation in the current that exists after a time T from an arbitrarily chosen time zero. The deviation in the current from the mean is calculated by averaging over all possible carrier distributions and transport histories weighed against the probability of realising each of them. The details of this theory have never been published previously, only its results (Moglestue, 1983c).

To break up the exposition of the material, this section will be divided into subsections as follows:

1) Calculations of the correlation for all possible transport histories
2) Calculations of all possible fluctuations in the electric field
3) Estimation of the noise spectrum and the exaggeration factor.

To avoid making matters more complicated than necessary to understand the theory, we shall restrict ourselves to calculating the noise of a current flowing through a rod of uniform doping and cross-section, thus through an ordinary resistor. This example should be sufficient to demonstrate the salient points of our approach. Our method can be applied to more complicated systems by rather straightforward adaptations.

12.3.1 The correlation current

We consider the current through a rod of semiconducting material of rectangular cross-section $N_x h_x N_z h_z$ where h_x and h_z represent the size of the mesh cell in the x and z directions and N_x and N_z the number of mesh cells in these two directions. A Cartesian coordinate system will be defined with its y axis in the direction of the current flow (see inset Fig. 12.3) and x and z axes perpendicular to it. For simplicity, we shall assume the material to extend to infinity in the z direction so that we can use a two-dimensional model in geometrical space. The suppression of one geometrical dimension makes no essential changes to our noise theory; we gain by making the mathematics more transparent. An extension to three dimensions is straightforward. Now our rod rather becomes a slab, $N_z = 1$.

The current flowing through the slab is

$$I_y = (n_e e h \langle k_y \rangle / m^* + \varepsilon \varepsilon_0 \partial \langle F_y \rangle / \partial t) A. \tag{12.12}$$

Here A represents the cross-sectional area perpendicular to the direction of the mean current, e the charge of the carrier, *not* the supercharge - it may be that of an electron or a hole. To simplify matters, only one type of carrier will be considered, and we assume all of them reside in the same minimum of the valence band or in only one conduction band minimum. The carriers have an effective mass m^*, which is independent of the particle's energy or direction of motion, k_y represents the wave vector in the y direction and F_y the electric field along the main flow of the current. ε_0 denotes the permittivity of vacuum and ε the dielectric constant of the material. The average, $\langle\ \rangle$, is taken over all possible distributions of carriers and wave vectors throughout the entire rod, weighed against the respective probabilities of particular values being realised, in other words, over all possible transport histories appropriately weighted.

The *noise spectrum* is calculated from the *Wiener–Khintchine theorem*:

$$C(\omega) = \int_{-\infty}^{\infty} \langle \Delta I_y(0) \Delta I_y(T) e^{-i\omega T} dT = \int_{-\infty}^{\infty} C(T) e^{-i\omega T} dT \tag{12.13}$$

where ΔI_y denotes the deviation from the mean current. The usual approach is to consider the *correlation* function

$$C(T) = \langle \Delta I_y(0) \Delta I_y(T) \rangle = \delta(T) \tag{12.14}$$

as a δ-function, justified by the assumption that T is much larger than the momentum or energy relaxation time. This means that we only work in very long time scales, when all information about the state in the beginning has been lost. We shall abandon this approach because, as we have seen in Chapter 4, often some amount of information on the state just prior to scattering is being preserved during scattering. Important high frequency information retained for short T is being lost through such a δ-function approach. Instead we have to look at all possible transport histories of the particles. We wish to gain information on the noise caused by scattering and the distribution of velocities or momenta of the particles. The terahertz part of the noise spectrum reveals information of the scattering.

The correlation in the noise current is independent of the choice of the initial time zero during stationary conditions assumed here. Inclusion of transient states complicate matters but not essentially; we shall therefore consider transients as being beyond the scope of this chapter.

During steady state $C(T) = C(-T)$. From Equation (12.14)

$$\begin{aligned} C(T) &= \langle \Delta I_y(0) \Delta I_y(T) \rangle = \langle I_y(0) I_y(T) \rangle - \langle I_y(0) \rangle \langle I_y(T) \rangle \\ &= (An_e e\hbar/m^*)^2 \langle \Delta k_y(0) \Delta k_y(T) \rangle + (\varepsilon\varepsilon_0 A)^2 \langle \partial F_y(0)/\partial t \, \partial F_y(T)/\partial t \rangle \end{aligned} \tag{12.15}$$

where k_y is the y component of the wave vector $\mathbf{k}$. Below we shall study the correlation $\langle \Delta\mathbf{k}(0) \bullet \Delta\mathbf{k}(T) \rangle$.

To calculate it we follow one particle in detail for a time interval T. At a chosen universal time 0 the particle was last scattered time t' ago, and its wave vector is now, Equation (5.42), without the magnetic field:

$$\mathbf{k}(T) = \mathbf{k}_0 + (e/\hbar) \int_{-t'}^{0} \mathbf{F}(T) dt \tag{12.16}$$

where $\mathbf{F}$ represents the electric field, $\mathbf{k}_0$ the wave vector of the particle just after the last scattering event, e the charge of it which is the elementary charge, *not* the charge of the superparticle, and t a dummy time variable. A time T later we find that the particle has been scattered n times, namely at the instances t_1, t_2, . . ., t_n where

$$0 < t_1 < t_2 < \ldots < t_n, \tag{12.17}$$

for $n = 0, 1, 2, \ldots, n$. We shall refer to a particle that has been scattered n times as an *n-particle*.

In particular *zero-particles* have not suffered scattering at all. At time T their wave vector is

$$\mathbf{k}(T) = \mathbf{k}_0 + (e/\hbar) \int_{-t'}^{T} \mathbf{F}(t)dt, \tag{12.18}$$

the deviation in the wave vector from the average is

$$\begin{aligned}\langle \Delta\mathbf{k}(0)\Delta\mathbf{k}(T)\rangle_0 &= \langle \mathbf{k}(0)\bullet\mathbf{k}(T)\rangle_0 - \langle \mathbf{k}(0)\rangle_0\bullet\langle \mathbf{k}(T)\rangle_0 \\ &= \langle \mathbf{k}_0^2\rangle_0 - \langle \mathbf{k}_0\rangle_0^2 - (e/\hbar)^2\left\{\left\langle \int_{-t'}^{0} \mathbf{F}(t)dt\bullet \int_{-t'}^{T} \mathbf{F}(t)dt\right\rangle_0 \right. \\ &\left. - \left\langle \int_{-t}^{0} \mathbf{F}(t)dt\right\rangle_0\bullet\left\langle \int_{-t'}^{T} \mathbf{F}(t)dt\right\rangle_0\right\}\end{aligned} \tag{12.19}$$

where $\langle\ \rangle_0$ denotes the average over all possible values of t', $\mathbf{k}_0$ and $\mathbf{F}$, for zero-particles weighted against the probability of realisation. All information about the state of the zero-particles at time $-t'$ remains intact at time T.

The wave vector of an n-particle at the same time is

$$\mathbf{k}(T) = \mathbf{k}_n' + (e/\hbar) \int_{t_n}^{T} \mathbf{F}(t)dt. \tag{12.20}$$

Much information about their original state has been lost during scattering. What remains is carried by $\mathbf{k}_n'$, the wave vector just after the last scattering event. The value of $\mathbf{k}_n'$ depends on the precise history of the particle. During the lth scattering event the wave vector changes from $\mathbf{k}_l$ to $\mathbf{k}_l'$ by an amount $\mathbf{K}_l$ such that

$$\mathbf{k}_l' = \mathbf{k}_l + \mathbf{K}_l. \tag{12.21}$$

We need to know $\mathbf{k}(0)\bullet\mathbf{k}(T)$ to calculate $\langle\Delta\mathbf{k}(0)\bullet\Delta\mathbf{k}(T)\rangle_n$. Here $\langle\ \rangle_n$ denotes the average over all t_n, $\mathbf{k}_n'$, $\mathbf{F}$, scattering processes and times, in other words, over all possible scattering histories of the n-particles weighed by the probability of realisation.

We wish to express $\langle\mathbf{k}(0)\bullet\mathbf{k}(T)\rangle_n$ in terms of $\langle\mathbf{k}_1'^2\rangle_n$ and $\langle\mathbf{k}_1\bullet\mathbf{k}_1'\rangle_n$ which can be obtained from scattering theory. We write

$$\mathbf{k}(0) = \mathbf{k}_l - (e/\hbar) \int_{-t'}^{t_1} \mathbf{F}(t)dt \tag{12.22}$$

to get

$$\begin{aligned}\mathbf{k}(0)\bullet\mathbf{k}(T) &= \mathbf{k}_l\bullet\mathbf{k}_n' - (e/\hbar)^2 \int_{-t'}^{t_1} \mathbf{F}(t)dt\bullet\int_{t_n}^{T} \mathbf{F}(t)dt \\ &- (e/\hbar)\left\{\mathbf{k}_n'\bullet \int_{-t'}^{t_1} \mathbf{F}(t)dt + \mathbf{k}_l\bullet\int_{t_n}^{T} \mathbf{F}(t)dt\right\}.\end{aligned} \tag{12.23}$$

The detailed scattering history of a particular n-particle is

$$\mathbf{k}_n' - \mathbf{k}_l = (e/\hbar) \int_{t_1}^{t_n} \mathbf{F}(t)\, dt + \sum_{l=1}^{n} \mathbf{K}_l . \tag{12.24}$$

Calculating the square of this, making use of the relation

$$\mathbf{k}_l = \mathbf{k}_{l-1} + (e/\hbar) \int_{t_{l-1}}^{t_l} \mathbf{F}(t)\, dt \tag{12.25}$$

and Equation (12.21), we get

$$\begin{aligned}\langle \mathbf{k}(0)\bullet\mathbf{k}(t)\rangle_n = {} & \sum_{l=1}^{n} \langle \mathbf{k}_l \bullet \mathbf{k}_l' \rangle_n - \sum_{l=1}^{n-1} \langle \mathbf{k}_l'^2 \rangle_n - \tfrac{1}{2} \sum_{l \neq j}^{n} \langle \mathbf{K}_j \bullet \mathbf{K}_l \rangle_n \\ & - (e/\hbar) \Bigg\langle \mathbf{k}_0 \bullet \int_0^{t_n} \mathbf{F}(t) dt + \mathbf{k} \bullet \int_{t_n}^{T} \mathbf{F}(t) dt \\ & + \sum_{l=1}^{n} \mathbf{K}_l \bullet \int_0^{t_n} \mathbf{F}(t) dt + \sum_{l=1}^{n-1} \left(\sum_{j=1}^{l} \mathbf{K}_j \right) \bullet \int_{t_1}^{t_{l+1}} \mathbf{F}(t) dt \Bigg\rangle_n \\ & - (e/\hbar)^2 \Bigg\langle \int_{-t'}^{0} \mathbf{F}(t) dt \bullet \int_{t_n}^{T} \mathbf{F}(t) dt - \int_{-t'}^{t_n} \mathbf{F}(t) dt \bullet \int_{0}^{t_n} \mathbf{F}(t) dt \Bigg\rangle_n .\end{aligned} \tag{12.26}$$

$\langle \mathbf{k}(0) \rangle \bullet \langle \mathbf{k}(T) \rangle_n$ is calculated in a similar way, the calculation of the latter differs in that the averaging is carried out first. Hence

$$\langle \Delta\mathbf{k}(0) \bullet \Delta\mathbf{k}(T) \rangle = J(T) + (e/\hbar) \sum_{n=0}^{\infty} F_n^c P_n \tag{12.27}$$

with

$$J(T) = (\langle \mathbf{k}_0^2 \rangle_0 - \langle \mathbf{k}_0 \rangle_0^2) P_0 + \sum_{n=1}^{\infty} \sum_{l=1}^{n} \{ \langle \mathbf{k}_l \bullet \mathbf{k}_l' \rangle_n - \langle \mathbf{k}_l \rangle_n \bullet \langle \mathbf{k}_l' \rangle_n \} P_n , \tag{12.28}$$

the Johnson noise functions

$$F_0^c \equiv \Bigg\langle \int_{-t'}^{0} \mathbf{F}(t) dt \bullet \int_{-t'}^{T} \mathbf{F}(t) dt \Bigg\rangle_0 \tag{12.29a}$$

for $n = 0$ and

$$F_n^c \equiv \left\langle \int_{-t'}^{t_n} \mathbf{F}(t)dt \bullet \int_0^{t_n} \mathbf{F}(t)dt \right\rangle_n - \left\langle \int_{-t'}^{0} \mathbf{F}(t)dt \bullet \int_{t_n}^{T} \mathbf{F}(t)dt \right\rangle_n \quad (12.29b)$$

for $n > 0$, and P_n represents the probability that a given particle is an n-particle which will be given by Equation (12.33) below.

The time of free flight, t'', of the particles is given by Equation (5.17) and is distributed with a density

$$f_t(t'') = \exp(-t''/\tau_s) \quad (12.30)$$

where τ_s depends on the energy of the particle and τ_s represents the total inverse scattering rate. Introducing self-scattering, τ_s becomes constant, but the actual distribution of the duration of the flights will still be given by Equation (12.30).

The set of scattering times

$$\{t\} = \{t_1, t_2, \ldots, t_n, \ldots\} \quad (12.31)$$

makes an ordered set as $t_l < t < t_{l+1}$, and t itself has a distribution over the interval $0 \leqslant t \leqslant T$. The probability that just n variables satisfy the inequality

$$0 \leqslant t_1 \leqslant t_2 \leqslant \ldots \leqslant t_n \leqslant T \quad (12.32)$$

is

$$P_n(T) = \exp(-T/\tau_s)\,(T/\tau_s)^n/n! \quad (12.33)$$

(Feller, 1971); this also applies when self-scattering is excluded. $P_n(T)$ expresses the probability that a given particle is an n-particle at the universal time T.

Feller also proves that the probability density for the n first variables of $\{t\}$ satisfying Equation (12.30) is:

$$h_{n,l}(t_l) = \frac{n}{T}\binom{n-l}{l-1}\left(\frac{t_1}{T}\right)^{l-1}\left(\frac{T-t_1}{T}\right)^{n-l}. \quad (12.34)$$

The local electric field F consists of the externally applied field $\mathbf{F}_0$ and the fluctuation field $\Delta\mathbf{F}$, defined solely by the presence of the ionic impurities (background charge) and the mobile charge carriers:

$$\mathbf{F} = \mathbf{F}_0 + \Delta\mathbf{F}. \quad (12.35)$$

The latter is calculated at regular intervals, δT, the field-adjusting time step. $\mathbf{F}$ stays fixed between each recalculation and is constant within each cell. This assumption gives rise to *model noise*. Using Equation (12.35) F_0^c can be split to read.

$$F_0^c = \left\langle \int_{-t'}^{0} \mathbf{F}_0 dt \bullet \int_{-t'}^{T} \mathbf{F}_0 dt \right\rangle_0 + \left\langle \int_{-t'}^{0} \Delta\mathbf{F}(t)dt \right\rangle_0 \bullet \left\langle \int_{-t'}^{T} \Delta\mathbf{F}(t)dt \right\rangle_0 \quad (12.36)$$

as terms containing products of $\int \mathbf{F}_0 dt$ and $\int \Delta\mathbf{F}(t)dt$ vanish because

$$\langle \Delta\mathbf{F} \rangle_n = 0 \tag{12.37}$$

for all n, as will be explained in Section 12.3.2. This equation says that the average in the fluctuation of the electric field cancels. As $\mathbf{F}_0$ is constant throughout the bulk

$$\left\langle \int_{-t'}^{0} \mathbf{F}_0 dt \bullet \int_{-t'}^{T} \mathbf{F}_0 dt \right\rangle_0 = \mathbf{F}_0^2\{T\langle t'\rangle_0 + \langle t'^2\rangle_0\} = \mathbf{F}_0^2(T\tau_s + 2\tau_s^2) \tag{12.38}$$

we obtain the latter noting that

$$\int_0^\infty t^j f_t(t)dt = j\tau_s^j \tag{12.39}$$

for $j = 1$ and 2.

In the next subsection we shall prove that the second term in F_0^c simplifies to $2\tau_s^2\langle \Delta F^2\rangle$ where $\langle \Delta\mathbf{F}^2\rangle$ represents the variance of the fluctuation field. Likewise, from Equation (12.29) and the next subsection,

$$\sum_{n=0}^{\infty} F_n^c P_n = \mathbf{F}_0^2[2\tau_s^2 + \tau_s^2 \mathcal{E}(T)]] + \langle \Delta\mathbf{F}^2\rangle \{(T\delta T - 2\tau_s \delta T + 2\tau_s^2) - (2\tau_s^2 + 2\tau_s\delta T)\frac{\exp(T/\tau_s) - \exp(\delta T/\tau_s)}{\exp(\delta T/\tau_s) - 1}\exp(-T/\tau_s)\}. \tag{12.40}$$

Here $\mathcal{E}(T)$ is a function of T defined by Equation (12.65) below. When the electric field $\mathbf{F} = 0$ $\langle \Delta\mathbf{k}(0)\bullet\Delta\mathbf{k}(T)\rangle$ represents *Johnson noise*. With $\mathbf{F} \neq 0$ the correlation $\mathbf{k}$ still represents such noise, in other words, Johnson noise represents the fluctuation in the thermal and drift velocity of their carriers. We may expect the Johnson noise to depend implicitly on the external electric field because it accelerates the particles.

Making use of Equation (12.21) we get

$$J(T) = \langle \mathbf{k}_0^2\rangle P_0 + \sum_{n=1}^{\infty}\sum_{l=1}^{n} \langle \mathbf{k}_l \bullet \mathbf{k}_l'\rangle_n P_n - \langle \mathbf{k}_0\rangle_0^2 P_0 - \sum_{n=1}^{\infty}\sum_{l=1}^{n} \langle \mathbf{k}_l'\rangle_n^2 P_n + \sum_{n=1}^{\infty}\sum_{l=1}^{n} \langle \mathbf{k}_l'\rangle_n \bullet \langle \mathbf{K}_l\rangle_n P_n. \tag{12.41}$$

In cubic semiconductors with no anisotropy there is no preferential direction of $\mathbf{K}_l$ so the term in $\mathbf{K}_l$ vanishes. In some single crystal semiconductors the anisotropy in $\mathbf{K}_l$ may become important for some scattering processes. Restricting ourselves to isotropic material this term can be omitted from the subsequent discussion. We turn our attention to $\langle \mathbf{k}_l \bullet \mathbf{k}_l'\rangle_n$.

The product $\mathbf{k}_l \bullet \mathbf{k}_l'$ depends on the type of scattering. There are a number of different scattering processes; the probability density that a particular one, type i, say, takes place is $S_i(\mathbf{k}, \mathbf{k}')$, which is proportional to the corresponding scattering rate, λ_i:

$$\int d^3\mathbf{k}\, S_i(\mathbf{k}, \mathbf{k}') = \lambda_i/\Gamma_0. \tag{12.42}$$

If self-scattering is not considered, Γ_0 is replaced by Γ, the total real event scattering rate.

$$\langle \mathbf{k}_l \bullet \mathbf{k}_l' \rangle = \sum_i \int d^3\mathbf{k}_l d^3\mathbf{k}_l' S_i(\mathbf{k}_l, \mathbf{k}_l') f_k(t)/n = \langle \mathbf{k} \bullet \mathbf{k} \rangle / n \tag{12.43}$$

where $f_k(t)$ represents the distribution of wave vectors, which has been discussed in Chapter 3. The factor $1/n$ stems from the fact that there are n scattering events which on the average will be spaced equidistant over the interval T. Under equilibrium conditions the integral of Equation (12.43) is independent of l and n, which is expressed by omitting the label l from the last part of it. This implies that

$$\langle \mathbf{k}_l' \rangle = \langle \mathbf{k}_0 \rangle \tag{12.44}$$

otherwise heating would take place. The Johnson noise function hence takes the form

$$J(T) = \langle \mathbf{k}_0^2 \rangle P_0 + \langle \mathbf{k} \bullet \mathbf{k}' \rangle (1 - P_0) - \langle \mathbf{k}_0 \rangle^2. \tag{12.45}$$

As $T \rightarrow \infty$, P_0 approaches zero so that it vanishes from the Johnson noise function.

We now turn our attention to the displacement current, the last term of Equation (12.15). Only the fluctuation in the electric field, $\Delta\mathbf{F}_y$ can enter because the time differential in F_0 vanishes. $d\mathbf{F}/dT$ is replaced by $\Delta\mathbf{F}/\delta T$ so that the correlation in the fluctuation field can be formulated as

$$\begin{aligned} \langle \Delta\mathbf{F}_y(0)/\delta T\, \Delta\mathbf{F}_y(T)/\delta T \rangle &= \langle \Delta\mathbf{F}_y(0) \Delta\mathbf{F}_y(T) \rangle / \delta T^2 \\ &= \langle \Delta\mathbf{F}_y(0) \rangle \langle \Delta\mathbf{F}_y(T) \rangle / \delta T^2 \end{aligned} \tag{12.46}$$

which vanishes for $T > \delta T$ as we shall see in the next subsection.

12.3.2 The fluctuation field

In Chapter 3 we learnt that the carriers move like free particles in a box but with an effective mass different from that of free electrons. Because the contacts inject and absorb carriers the charge probability distribution has to be the same everywhere provided the externally applied electric field is uniform. Plots showing the positions of the charge carriers obtained during simulation of uniformly doped resistors support this assumption. During the flow of the steady state current the absorbed particles will be replaced by new ones as explained in Chapter 8. Let us use the Nearest-Grid-point charge assignment scheme (Section 7.3) to calculate the local fluctuation field defined by the presence of all the other particles. The simulation area is divided into cells numbered from 0 to N_x in the x direction and from 1 to N_y in the y direction. By cell (i, j) we mean the cell in column i in the x direction and row j in the y direction. In the two-dimensional model the charged particles are considered rods of zero thickness and uniform charge density extending over the entire width, L_z, i.e. the z direction, for the purpose of solving the field equation. The ionised donors or acceptors, representing the background charge, can be regarded as being distributed uniformly, with ν_s charges in each cell: ν_s is also the average number of carriers per cell. With N_xN_y cells and n_{kl} particles in cell (k, l) the particles can be distributed over the resistor according to:

$$P_\phi^{ji}(\{n_{kl}\}) = \frac{(N_xN_y\nu_s)!}{\prod_{k=0}^{N_x} \prod_{l=1}^{N_y} n_{kl}!} \left(\frac{1}{N_xN_y}\right)^{N_xN_y} \tag{12.47}$$

which also expresses the probability that the field in cell (i, j) is

$$\Phi^{ji} = \sum_{k=0}^{N_x} \sum_{l=1}^{N_y} (n_{kl} - \nu_s)\Phi_{kl}^{ji} + \Phi_{\text{ext}} \tag{12.48}$$

with

$$\begin{aligned} \Phi_{kl}^{ji} &= \frac{e}{4\pi\varepsilon\varepsilon_0 L_z} \int_{-L_z/2}^{L_z/2} [(i-k)^2h_x^2 + (j-l)^2h_y^2 + z^2]^{-\frac{1}{2}}\, dz \\ &= \frac{e}{4\pi\varepsilon\varepsilon_0 L_z} \log \frac{[L_z^2/4 + (i-k)^2h_x^2 + (j-l)^2h_y^2]^{\frac{1}{2}} + L_z/2}{[L_z^2/4 + (i-k)^2h_x^2 + (j-l)^2h_y^2]^{\frac{1}{2}} - L_z/2} \end{aligned} \tag{12.49}$$

for our particular representation of the superparticles. Φ_{kl}^{ji} expresses the potential in cell (i, j) due to one superparticle in cell (k, l). The electric field $\Delta\mathbf{F}^{ji} = -\Delta\Phi^{ji}$ in cell (i, j) has y component

$$\Delta\mathbf{F}_y^{ij} = -\partial\Phi_{kl}^{ji}/\partial y \simeq (\Phi_{kl}^{j+1,i} - \Phi_{kl}^{j-1,i})/(2h_y) \tag{12.50}$$

Within one cell the acceleration of a particle is constant. During simulation we leave the acceleration unchanged when the particle crosses the boundary to the neighbouring cell, and let it remain unchanged until the next scatter event or recalculation of the electric field at the end of the field-adjusting time step. Of course, the field-adjusting time step has been chosen sufficiently small that the particle has not moved too far into the next cell.

The average fluctuation field in cell (i, j) is given by

$$\langle \Delta\mathbf{F}_y \rangle = \frac{1}{N_xN_y} \sum_{\{n_{kl}=0\}} P\phi^{ji}(\{n_{kl}\})\Delta\mathbf{F}_y^{ji} = 0 \tag{12.51}$$

and its second moment is

$$\langle \Delta \mathbf{F}_y^2 \rangle = \frac{1}{N_x N_y} \sum_{\{n_{ij}\}} \sum_{i=0}^{N_x} \sum_{j=1}^{N_y} P_{\Phi}^{ji}(\{n_{kl}\}) (\Delta \mathbf{F}_y^{ji})^2$$

$$= \left(\frac{e}{4\pi \varepsilon \varepsilon_0 L_z} \right)^2 \frac{\nu_s B_f}{4h_x^2 N_x} \left(1 - \frac{1}{N_x N_y} \right) \tag{12.52}$$

where $\sum_{\{n_{ij}\}}$ denotes summation over all n_{ij}, subject to the condition

$$\sum_{i=0}^{N_x} \sum_{j=1}^{N_y} n_{ij} = N_x N_y \nu_s. \tag{12.53}$$

B_f is a *bulk factor* defined by

$$B_f \equiv \frac{1}{N_x N_y} \sum_{i=0}^{N_x} \sum_{j=1}^{N_y} \sum_{k=0}^{N_x} \sum_{l=1}^{N_y} \log^2 \frac{(j-l-1)^2 h_x^2 + (i-k)^2 h_y^2}{(j-l+1)^2 h_x^2 + (i-k)^2 h_y^2} \tag{12.54}$$

where $L \gg h_x$ and $L \gg h_y$ which is the case for most applications. The bulk factor reflects the shape of the bulk. Table 12.1 shows B_f for a block of $N_x N_y = (2^u + 1)2^{u+2}$ cells. B_f first increases with u then levels off, suggesting that the effect on $\langle \Delta E^2 \rangle$ of dividing the bulk into cells decreases with an increasing number of cells. The index y has no other function than to remind the reader that the steady state average current travels in the y direction.

Note we have not approximated the factorials with Sterlings' formula, as is usual in statistical mechanics. We allow N_x and N_y to be relatively small, so we prefer to work with exact expressions. Evaluating Equation (12.19) with a field given by Equation (12.35) we shall find, because of Equation (12.37) or (12.51), that only the terms of the order ΔF^2 will remain.

To compute the integral $\int \Delta \mathbf{F}(t) dt$ let $T = K\delta T$, where K represents an integer, δT the field-adjusting time step and T the duration of the simulation. In the expressions for F_0^c and F_n^c, Equations (12.29), it is necessary to evaluate the integrals of $\mathbf{F}(t)$ over time intervals which generally extend over several field-adjusting time steps. $\Delta \mathbf{F}$ is constant within each time step for a particular charge carrier, so that in the expression for F_n^c we need to calculate

$$G_n = \int_{-t'}^{t_n} \Delta \mathbf{F}(t) dt \bullet \int_0^{t_n} \Delta \mathbf{F}(t) dt - \int_{-t'}^{0} \Delta \mathbf{F}(t) dt \bullet \int_{t_n}^{T} \Delta \mathbf{F}(d) dt \tag{12.55}$$

Table 12.1 Bulk factor B_f against u. Bulk factor for a mesh with 2^{u+2} cells in the directions of the current; 2^{u+1} in the direction perpendicular to it

u	B_f
1	24.9
2	50.6
3	80.1
4	101.5
5	101.6

which deals with the fluctuation fields only.

The last scattering event falls within the $(p+1)^{\text{st}}$ time step, $p\delta T \leqslant t_n \leqslant (p+1)\delta T$, $(p = 0, \ldots, K-1)$. The probability distribution of the particles is always uniform during equilibrium, we can therefore prove there is no correlation in $\Delta\mathbf{F}$ from one time-step to the next. Hence

$$\langle G_n \rangle = \left\langle \int_{-t'}^{0} \Delta\mathbf{F}(t)dt \right\rangle_n \bullet \left\langle \int_{0}^{t_n} \Delta\mathbf{F}(t)dt \right\rangle_n + \left\langle \left(\int_{0}^{t_n} \Delta\mathbf{F}(t)dt \right)^2 \right\rangle_n$$
$$- \left\langle \int_{-t'}^{0} \Delta\mathbf{F}(t)dt \right\rangle_n \bullet \left\langle \int_{t_n}^{T} \Delta\mathbf{F}(t)dt \right\rangle_n . \tag{12.56}$$

(The range of integration in the first term of Equation (12.55) has been split in two.) Because of Equation (12.51) only the middle term will remain. It is therefore sufficient to consider only that term further:

$$G'_n = \int_0^{t_n} \Delta\mathbf{F}(t)dt \bullet \int_0^{t_n} \Delta\mathbf{F}(t)dt$$
$$= \sum_{i=0}^{p-1} \int_{i\delta T}^{(i+1)\delta T} \Delta\mathbf{F}(t)dt \bullet \int_{i\delta T}^{(i+1)\delta T} \Delta\mathbf{F}(t)dt + \int_{p\delta T}^{t_n} \Delta\mathbf{F}(t)dt \bullet \int_{p\delta T}^{t_n} \Delta\mathbf{F}(t)dt$$
$$+ 2\sum_{i=0}^{p-1} \int_{i\delta T}^{(i-1)\delta T} \Delta\mathbf{F}(t)dt \bullet \int_{p\delta T}^{t_n} \Delta\mathbf{F}(t)dt$$
$$+ 2\sum_{i<j}^{p-1} \int_{i\delta T}^{(i+1)\delta T} \Delta\mathbf{F}(t)dt \bullet \int_{j\delta T}^{(j+1)\delta T} \Delta\mathbf{F}(t)dt \tag{12.57}$$

Here, again, only the terms without the factor 2 before them survive on forming $\langle G' \rangle_n$ so it suffices to consider the expression

$$G''_n(p) = \sum_{i=0}^{p-1} \left(\int_{i\delta T}^{(i+1)\delta T} \Delta\mathbf{F}(t)dt \right)^2 + \left(\int_{p\delta T}^{t_n} \Delta\mathbf{F}(t)dt \right)^2$$
$$= \delta T^2 \sum_{i=0}^{p} \Delta\mathbf{F}_i^2 + (t_n - p\delta T)^2 \Delta\mathbf{F}_p^2 \tag{12.58}$$

where $\Delta\mathbf{F}_i$ is the local field deviation a particular particle experiences during the ith time-step. Using Equation (12.49)

$$\langle G_n \rangle = \langle G''_n(p) \rangle_n = \langle \Delta F^2 \rangle \left[p(p+1)\delta T^2 + \langle t_n^2 \rangle_n - 2p\delta T \langle t_n \rangle_n \right]. \tag{12.59}$$

To obtain $\Sigma \langle G_n \rangle P_n$, it is necessary to sum over all p:

$$\sum_{n=1}^{\infty} \langle G_n \rangle P_n = \sum_{n=1}^{\infty} \sum_{p=0}^{K-1} \langle G''_n \rangle_n = \sum_{n=1}^{\infty} \frac{\delta T^2 \exp(-T/\tau_s)}{n!}$$

$$\times \sum_{p=0}^{k-1} \int_{p\delta T}^{(p+1)\delta T} [p(p+1)\delta T^2 + t_n^2 - 2pt_n\delta T] n t_n^{n-1} dt_n \tag{12.60}$$

Evaluating the integrals over t_n and then writing out terms in p this can be written

$$\sum_{n=1}^{\infty} \langle G_n \rangle_n P_n = \Delta \mathbf{F}^2 \exp(-T/\tau_s) \sum_{n=1}^{\infty} \frac{\delta T^{n+2}}{\tau_s^n} \left\{ \frac{K^{n+1}}{(n-1)!} \left[\frac{(K-1)(1-n)}{n+1} + \frac{K}{n+2} \right] - \sum_{p=1}^{k-1} \frac{p^{n+1}}{(n+1)!} \right\} \tag{12.61}$$

Because the sums over both p and n converge, the order of summation can be interchanged. Furthermore, the simulation runs over a large interval of time, so that $K \gg 1$. Hence

$$\sum_{n=1}^{\infty} \langle G_n \rangle_n P_n = 2\Delta \mathbf{F}^2 \left\{ \tau_s^2 \exp(T/\tau_s) - \tau_s^2 + \tfrac{1}{2} T\delta T - \tau_s \delta T \exp(\delta T/\tau_s) \frac{\exp(T/\tau_s) - 1}{\exp(\delta T/\tau_s) - 1} \right\} \exp(-T/\tau_s) \tag{12.62}$$

In the expression for F_0 we have, dealing with fluctuation fields,

$$G_0 \equiv \left\langle \int_0^{t'} \Delta \mathbf{F}(t) dt \bullet \int_0^{t'+T} \Delta \mathbf{F}(t) dt \right\rangle_0 = \left\langle \left(\int_0^{t'} \Delta \mathbf{F}(t) dt \right)^2 \right\rangle_0 = 2\Delta \mathbf{F}^2 \tau_s^2. \tag{12.63}$$

As $T \rightarrow \infty$, $\sum_{n=1}^{\infty} \langle G_n \rangle_n P_0$ approaches

$$2\Delta \mathbf{F}^2 \left\{ t_s^2 - \frac{\tau_s \delta T \exp(\delta T/\tau_s)}{\exp(\delta T/\tau_s) - 1} \right\} \tag{12.64}$$

The function

$$\mathcal{E}(T) \equiv T^2 \sum \frac{n(T/\tau_s)^n}{(n+2)(n+1)!} \tag{12.65}$$

cannot be expressed analytically. To get an idea of its behaviour, we note that the lower limit is 0 as $T \rightarrow 0$, and $\mathcal{E}(T > 0) > 0$. Furthermore

$$\frac{n}{(n+1)(n+2)!} < \frac{n+1}{(n+1)(n+2)!} = \frac{1}{(n+2)!} \tag{12.66}$$

so that $0 < \mathcal{E}(T) < \mathcal{E}'(T)$ with $\mathcal{E}'(T)$ given below. For large n this inequality represents almost equality so that we may replace $\mathcal{E}'$ by $\mathcal{E}$. With

$$\mathcal{E}'(T) \equiv T^2 \sum_{n=0}^{\infty} \frac{(T/\tau_s)^n}{(n+2)!} = \tau_s^2 \left[1 - \exp(T/\tau_s) \left(1 + \frac{T}{\tau_s} + \frac{T^2}{2\tau_s^2} \right) \right]. \tag{12.67}$$

The Fourier transform of $\mathcal{E}'(T)$

$$\int_{-\infty}^{\infty} \mathcal{E}(T) e^{-i\omega T} dT < \mathcal{E}'(T) \equiv \int_{-\infty}^{\infty} \varepsilon'(T) e^{-i\omega T} dT \tag{12.68}$$

12.3.3 Summary of noise theory

We have thus far considered only one type of carrier. It is straightforward to generalise to ensembles with different particles. In this context particles belonging to different extrema of the valence or conduction bands are considered to be different. The correlation in the current through a block of uniformly doped semiconducting material is expressed as

$$\begin{aligned} C(T) = \langle \Delta I_y(0) \Delta I_y(T) \rangle &= \sum r_\alpha (n_\alpha e_\alpha / m^*{}_\alpha)^2 \\ &\times \Big\{ (A\hbar)^2 J(T) + (A e_\alpha F_0)^2 [2\tau_s^2 + \mathcal{E}(T)] \\ &+ \left(\frac{n_\alpha e_\alpha^2 L_z}{8\pi\varepsilon\varepsilon_0} \right)^2 \frac{(h_x h_y)^2}{\nu_s} N_x B_f f(T) \Big\} \end{aligned} \tag{12.69}$$

where $J(T)$ and $\mathcal{E}(T)$ are given by Equations (12.28) and (12.65) respectively and

$$f(T) \equiv T\delta T \exp(-T/\tau_s) + 2\tau_s^2 - 2\tau_s \delta T \frac{1 - \exp(-T/\tau_s)}{\exp(\delta T/t_s) - 1} \exp(\delta T/\tau_s) \tag{12.70}$$

approaches

$$f(T) = 2\tau_s^2 - \frac{2\tau_s \delta T}{\exp(\delta T/\tau_s) - 1} \simeq 0 \tag{12.71}$$

as $T \to \infty$. This has a value near zero for small $\delta T/\tau_s$. The bulk factor, B_y, is defined by Equation (12.54). Furthermore, r_α represents the relative abundance of carrier type α which has an effective mass m_α^* and a charge e_α.

The *power spectrum* is given by

$$C(\omega) = \sum r_\alpha (n_\alpha e_\alpha / m_\alpha^*)^2 \Big\{ 2(A\hbar)^2 [\langle \mathbf{k}_0 \rangle_0^2 - \langle \mathbf{k} \bullet \mathbf{k}' \rangle_0]$$

$$\times \frac{\tau_s}{1 + (\omega\tau_s)^2} + (Ae_\alpha \mathbf{F}_0)^2(\omega) + \left(\frac{n_\alpha e_\alpha L_z}{8\pi\varepsilon\varepsilon_0}\right) \frac{(h_x h_y)^2}{\nu_s} N_x B_y$$

$$\times \left[\frac{\tau_s \delta T}{1 + (\omega\tau_s)^2} \left(\frac{2\exp(\delta T/\tau_s)}{\exp(\delta T/\tau_s) - 1} + \frac{1 - (\omega\tau_s)^2}{1 + (\omega\tau_s)^2} \right) \right.$$

$$\left. + \left(\tau_s^2 - \frac{\tau_s \delta T \exp(\delta T/\tau_s)}{\exp(\delta T/\tau_s) - 1} \right) \delta(\omega) \right] \Bigg\}. \tag{12.72}$$

Although the autocorrelation was evaluated for $T \geqslant 0$, the range $T < 0$ needed to get this is obtained using the relation $C(T) = C(-T)$.

Figure 12.2 shows the noise spectrum, the correlation current versus frequency. We see that for very low frequencies the noise spectrum tends to level, the noise is almost white. When $\omega\tau_s = 1$ the spectrum is *1/f noise*-like; for higher frequencies it falls off faster than $1/f$ and is then of a Lorenzian type.

Equation (12.69) and its Fourier transform, the *noise spectrum*, Equation (12.72), have been calculated for one particle. The total number of super- or real particles, N_s, does fluctuate a bit during a simulation. Disregarding this, the total correlation current spectrum is therefore the average over all the superparticles because each particle is equally likely to experience the same flight history:

$$I_c = C(\omega)/N_s \tag{12.73}$$

We could just as well calculate the correlation current for all the N real particles; the average is

$$I_c' = \lim_{\delta T \to 0} C(\omega)/N \tag{12.73'}$$

where lim $C(\omega)$ represents the limit of $C(\omega)$ when $\delta T \to 0$, and $N_x \to \infty$ and $N_y \to \infty$ but in such a manner that $N_x h_x$ and $N_y h_y$ remain unchanged. This eliminates the model noise, the noise caused by the selection of a finite time and space grid to solve Poisson's equation. The limiting value of $C(\omega)$ is

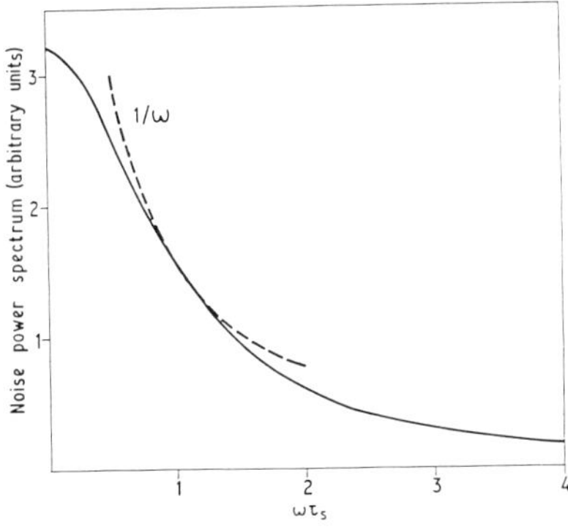

Figure 12.2 *Power spectrum of the current through a uniformly doped resistor. The dashed curve represents 1/f (power against inverse frequency) behaviour*

$$\lim_{\delta T \to 0} C(\omega) = \sum_{\alpha} r_{\alpha} (n_{\alpha} e_{\alpha}/m_{\alpha}^{*})^2 \{2(A\hbar)^2 [\langle \mathbf{k}_0 \rangle_0^2 - \langle \mathbf{k} \bullet \mathbf{k}' \rangle_0]$$

$$\times \frac{\tau_s}{1 + (\omega \tau_s)^2} + (A e_{\alpha} F_0)^2 \mathcal{E}(\omega) + \frac{2\tau_s}{1 + (\omega \tau_s)^2} . \quad (12.74)$$

The ratio between the simulated and the real noise current spectrum is:

$$\frac{I_c}{I_c'} = \frac{NC(\omega)}{N_s \lim_{\delta T \to 0} C(\omega)} \quad (12.75)$$

When the grid is fine enough and the field-adjusting time step small enough

$$C(\omega) \simeq \lim_{\delta T \to 0} C(\omega) \quad (12.76)$$

which means that the simulated correlation current power spectrum is exaggerated by a factor N/N_s. The noise of the simulated current itself augments by a factor

$$E_{xg} = \sqrt{\frac{N_s C(\omega)}{N \lim C(\omega)}} \quad (12.77)$$

the exaggeration factor. When the Relation (12.76) is sufficiently accurate, the exaggeration factor reduces to $\sqrt{N/N_s}$.

12.4 COMPUTER EXPERIMENTAL VERIFICATION

In order to check the validity of our theoretical description, we have compared our predicted correlation current with the one obtained by simulating a current through a rectangular block of semiconducting n-type GaAs. The material has an impurity concentration of $10^{23}\,\mathrm{m}^{-3}$, each impurity atom yielding one conduction electron. Where nothing else is indicated, the field-adjusting time-step is 50 fs. Due to the way Poisson's equation is solved, Chapter 7, the block is divided into

$$N = (2^u + 1)2^{u+2} \quad (12.78)$$

cells of square cross-section h^2. There are 2^{u+2} cells in the direction of the current. (Inset Fig. 12.3.) The slope of Fig. 12.3 represents the particle current, the deviation from the straight line represents particle noise. During the lth field-adjusting time step N_l superparticles are absorbed, the fluctuation in the number of absorbed particles represents flicker noise. Both N_l and the electric field $\mathbf{F}_l$ are recorded for each field-adjusting time step and the current I_l is obtained as the sum of the particle and the displacement current. If the current for time step 1 and $1 + \alpha$ *is* I_1 and $I_{1+\alpha}$ respectively the autocorrelation current is given by

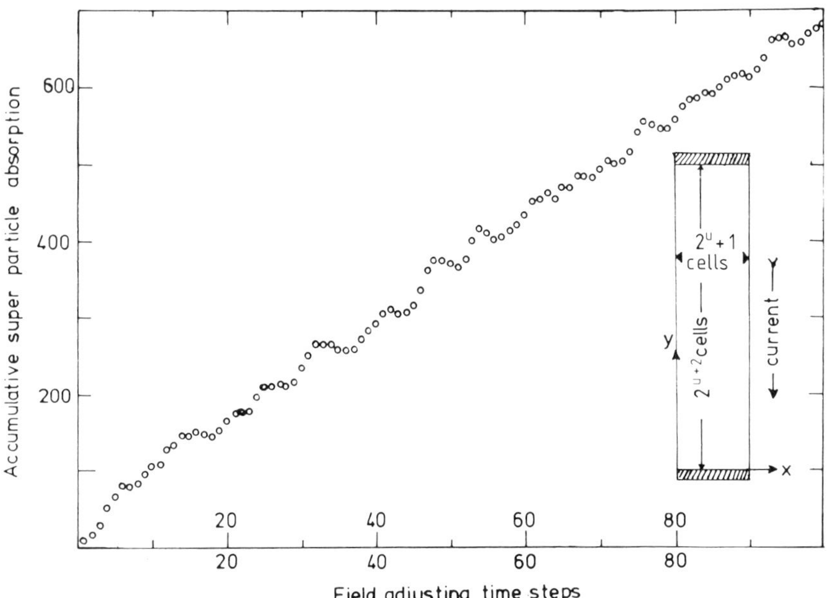

Figure 12.3 *Typical count of superparticles absorbed by the anode. The count starts at the beginning of the simulation and only the first 100 field-adjusting time steps are shown. One time step represents 50 fs, one superparticle represents 2 × 10⁷ electrons per metre. Inset shows a cross-section in the (x, y) plane of the simulated block. The current flows in the y direction, the bottom electrode serves as anode. The externally applied voltage is 0.05 V*

$$C(\alpha\delta T) = \langle I_y(0)I_y(T)\rangle = \frac{1}{N' - \alpha}\sum_{l=1}^{N'-\alpha} I_l I_{l+\alpha}, \tag{12.79}$$

where N' denotes the number of time steps in the simulation ($\alpha < N'$),

$$I_y(0) = \langle I_y\rangle + \Delta I_y(0) \tag{12.80}$$

and

$$I_y(\alpha\delta T) = \langle I_y\rangle + \Delta I_y(\alpha\delta T). \tag{12.81}$$

The 'measured' correlation current can be written

$$C(\alpha\delta T) = \langle I_y(0)\rangle^2 + \langle \Delta I_y(\alpha\delta T)\rangle + \langle I_y(0)\rangle[\langle \Delta I_y(0) + \Delta I_y(\alpha\delta T)\rangle]. \tag{12.82}$$

A plot of $C(\alpha\delta T)$ against α (i.e. time) yields $\langle I_y\rangle^2$. Furthermore

$$C(0) = \langle I_y(0)I_y(0)\rangle = \frac{1}{N}\sum_{j=1}^{N} I_j I_j = \langle I_y\rangle^2 + \langle \Delta I_y(t)\rangle^2 = \langle I_y\rangle^2 + \Delta I^2. \tag{12.83}$$

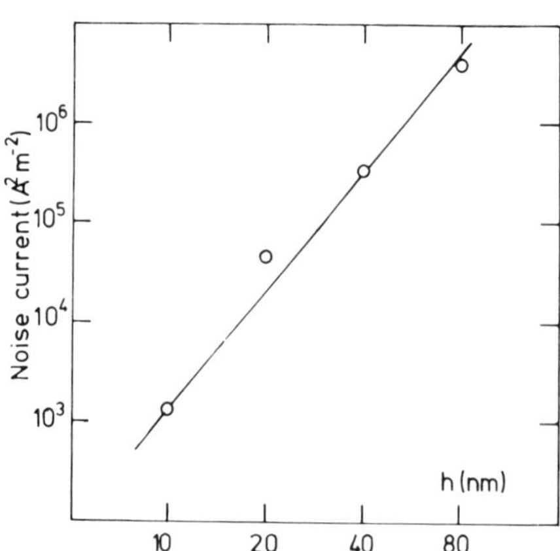

Figure 12.4 *Noise power current against cell width for a bulk of 17 × 64 cells with eight superparticles per cell. Circles show simulated results, full curve theory. The full line in this and the following figures have been drawn only as a guide for the eye*

The term in $J(T)$ of Equation (12.69) is due to deviation in the thermal velocity of the carriers from their average and is therefore attributed to Johnson noise. The term is explicitly model independent as $n_\alpha e_\alpha$ is the same for superparticles as it is for real carriers. The term in $\mathbf{F}_0$ is also explicitly model independent and can be attributed to *shot noise*. It represents the acceleration of the particles caused by the external field.

The last term of Equation (12.69) contains the effects of the internal fluctuation fields. The factor containing $f(T)$ also reflects the chosen model parameters as the fluctuation field correlation is proportional to the square of the cross-sectional area of the cell, and inversely proportional to the number of superparticles per cell; this has been confirmed in Figs 12.4 and 12.5, which show, in a double logarithmic plot, how ΔI^2 varies with h, the cell width, and ν_s, the superparticle factor. In these and subsequent figures, the results of simulation

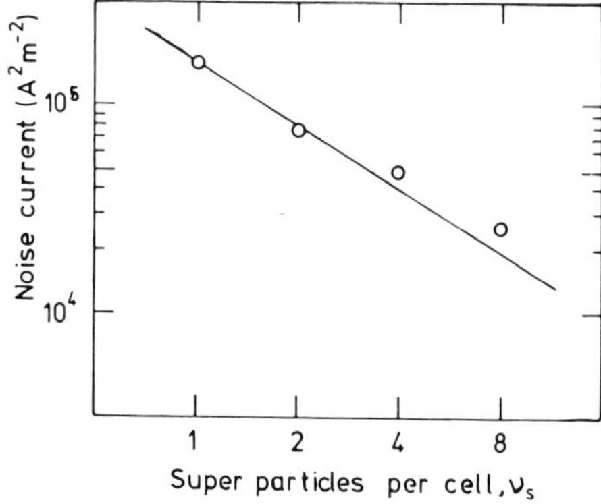

Figure 12.5 *Noise power current against number of superparticles per cell.* ○, *simulated results, full curve theory*

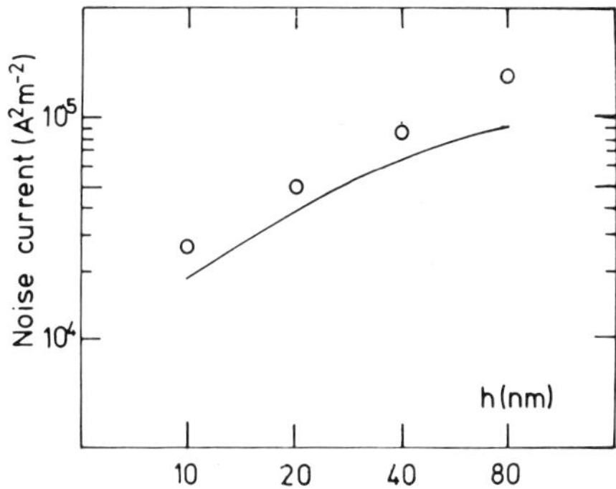

Figure 12.6 *Noise power current against cell size for a bulk of cross-section 340 × 1280 nm. The number of cells is* $(2^u + 1) \times 2^{u+2}$, *the cell width* 320×2^{-u} *nm,* $u = 2, 3, 4, 5$. ○ *show simulated results, full curve theory*

are represented by circles, while expected values are indicated by a solid curve. The inverse proportionality with ν_s agrees with Barnes and Dunn (1967).

To obtain the results of Fig. 12.4, bulks of different sizes, but with the same number of cells, and hence with the same bulk factors, were used. The choice $h = h_x = h_y$ ensured that the bulk factor was constant (Equation 12.54).

Keeping the size of the bulk fixed, $h_x h_y / \nu_s$ remains constant, and the noise current becomes proportional to $h_x h_y$, or h^2 in our case. However, $N_y = 2^{u+2}$ is no longer constant; N_x halves as h doubles, making an h dependency rather than an h^2 one. In addition, the bulk factor B_f also comes into the picture here. Figure 12.6 shows the noise current against h. In this case the simulated results agree fairly well with the expected h dependency; the points fall on a straight line. This suggests a constant B_f. The discrepancy between simulation and expectation is largest for bulks with the fewest cells (those with $h = 80$ nm).

Table 12.1 shows that the bulk factor levels off with rising u. This indicates that the effect of the surface cells reduces with bulks divided into a larger num-

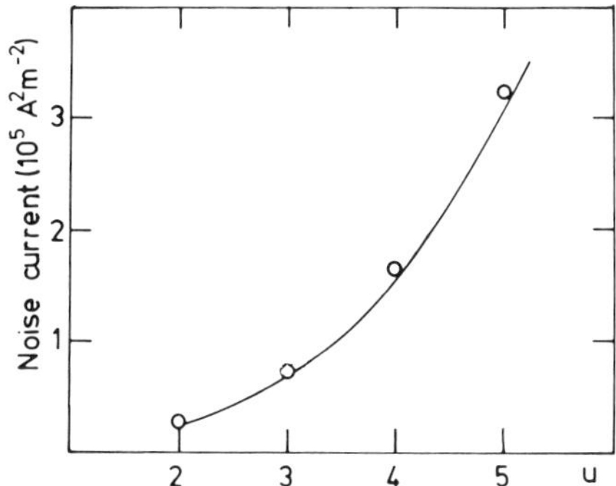

Figure 12.7 *Noise power current against u for a bulk of cross-section* $(2^u + 1) \times 2^{u+2}$ *cells, the cell width is 20 nm, the number of superparticles per cell is 1.* ○, *simulation results, full curve theory*

ber of cells. Figure 12.7 shows the noise current against u for bulks with h and ν_s constant. (These bulks vary of course in size.) The solid line represents $B_f N_x = B_f 2^2$ and agrees well with the results of simulation. If B_f were constant, the slope would be less steep and the agreement with the simulation would be worse; this indicates that the bulk factor does have a significance which Fig. 12.6 did not point out that clearly.

We have also found that the noise current does not change significantly for a mesh of 2^{u+3} cells in the y direction (simulated with the same external F_0); this indeed agrees with the prediction made by Equation (12.69) namely that N_y does not alter the most significant term explicitly.

We now turn to the expression for $J(T)$ in Equation (12.41). In the case of equilibrium a Maxwellian distribution will remain Maxwellian (Lenard, 1960), so that $\sum_{l=0}^{n} \langle \mathbf{k}_l \rangle_n = \langle \mathbf{k}_0 \rangle_0$. Using Equation (12.21)

$$\langle \mathbf{k}_l \bullet \mathbf{k}_l \rangle = \langle \mathbf{k}_l 2 \rangle - \langle \mathbf{k}_l \bullet \mathbf{k}_l \rangle. \tag{12.84}$$

If the latter term vanishes, J reduces to

$$J(T) = \langle \mathbf{k}^2 \rangle - \langle \mathbf{k}_0 \rangle^2 \tag{12.85}$$

which is the familiar expression for Johnson noise. The condition $\langle \mathbf{k}_l \bullet \mathbf{k}_l' \rangle = 0$ is valid when the scattering process is randomising, e.g. acoustic phonon scattering. However, there is a fraction f of processes, e.g. scattering from polar optical phonons, for which $\langle \mathbf{k}_l \bullet \mathbf{k}_l \rangle = C\langle \mathbf{k}^2 \rangle$ so that

$$J(T) = \langle \mathbf{k}_0^2 \rangle \{ (1 - C)P_0 + C \} - \langle \mathbf{k}_0 \rangle^2, \tag{12.86}$$

which reduces to $C\langle \mathbf{k}_0^2 \rangle - \langle \mathbf{k}_0 \rangle^2$ for very large T. This means that when $\langle k_0 \rangle = 0$ (which is the case for a Maxwellian gas) there will always be some correlation.

The power spectrum of $J(T)$ is

$$J(\omega) = \delta(\omega) \{ \langle k_0^2 \rangle C - \langle k_0 \rangle^2 \} + 2\tau_s (1 - C) \langle k_0^2 \rangle [1 + (\omega\tau_s)^2]^{-1}, \tag{12.87}$$

which represents a white spectrum when $\omega\tau_s \ll 1$, and becomes $(\omega\tau_s)^{-2}$ for $\omega\tau_s \gg 1$. Recalling that $J(\omega)$ is the power spectrum of ΔI^2, $\Delta I_y(\omega)$ varies like ω^{-1} for these frequencies.

From Equation (12.69) it is clear that the significant part of $C(0) = \langle \Delta I_y(0)^2 \rangle$ does not depend on ΔT. To see the dependency on δT it is necessary to turn to the power spectrum; the discrete Fourier transform of the most significant part of Equation (12.69) is, for $\omega = 0$,

$$\begin{aligned} C_{\text{disc}}(0) = {} & 2N\delta T\tau_s \{ \tau_s - \delta T \exp(\delta T/\tau_s) [\exp(\delta T/\tau_s) - 1]^{-1} \} \\ & - \Delta T^3 \{ [1 - \exp(\delta T/\tau_s)]^{-1} - [\exp(-\delta T/\tau_s)^{-2} - 1]^{-2} \} \\ & - 2\tau_s \delta T^2 \exp(\delta T/\tau_s) [1 - \cosh(\delta T/\tau_s)]^{-1} \end{aligned} \tag{12.88}$$

when the number of time steps $N \gg 1$.

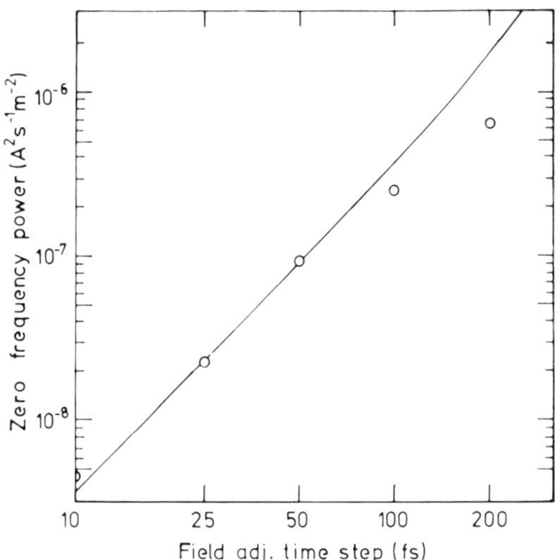

Figure 12.8 *Power spectrum at zero frequency against the field-adjusting time-step for a bulk of 17 × 64 cells with one superparticle per cell; cell width 20 nm. Both scales are logarithmic. Circles show simulated results, full curve theory*

Here the first term, corresponding to the coefficient of the δ-function in Equation (12.72), dominates. But the correlation has been obtained for a finite interval; we have plotted $C_{\text{disc}}(0)$ against the 'experimental' values in Fig. 12.8. The agreement is good, but for $\delta T \geqslant 100$ fs the theory overestimates $C(0)$. However, this region corresponds to the conditions when the simulated plasma starts to become numerically unstable (Hockney, 1971); these instabilities may affect the simulated value for $C(0)$. This also reflects itself in the lower values obtained for the steady state current when $\delta T = 200$ fs.

We may express the noise current in terms of τ_s as

$$C(0) = [\Delta I_y(t)]^2 = a\tau_s^2 \tag{12.89}$$

where τ_s can be interpreted as the average free flight time between two successive real scattering events. A simulation including a significantly larger proportion of self-scattering than used previously in this section yields the same ratio $(\Delta I^2 + I^2)/I^2$, which indicates that τ_s does not depend on the Rees self-scattering. To obtain the best fit to simulated results, Fig. 12.8, a value of $\tau_s = 7.1 \times 10^{-14}$ s has been obtained.

12.5 TURBULENCE AND CHAOS

In the previous section we have studied the noise power spectrum from the current correlation for individual particles in order to derive our noise theory. The noise originates from stochastic fluctuations of the particle distribution which causes statistical distribution in configuration (i.e. momentum and geometrical) space. We shall refer to this as *genuine noise*. We found that using N_s superparticles instead of the N real ones the noise gets exaggerated by $\sqrt{N/N_s}$.

In Section 10.8 we studied Gunn oscillations in a transistor. Gunn domains nucleated at the drain end of the gate, moved across to the drain in order to become absorbed there while the next one developed at the gate. From Equation (8.10) we found that the condition for Gunn domain formation was fulfilled for that transistor. The phenomenon has a hydrodynamic equivalent but can only be described properly by solving Poisson's field and Boltzmann's transport equation self-consistently, which is taken care of by the Monte Carlo particle model. We found that the Gunn domains arrived at the drain with a frequency of 100 GHz; in a real device such fluctuations in the drain current would probably be interpreted as noise when they could not be observed directly.

Many modern devices have highly doped layers in which the condition for wandering Gunn domain formation is fulfilled so we can expect a turbulent or chaotic current flow. The distances these domains have to travel is often smaller than in the above-mentioned transistor, the oscillations in the terminal currents can easily reach the terahertz range. These fluctuations may often be caused by carriers scattered into another band extremum where their effective mass changes. In proper analysis we should be able to distinguish genuine noise from chaotic transport; this adds a new dimension to the interpretation of measured noise data.

In addition to this there is another complication caused by plasma oscillations which genuine noise theory has not coped with. Equation (3.117) says that the plasma oscillation frequency depends on the local carrier density, and as this varies throughout the device there will be oscillations of many different frequencies.

When fluctuations in the currents out of the device cannot be observed directly we attribute them to noise. This noise should be decomposed into components of genuine noise, turbulent flow and plasma oscillations. The former produces frequencies related to the scattering rates and vary in the terahertz range for phonon and ionised impurity scattering down to kilohertz or even further for the trapping and release of carrier and multiphonon effects. Plasma oscillations will also manifest themselves in the high frequency range; the origin of the turbulences lies in the geometry of the device and in the capacitive coupling to the environment.

13

Computers: Scope of Modelling

13.1 INTRODUCTION: ASPECTS OF MODELLING

From the knowledge gathered from the first 12 chapters we should have a physical understanding of transport in semiconductor devices, although there is always room to add to it. The reader should be able to write their own software being capable of simulating individual devices of any complexity both under a transient and a steady state regime. The model is based on first physical principles so that we should be able to obtain a deep insight into the physics of the problem. This is of great value for the researcher who aims at understanding how the devices work, why they exhibit their properties and how to improve on them or design new ones.

A discussion of the relative importance of the various physical processes taking place has been given. The writer of the model has to judge which processes and how much physics to include in it. The inclusion of additional scattering processes does not necessarily mean a drastic decrease in the speed of the computation. But the model, however, is never better than the physics put into it; the accuracy of the results of the simulation can only be judged by comparison with experimental results. This, however, is not always straightforward as there is a spread in the characteristics of nominally identical devices. To facilitate the comparison the cause of the spread should be understood. The improvement in the uniformity of nominally equal devices has improved drastically during the last year and we are confident this trend will continue. We have seen that it has been possible to make a comparison within a few per cent thanks to improved manufacturing technology. An understanding of the measurements is just as important as the physical model; we have to make sure that we really measure what we want to know, e.g. that the results have not been marred by stray effects due to the shape of the electrodes or of the gate recess. But the Monte Carlo modeller also has to ensure as far as possible that they study the right geometry with the right boundary conditions.

In many of the conventional modelling techniques, device-dependent parameters have been introduced to get agreement with observed data. This limits the model to those kinds of devices without deepening the physical insight into them. The idea behind Monte Carlo particle modelling is that such fitting parameters should not be introduced. Every parameter we have used, e.g. deformation potentials, non-parabolicities, dielectric constants, effective masses, energy band gaps etc. have a physical justification independent of the device we are interested in. Modified effective masses and deformation potentials in pseudomorphic hot electron emission transistors (HEMTs) should be introduced on the basis of the strained lattice which is an effect of lattice mismatch, e.g. $In_xGa_{1-x}As$ to GaAs, and not on the geometry of the transistor.

We discussed comparison between calculated and simulated characteristics for modulation doped transistors (MODFETs). To get agreement we introduced additional physical processes like tunnelling, quantum transport and through alloyed Ohmic contacts. The extent of alloyed Ohmic contacts depends primarily on the manufacturing history and on the quality of the crystal. These processes were irrelevant for the field-effect transistors discussed earlier in the book. The inclusion of the additional physics has been defended on their introduction and is, in principle, except for the actual choice of height and width of the tunnel barrier and the width and depth of the quantum well, independent of the device.

In this last chapter we shall discuss some of the limitations to the model, some of the computational requirements, choice of best computer architecture, and the future of Monte Carlo modelling. We shall also present a few remarks on the development of our model which was aimed at complete devices from the start rather than bulk transport like many other modellers. Finally a note on the term Monte Carlo closes the book.

13.2 LIMITATIONS TO THE MODEL

Having read the 12 previous chapters the reader may have an impression that the Monte Carlo particle model is a wonder method capable of everything, but is very computer intensive. However, it has its limitations and disadvantages.

Monte Carlo modelling is nothing new having started in the 1960s studying material properties. It was not until the late 1970s that the first simulations of entire devices was reported (Warriner, 1977). The author of this book has since published results of Monte Carlo particle modelling of devices, but the world received the method as an interesting scientific curiosity without paying any more attention to it. Researchers at industrial laboratories have disdained it as being unnecessarily complex and time-consuming and argued that the conventional drift-diffusion model is sufficiently accurate, encouraging its development beyond the scope of its validity. A pure parameter fitting has also been a favoured method of device description among electrical engineers. The resistance towards lengthy calculations originates from industrialists who thought that particle modelling would be too expensive. This picture changed in the middle of the 1980s and now Monte Carlo particle modelling has been accepted with an increasing number of research laboratories around the industrialised part of the world becoming interested in it. With today's computers, however, it is not feasible to follow all the particles. This is why the superparticle was introduced in Section 8.5. A sample of N_s particles out of the total of N has been chosen to be followed. As a consequence of this, any noisy fluctuations are exaggerated approximately by a factor $\sqrt{N/N_s}$. Any signal weaker than this factor will therefore be overwhelmed by the exaggerated noise, and can therefore not easily be discerned. This is the most serious essential limitation of the Monte Carlo particle modelling. The superparticle introduces some 'graininess' in the model which may conceal some details. Effects that are due to processes of a low rate, e.g. the trapping and release of carriers, will require a long computation run before it shows. The computation time it requires may also be considered as a limitation to its applicability, but this is purely a personal opinion of the user. It is true that interactive use, which the developer of new components would like to see, cannot be made today.

However, as the geometrical size of components reduces, and computers become faster and cheaper it is only a matter of time before the particle model software can be run interactively. The future will also bring a computer with sufficient capacity and speed that we may choose $N_s = N$, and simulate all the particles in a three-dimensional geometrical model. If the signal is weaker than the noise when $N_s = N$, then this is a true result: the signal will also be overwhelmed by the noise in the corresponding real device in this case.

A model which is fully three-dimensional in geometrical space will take longer to run; the increase in computer time lies in the solution of Poisson's equation which will take about N_z times as long, where N_z represents the number of cells along the third dimension. The particle pushing part does not need to take any more time. The simulations described in this book have been carried out with a model which is two-dimensional in geometrical space. This is sufficiently accurate for devices which are long in the third dimension. When this is not the case, corner and end effects start to play a role – three geometrical dimensions therefore have to be considered. This extra dimension does not make any essential change to the method of particle modelling.

Still, for many years to come the Monte Carlo particle model will use much computer time. The alternative to such modelling is to use hydrodynamic modelling, which is faster, but relies on parameters obtained from Monte Carlo particle modelling. We may gain processing time by combining with simpler models. Monte Carlo particle modelling is used in areas where the local electric field varies greatly over short distances etc., and use a simpler model in areas where such large charges do not take place, as mentioned in Section 1.4.

As we have stated several times, the model is purely a physical one; it can be no better than the physics introduced into it. Any addition of physical processes will not make any essential changes to the modelling technique. Inaccurate physics of course implies a limitation of the applicability of the model.

13.3 COMPUTER REQUIREMENTS

What sort of computer is most suitable for Monte Carlo particle modelling? Often the reader has no choice when setting up a model and has to accept what is available. However, if large scale Monte Carlo particle modelling is envisaged, we may reflect on what computer architecture would be most efficient in terms of the speed of calculation.

An efficient code takes advantage of the computer architecture. There may be conflicting issues concerning this when we want portable software. The parts of the code most often used, such as the particle pushing and the solution of Poisson's equation, should be written with efficiency of execution in mind. The fastest algorithms to solve mathematical problems should be chosen. Often an ancient algorithm originally developed for manual computing or calculating machines is the best one to use.

After the necessary initialisation, i.e. the computer reading the required material constants and definition of the device geometry doping and biases, the computer can start processing. The simulation proceeds alternately by solution of the field equation and pushing the particles.

Poisson's field equation can be solved by fast Fourier transforms. The algorithm explained in Chapter 7 used to be the fastest ever written (Hockney, 1980). After the Fourier transform we have obtained $(N_x + 1)(N_z + 1)$ one-dimensional field equations in Fourier space, all of which are solved according to exactly the same sequence of instructions. This problem therefore seems ideal for a transputer or connection machine with a large number – i.e. thousands – of processors which all receive the same sequence of executable instructions from a central processor and work in parallel. Each processor works with its own data and finishes its calculations at the same time. The processing time is proportional to $(N_x + 1)(N_z + 1)$ divided by the number of processors, rounded off upwards to the nearest integer. Even if the individual transputers work slowly compared to modern sequential machines, the total computing time can be considerably shorter than that required by the most powerful vector or sequential machine. The processing speed could be increased further by using hardware specifically designed or constructed for this very task. We have so far only considered one field equation, namely Poisson's equation. In principle we could include electromagnetic induction introducing the other Maxwell's equations. We have only considered static magnetic fields in this book.

The particle pushing is more complex. Although the same set of instructions (program) is used for all the particles, each particle has its individual transport history and therefore its own path through the code. In the *transputer machine* each processor works independently but there seems to be some unsolved problems in connection with the communication between the individual processors making the machine less efficient than we would expect *a priori*. However, these will certainly be eliminated in the future. Ordinary sequential machines with a few processors working independently of one another would accelerate the simulation.

The *connection machine* consists of several bit processors which all have to do the same thing at the same time. They are directed from a central processor; turning on and off individual processors according to which path the individual particles take through the program does not seem to be possible. The way to program such a machine is to compile tables of all possible outcomes of scattering events, including reflection from boundaries and barriers and transmission through tunnel barriers, and then simulate the transport histories by selecting entries from these tables by means of random numbers. These tables may look different for the different parts of the simulated device. Rieger and Vogl (1989) have discussed the aspects of such programming.

Vector or *pipe lining* machines may be faster than conventional sequential machines. Each instruction consists of a sequence of elementary instructions, e.g. multiplications consist of binary additions and shifts; each addition and shift is an elementary instruction. As soon as one such instruction has been carried out it can be repeated for the next datum. Although each multiplication takes just as long as in a sequential machine, several multiplications can be processed simultaneously in stages so that one result will be delivered for each elementary instruction. This process is analogous to the manufacture of goods on a conveyor belt. A car factory may deliver one car an hour, but it takes several days to build an individual vehicle.

Pipe lining works well until a decision comes up as to which path through the instruction set has to be followed. In the particle pushing code there are many such crossroads. The pipe line has to be empty before the decision can be made; i.e. the last elementary instruction must have been carried out because the next instruction depends on the outcome of the previous one – the processing is therefore held up for a while. Gain on the sequential computer can only be obtained when going through that part of the code where the next instruction does not depend on the result of the previous one.

The tendency in the past has been to buy larger and larger machines, *mainframes*. These are expensive and have to be shared between several users connected to it via terminals. The user submits the job which is placed in a queue until it can be processed. Though the job may take a few hours or less to run, the user may have to wait much longer for the results.

Today the trend is to share smaller machines between fewer users and heavy users may even have a machine to themselves. Each machine is slower than a modern mainframe but one or two orders of magnitude cheaper to purchase. Such a machine, often a *work station*, delivers the results sooner: it is often of sequential type, but can also be a reduced instruction set (RISC) machine which is faster than the conventional serial one. The newest machines on the market can compete with mainframes in speed and price. The computer manufacturers are continually launching faster and cheaper machines. In the near future it may even be possible to run the Monte Carlo particle model interactively on a work station for the purpose of developing new microelectronic components.

The author's software was originally written for an IBM 350/195 which is a sequential machine. Compared to modern mainframes this machine is small; the code was written with both memory saving and processing speed in mind. Later the program was moved to the VAX. The VAX II had a speed of 0.9 mips (million instructions per second). Without any modification to the source code it has also been recompiled for the CRAY II X-MP/216 and found to run only 40 times faster, although the CRAY is supposed to deliver 100 mips. The CRAY, however, is a vector processing machine; the observed reduction in the expected speed performance was due to the number of decisions that had to be made during processing.

13.4 A SHORT HISTORY OF MONTE CARLO MODELLING

Probably the first attempt to calculate electric current flow by means of random numbers was reported by Lüthi and Wyder (1960). Kurosawa (1966) applied this technique to the transport of hot electrons. The earliest modellers, however, were only interested in the transport of particles in a uniform electric field. Boardman *et al.* (1968) calculated the drift velocity of the electrons in gallium arsenide against electric field. Lebwohl and Price (1971b) developed a hybrid model where the Monte Carlo technique was used to simulate the hot electrons, while they used the drift-diffusion model to describe the overwhelmingly larger amount of thermal electrons.

The first Monte Carlo models were known as one particle models because there was no interaction between moving charges. Only one particle was considered; this was followed long enough to gain the necessary information. The calculation was only carried out in k-space. Later the interest for entire devices developed. This has been reviewed very briefly in the introductory chapter.

During the 1970s an interdisciplinary group led by Professor Hockney in Reading (England) developed a model which aimed at self-consistent device simulation. This model, which forms the basis of the author's model, combines the transport model with an algorithm solving Poisson's equation by means of fast Fourier transforms and Green's functions (Hockney, 1965). The particle pushing code is based on a previously developed model to simulate the dynamics of a plasma of electrically charged particles (Langdon and Birdsall, 1970). The plasma was collisionless, the particles only interacting with one another through Poisson's equation. Reinterpreting this field equation as the gravitational field equation this model could successfully simulate the evolution of galaxies from a uniform disc of stars (Berman *et al.*, 1978; Brownrigg, 1975). Warriner (1977) introduced interaction with phonons so the model could be adapted to transistors. The first device simulation was published by Hockney *et al.* (1974b); the contributions from this group were published a few years later by Warriner (1977) and Moglestue and Beard (1979). Much of this work has been described by Hockney and Eastwood (1988) in their book on the computer simulation of particles.

Self-consistent Monte Carlo particle modelling like this was considered a scientific curiosity for many years because of the expensive computer time. In the middle of the 1980s the method finally started to gain popularity as mentioned in Section 1.1. The method is going to be an important research tool and will undoubtedly develop and improve by adding more physics to it.

To simulate devices which only measure a few tens of nanometres we need a fully quantum mechanical model with quantised energy levels everywhere and with scattering of finite duration. The development of such a model was started by introducing the quantum wells and tunnel barriers in Sections 10.5 and 10.4, while delayed scattering has been hinted at in Section 4.9. Fully quantum mechanical models for fixed fields are in development now and the future will certainly see such models made self-consistent with the field equation and applied to entire devices.

13.5 WHY CALL IT THE MONTE CARLO PARTICLE MODEL?

The reader may or may not have reflected on why our method is called the Monte Carlo model. This name arises from the association with the famous casino in *Monte Carlo*, Fig. 13.1, the capital city of the principality of Monaco, situated on the Mediterranean coast. Here gamblers from all over the world have won and lost fortunes by playing roulette. The *roulette* consists of a shallow bowl enclosing a rotating disc that has numbered sectors alternatively coloured red and black, and a small steel ball. (Figure 13.2.) The players bet on which sector the ball will come to rest.

To make the outcome of the game as unpredictable as possible, the wheel is set in motion by a special handle while the ball is thrown in a direction opposite to that of rotation. Mathematically seen, the roulette wheel is as perfect a random number generator as can be constructed by mechanical means. Our simulation technique also requires random numbers, hence the name; given to it by the Italian physicist Enrico Fermi.

But why is it called after Monte Carlo? What does Monte Carlo have that the gambling casinos in Las Vegas, Knokke (Belgium) or Baden Baden (Germany) do not? People have gambled with stakes since prehistoric times and have put great efforts into making the outcome of the game as unpredictable as possible to avoid quarrelling over unfair procedures. Card games and dice are also examples of such games.

Perhaps a small excursion into history will shed some light on the choice of name. The Phoenicians founded a colony in the territory of modern Monaco. The

Figure 13.1 *The entrance to the casino in Monte Carlo (Photo: J. Rosenzweig). Note, this figure is reproduced in colour as Plate 11*

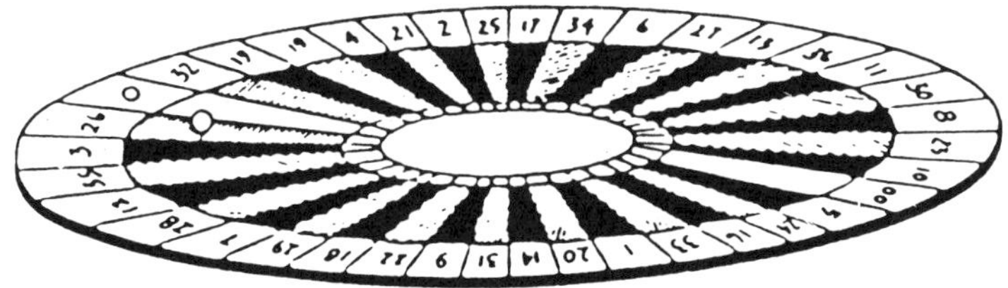

Figure 13.2 *Roulette wheel. It is placed in a shallow bowl and set in rotation by means of a handle (not shown) in the centre. A steel ball (here crossing into the sector marked 0) is thrown in the opposite direction*

harbour was important both for the navy and merchant ships, but later it lost importance. The Genoese family Grimaldi were given Monaco as a present in 980 as a reward for supporting the count of Arles in his fight against the Saracens. However, both the French kings and the rival Genoese families Doria and Spinola contested the right of the Grimaldis to the throne of Monaco.

Slowly the connection between the Grimaldis and Genoa weakened on account of the French influence. In 1641 the ruling Grimaldis were forced to accept a French garrison on their territory but were allowed to continue governing. In 1815 Monaco became a Sardinian protectorate. A popular rising in 1848 gave Sardinia the pretext to occupy the provinces of Mentane and Roccabruna, which where ceded to France in 1861. This deprived the principality of an important economic basis for its existence.

In 1857 Prince Charles III gave consent to the establishment of a bathing resort and a casino. After a few unsuccessful attempts the casino soon gained fame and earned sufficient money to support the principality.

Today, however, contrary to popular belief, the main source of revenue is derived from trade, tourism and industry, while the proceeds from the casino contribute only about 3% to the national income.

Appendix

Useful Constants

e	e	2.712
pi	π	3.14159265
Velocity of light	c	2.997×10^{8} m s^{-1}
Planck's constant	h	6.655×10^{-34} J s
Reduced Planck's constant	h	1.054×10^{-34} J s
Electronic charge	e	1.602×10^{-19} C
Free electron rest mass	m_0	9.1083×10^{-31} kg
Permittivity of vacuum	ε_0	8.85416×10^{-12} F m^{-1}
Permeability of vacuum	μ_0	1.25664×10^{-6} H m^{-1}
Boltzmann's constant	k_B	1.38044×10^{-23} J K^{-1}

References

Adachi, S. (1985). GaAs, AlAs, and $Al_xGa_{1-x}As$: material parameters for use in research and device applications. *J Appl Phys* **58**, R1–R29.

Agrawal, G.P. and Dutta, N.K. (1986). *Long-wavelength Semiconductor Lasers*, Van Nostrand, New York.

Albright, J.R. (1977). Integrals of products of Airy functions. *J Phys A: Math Gen* **10**, 485–90.

Alferov, Zh, I., Andreev, V.M., Garbuzov, D.Z. and Trukan, M.K. (1974). High-efficiency injection luminescence of an electron–hole plasma in double-heterojunction structures. *Sov Phys Semicond* **8**, 358–60.

Anastassakis, E. (1983). Ionic photoelasticity of GaAs. *J Phys C: Solid State Phys* **16**, 3329–48.

Anderson, R.W. (1967). S-parameter techniques for faster, more accurate network design. *Hewlett-Packard Journal* 18 Feb. 1967. Reprinted in Chapter 3 of *Hewlett-Packard Application Note 95: S-parameters . . . Circuit Analysis and Design*, Sept. 1968.

Ando, T., Fowler, A.B. and Stern, F. (1982). Electronic properties of two-dimensional systems. *Rev Mod Phys* **54**, 437–672.

Anwar, A.F.M., Khondker, A.N. and Khan, M.R. (1989). Calculation of the traversal time in resonant tunnelling devices. *J Appl Phys* **65**, 2761–5.

Arora, V.K. and Awad, F.G. (1981). Quantum size effect in semiconductor transport. *Phys Rev* **B23**, 5570–5.

Asche, M. and Sarbei, O.G. (1981). Electron–phonon interaction in n-Si. *Phys Stat Sol* **(b)103**, 11–50.

Aspnes, D.E., Kelso, S.M., Logan, R.A. and Bhat, R. (1986) Optical properties of $Al_xGa_{1-x}As$. *J Appl Phys* **60**, 754–67.

d'Avanzo, D.C. (1982). Proton isolation for GaAs integrated circuits. *IEEE Trans Electron Devices* **29**, 1051–9.

Azimov, S.A., Katalevskii, Yu. A., Muminov, R.A. and Rakhimova, A.G. (1980). Spatial distribution of a highly degenerate electron–hole plasma in the base of a heterostructure. *Sov Phys Semicond* **14**, 678–81.

Bächtold, W. and Strutt, M.J.O. (1968). Optimum source admittance for minimum noise figure of microwave transistors. *Electr Lett* **4**, 346–8.

Bahder, T.B. (1990). Eight-band **k•p** model of strained zinc-blende crystals. *Phys Rev* **B41**, 11992–12001.

Bandyopadhyay, S., Klausmeier-Brown, M.E., Maziar, C.M., Datta, S. and Lundstrom, M.S. (1987). A rigorous technique to couple Monte Carlo and drift-diffusion models for computationally efficient device simulation. *IEEE Trans Electron Devices* **34**, 392–9.

Bannov, N.A., Ryzhii, V.I. and Svyatchenko, A.A. (1984). Numerical modelling of high-frequency pulsations of the current in diodes with ambipolar injection and quasiballistic motion of electrons and holes. *Sov Phys Semicond* **18**, 1389–90.

Barker, J.R. (1973). Quantum transport theory of high-field conduction in semiconductors. *J Phys C: Solid State Phys* **6**, 2663–84.

Barker, J.R. and Ferry, D.K. (1979). Self-scattering path-variable formulation of high-field, time-dependent, quantum kinetic equations for semiconductor transport in the finite-collision-duration regime. *Phys Rev Lett* **26**, 1779–81.

Barnes, C. and Dunn, D.A. (1967). *Symposium on Computer Simulation of Plasma and Many-body Problems*, College of William and Mary, Williamsburg, Virginia, April 1967 (NASA SP–153).

Basu, P.K. (1977). High-field drift velocity of silicon inversion layers: a Monte Carlo calculation. *J Appl Phys* **48**, 350–3.

Beneking, H. (1982a). On the response behaviour of fast photoconductive optical planar and coaxial semiconductor detectors. *IEEE Trans Electron Devices* **29**, 1431–41.

Beneking, H. (1982b). Gain and bandwidth of fast near-infrared photodetectors: a comparison of diodes, phototransistors and photoconductive devices. *IEEE Trans Electron Devices* **29**, 1420–31.

Berman, R.H., Brownrigg, D.R.K. and Hockney, R.W. (1978). Numerical models of galaxies: I. The variability of spiral structure. *Mon Not R Astron Soc* **185**, 861–75.

Berroth, M. and Bosch, R. (1990). Broad-band determination of the FET small-signal equivalent circuit. *IEEE Trans Microwave and Techniques* **38**, 891–5.

Berroth, M., Nowotny, U., Hurm, V., Köhler, K., Kaufel, U. and Hülsmann, A. (1990). Extreme low power 1:4 demultiplexer using double delta quantum well GaAs/AlGaAs transistors. *22nd Int Conf Solid State Devices and Materials*, Sendai, Japan, August 1990.

Bertoncini, R., Kriman, A.M., Ferry, D.K., Reggiani, L., Rota, L., Poli, P. and Jauho, A.P. (1989). Position broadening effect in hot-electron transport. *Solid State Electronics* **32**, 1167–71.

Birdsall, C.K. and Fuss, D. (1969). Clouds in clouds, clouds in cells, physics for many-body plasma simulation. *J Comp Phys* **3**, 494–511.

Blakemore, J.S. (1982). Semiconducting and other major properties of gallium arsenide. *J Appl Phys* **53**, R123–R181.

Bloch, F. (1929). On the quantum mechanics of electrons in crystal lattices (in German). *Zeitschrift für Physik* **52**, 555–600.

Boardman, A.D., Fawcett, W. and Rees, H.D. (1968). Monte Carlo calculation of the velocity-field relationship for gallium arsenide. *Solid State Communications* **6**, 305–7.

Bockelmann, U. and Bastard, G. (1990). Phonon scattering and energy relaxation in two-, one- and zero-dimensional electron gases. *Phys Rev* **B42**, 8947–51.

Bosi, S. and Jacoboni, C. (1976). Monte Carlo high-field transport in degenerate GaAs. *J Phys C: Solid State Physics* **9**, 315–9.

Boudville, W.J. and McGill, T.C. (1985). Ohmic contacts to n-type GaAs. *J Vac Sci Technol* **B3**, 1192–6.

Brennan, K. and Hess, K. (1984). Theory of high field transport of holes in GaAs and InP. *Phys Rev* **B29**, 5581–90.

Brennan, K.F. and Park, D.H. (1989). Theoretical comparison of electron real-space transfer in classical and quantum two-dimensional heterostructure systems. *J Appl Phys* **65**, 1156–63.

Brennan, K.F., Park, D.H., Hess, K. and Littlejohn, M.A. (1988). Theory of the velocity-field relation in AlGaAs. *J Appl Phys* **63**, 5004–8.

Brey, L. and Tejedor, C. (1985). Effect of the electron–electron interaction on the band structure of semiconductors. *Solid State Communications* **55**, 1093–6.

Brockhouse, B.N. (1959). Lattice vibrations in silicon and germanium. *Phys Rev Lett* **2**, 256–8.

Brown, R.A. (1977). The 'golden rule' for dislocation scattering. *J Phys F: Metal Physics* **7**, L155–L158.

Brownrigg, D.R.K. (1975). *Computer Modelling of Spiral Structure in Galaxies*. PhD thesis, University of Reading.

Brust, D. (1964). Electronic spectra of crystalline germanium and silicon. *Phys Rev* **134**, A1337–53.

Buot, F.A., Anderson, W.T., Christou, A., Campbell, A.B. and Knudson, A.R. (1985). A mechanism for radiation-induced degradation in GaAs field-effect transistors. *J Appl Phys* **57**, 581–90.

Buot, F.A., Anderson, W.T., Christou, A., Sleger, K.J. and Chase, E.W. (1987). Theoretical and experimental study of subsurface burnout and ESD in GaAs FETs and HEMTs. *25th Annual Proceedings Reliability Physics*, San Diego, California, 7–9 April 1987, pp 181–90.

Buot, F.A. and Jensen, K.L. (1990). Lattice Weyl-Wigner formulation of exact many-body quantum-transport theory and applications to novel solid state quantum-based devices. *Phys Rev* **B42**, 9429–57.

Canali, C., Jacoboni, C., Nava, F., Ottaviani, G. and Alberigi-Quaranta, A. (1975). Electron drift velocity in silicon. *Phys Rev* **B12**, 2265–84.

Casey, H.C. and Panish, M.B. (1978). *Heterostructure Lasers, Part B: Materials and Operation Characteristics*, Academic Press, London.

Chandra, A., Wood, C.E.C., Woodard, D.W. and Eastman L.F. (1979). Surface and interface depletion corrections to free carrier-density determinations by Hall measurements. *Solid State Electronics* **22**, 645–50.

Chang, Y.C. and Schulman, J.N. (1982). Complex band structure of crystalline solids: an eigenvalue method. *Phys Rev* **B25**, 3975–86.

Chattopadhyay, D. and Queisser, H.J. (1981). Electron scattering by ionised impurities in semiconductors. *Rev Mod Phys* **53**, 745–68.

Chelikowsky, J.R. and Cohen, M.L. (1974). Electronic structure of silicon. *Phys Rev* **B10**, 5095–107.

Cheng, D.Y., Wu, K., Hwang, C.G. and Dutton, R.W. (1988). Drain contact boundary specification in windowed Monte Carlo device analysis. *IEEE Electr Dev Lett* **9**, 503–5.

Choi, K.K., Tsui, D.C. and Palmateer, S.C. (1986). Electron–electron interaction in GaAs-$Al_xGa_{1-x}As$ heterostructures. *Phys Rev* **B12**, 8216–27.

Christensen, N.E. (1984). Deformation potentials and internal strain parameter of silicon. *Solid State Communications* **50**, 177–80.

Chu-Hao, Zimmermann, J., Charef, M., Fauquembergue, R. and Constant, E. (1985). Monte Carlo study of two-dimensional electron gas transport in Si-MOS devices. *Solid State Electronics* **28**, 733–40.

Cohen, M.L. and Bergstresser, T.K. (1966). Band structures and pseudopotential form factors for 14 semiconductors of the diamond and zinc-blende structures. *Phys Rev* **141**, 789–96.

Cohen, M.L. and Heine, V. (1970). The fitting of pseudopotentials to experimental data and their subsequent application. *Solid State Physics* **24**, 37–248.

Collet, J.H. and Amand, T. (1984). Electron–hole interaction in the presence of excitons. *Solid State Communications* **52**, 53–6.

Collins, C.L. and Yu, P.Y. (1984). Generation of nonequilibrium optical phonons in GaAs and their application in studying intervalley electron–phonon scattering. *Phys Rev* **B30**, 4501–15.

Constant, E. (1980). Modelling of sub-micron devices. In Caroll, J.E. (ed.), *Solid State Devices: IOP Conf Series* **57**, 141–68.

Constant, E. (1985). Non-steady state carrier transport in semiconductors in perspective with sub-micrometre devices. In Reggiani, L. (ed.), *Hot-Electron Transport in Semiconductors: Topics in Applied Physics*, Springer Verlag, Berlin, **58**, 227–61.

Costa, J., Peczalski, A. and Shur, M. (1989). Monte Carlo studies of electronic transport in compensated InP. *J Appl Phys* **66**, 674–9.

Coulson, C.A. (1961). *Valence*, 2nd edn, Oxford University Press, Oxford.

Curtice, W.R. and Yun, Y.-H. (1981). A temperature model for the GaAs MESFET. *IEEE Trans Electron Devices* **28**, 954–62.

Dalacu, N. and Kitai, A.H. (1991). Semiconductor hot-electron alternating current cold cathode. *Appl Phys Lett* **58**, 613–5.

Dalal, V.L. (1970). Hole velocity in p-GaAs. *Appl Phys Lett* **12**, 489–90.

Dasgupta, S. and Sengupta, S. (1985). Homogeneous deformation theory for piezoelectric crystals. *J Phys C: Solid State Physics* **18**, 2209–15.

Das Sarma, S. and Vinter, B. (1981). Effect of impurity scattering on the distribution function in two-dimensional Fermi systems. *Phys Rev* **B24**, 549–53.

Dingle, R.B. (1955). Scattering of electrons and holes by charged donors and acceptors in semiconductors. *Phil Mag* **46**, 831–40.

Dingle, R., St Börmer, H.L., Gossard, A.C. and Wiegmann, W. (1978). Electron mobilities in modulation-doped semiconductor heterojunction superlattices. *Appl Phys Lett* **33**, 665–7.

Dunstant, D.J. (1982). Kinetics of distant-pair recombination: amorphous silicon luminescence at low temperature. *Phil Mag* **46**, 679–94.

Eastwood, J.W. and Hockney, R.W. (1974). Shaping the force law in 2D particle-mesh models. *J Comp Phys* **16**, 342–59.

Eisele, S. (1978). Stress and intersubband correlation in the silicon inversion layer. *Surface Science* **73**, 315–37.

Engl, W.L., Dirks, H.K. and Meinerzhagen, B. (1983). Device modelling. *Proc IEEE* **71**, 10–33.

Erginsoy, C. (1950). Neutral impurity scattering in semiconductors. *Phys Rev* **79**, 1013–14.

Ezawa, H., Kawaji, S. and Nakamura, K. (1974). Surfons and the electron mobility in silicon inversion layers. *Jap J Appl Phys* **13**, 126–55. Errata, *ibid* **13**, 921–2.

Fauquembergue, R., Zimmermann, J., Kaszynski, A., Constant, E. and Microondes, G. (1980). Diffusion and the power spectral density and correlation function of velocity fluctuation for electrons in Si and GaAs by Monte Carlo methods. *J Appl Phys* **51**, 1065–71.

Fawcett, W., Boardman, A.D. and Swain, S. (1970). Monte Carlo determination of electron transport properties in gallium arsenide. *J Phys Chem Sol* **31**, 1963–90.

Fedders, P.A. (1983). Strain scattering of electrons in piezoelectric semiconductors. *J Appl Phys* **54**, 1804–7.

Feenstra, R.M., Stroscio, J.A., Tersoff, J. and Fein, A.P. (1987). Atom-selective imaging of the GaAs (110) surface. *Phys Rev Lett* **58**, 1192–5.

Feller, W. (1961). *An Introduction to Probability Theory and its Applications.* Vol. 1, John Wiley, New York.

Feller, W. (1971). *An Introduction to Probability Theory and its Applications.* Vol. 2, 2nd edn, John Wiley, New York.

Ferry, D.K. (1976a). First order optical and intervalley scattering in semiconductors. *Phys Rev* **B14**, 1605–9.

Ferry, D.K. (1976b). Optical and intervalley scattering in quantised inversion layers in semiconductors. *Surface Science* **57**, 218–28.

Ferry, D.K. (1976c). Hot-electron effects in silicon quantised inversion layers. *Phys Rev* **B14**, 5364–71.

Ferry, D.K. (1978). Scattering by polar-optical phonons in a quasi-two-dimensional semiconductor. *Surface Science* **75**, 86–91.

Ferry, D.K. and Barker, J.R. (1979). Balance equations for high-field transport in the finite collision duration regime. *Solid State Communications* **30**, 361–3.

Fischetti, M.V. (1991). Monte Carlo simulation of transport in technologically significant semiconductors of the diamond and zinc-blende structures. Part I: Homogeneous transport. *IEEE Trans Electron Devices* **38**, 634–49.

Fischetti, M.V. and Higman, J.M. (1991). Theory and calculation of the deformation potential electron–phonon scattering rates in semiconductors. In Hess, K. (ed.) *Monte Carlo Device Simulation, Full Band and Beyond* (to be published).

Fischetti, M.V. and Laux, S.E. (1988). Monte Carlo analysis of electron transport in small semiconductor devices including band-structure and space-charge effects. *Phys Rev* **B38**, 9721–45.

Foulon, Y., Priester, C., Allan, G. and Lannoo, M. (1990). Strained heterojunctions: band-offsets and effective masses, a theoretical approach. In Anastassakis, E.M. and Joannopoulos, J.D. (eds) *20th Internat Conf on The Physics of Semiconductors*, World Scientific, Singapore, 977–80.

Fowler, R.H. and Nordheim, L. (1928). Electron emission in intense electric fields. *Proc Royal Society (London)* **119**, 173–81.

Fukui, H. (1979). Optimal noise figure of microwave GaAs MESFET's. *IEEE Trans Electron Devices* **26**, 1032–7.

Ganguly, A.K., Hui, B.H. and Chu, K.R. (1982). Nonlinear analysis of the solid-state gyrotron oscillator by the Monte Carlo method. *IEEE Trans Electron Devices* **29**, 1197–1209.

Giner, C.T. and Comas, F. (1988). Electron–LO-phonon interaction in semiconductor double heterostructures. *Phys Rev* **B37**, 4583–8.

Goldstein, H. (1959). *Classical Mechanics*, Addison-Wesley, Reading, Massachusetts.

Gram, N.O. and Jørgensen, H.M. (1973). Influence of quadrupole and octupole electron–phonon coupling on the low-field transport properties of n-type silicon. *Phys Rev* **B8**, 3902–7.

Gray, H.F. (1989a). Vacuum microelectronics: the electronics of the 21st century. *Electronics Show and Convention April 11–13, 1989 at the Jacob K. Javits Convention Centre*, New York Paper 8, Nanoelectronics.

Gray, H.F. (1989b). Field-emitter arrays: more than a scientific curiosity? *Colloque de Physique* **50**, Supplément, C8, 67–72.

Gray, H.F. (1989c). Vacuum microelectronics: it promises subpicosecond switching and terahertz amplification. In Pierro, J. (ed.), *Proceedings of the New York Long Island MTT Chapter Symposium on Emerging Technologies*.

Gray, H.F., Campisi, G.J. and Greene, R.F. (1986). A vacuum field-effect transistor using silicon field-emitter arrays. *IEDM 86*, 776–9.

Gribnikov, Z.S. and Zheleznyak, V.B. (1991). Two-dimensional Gunn domains in structures with a gate electrode. *Sov Phys Semicond* **25**, 9–15.

Gružinskis, V., Kersulis, S. and Reklaitis, A. (1991). An efficient Monte Carlo particle technique for two-dimensional transistor modelling. *Semicond Sci Technol* **6**, 602–6.

Gubernatis, J.E. (1987). Quantum Monte Carlo simulations of many-body phenomena in a single-impurity Anderson model. *Phys Rev* **B36**, 394–400.

Guckel, H., Demirkol, A., Thomas, D. and Iyengar, S. (1982). The forward biased, abrupt p–n junction. *Solid State Electronics* **25**, 105–13.

Guha, S. and Ghosh, S. (1979). Longitudinal phonon–plasmon interaction in inhomogeneous semiconductors. *Phys Stat Sol* **(b)92**, K95–K97.

Hackenberg, W. and Fasol, G. (1989). Determination of the LO-phonon and $\Gamma \rightarrow L$ intervalley scattering time in GaAs from hot electron luminescence spectroscopy. *Solid State Electronics* **32**, 1247–51.

Hall, A. (1873). On the experimental determination of π. *Messeng Math* **2**, 113–14.

Hammersley, J.M. and Handscomb, D.C. (1964). *Monte Carlo Methods*. Chapman and Hall, London.

Harrison, J.W. and Hauser, J.R. (1976). Alloy scattering in ternary III–V compounds. *Phys Rev* **B13**, 5347–50.

Harrison, W.A. (1980). *Electronic Structure and the Properties of Solids. The Physics of the Chemical Bond.* Freeman and Company, San Francisco.

Herman, F. (1959). Lattice vibrational spectrum of germanium. *J Phys Chem Sol* **8**, 405–18.

Herring, C. and Vogt, E. (1956). Transport and deformation-potential theory for many-valley semiconductors with anisotropic scattering. *Phys Rev* **101**, 944–61.

Herzog, M., Schels, M., Koch, F., Moglestue, C. and Rosenzweig J. (1989). Electromagnetic radiation from hot carriers in FET devices. *Solid State Electronics* **32**, 1065–9.

Hess, K. (1979). Impurity and phonon scattering in layered structures. *Appl Phys Lett* **35**, 484–6.

Hess, K. and Vogl, P. (1979). Remote polar phonon scattering in silicon inversion layer. *Solid State Communications* **30**, 807–9.

Hesto, F., Pelouard, J.-L., Castagné, R. and Pône, J.-F. (1984). Nonelastic acoustic-phonon–electron interactions in Monte Carlo simulations at low fields. *Appl Phys Lett* **45**, 641–3.

Higman, J.M., Hess, K., Hwang, C.G. and Dutton, R.W. (1989). Coupled Monte Carlo drift-diffusion analysis of hot electron effects in MOSFET's. *IEEE Trans Electron Devices* **36**, 930–7.

Hirakawa, K. and Sakaki, H. (1988). Hot-electron transport in selectively doped n-type AlGaAs/GaAs heterojunctions. *J Appl Phys* **63**, 803–8.

Hockney, R.W. (1965). A fast direct solution of Poisson's equation using Fourier analysis. *J Assoc Comput Machinery* **12**, 95–113.

Hockney, R.W. (1968). Formation and stability of virtual electrodes in a cylinder. *J Appl Phys* **39**, 4166–70.

Hockney, R.W. (1971). Measurements of collision and heating times in a two-dimensional thermal computer plasma. *J Comp Phys* **8**, 19–44.

Hockney, R.W. (1980). Rapid elliptic solvers. In Hunt, B. (ed.), *Numerical Methods in Applied Fluid Dynamics*, Academic Press, London, 1–48.

Hockney, R.W. and Eastwood, J.W. (1988). *Computer Simulation Using Particles*, Adam Hilger, Bristol.

Hockney, R.W., Goel, S.P. and Eastwood, J.W. (1974a). Quiet high resolution computer models of a plasma. *J Comp Physics* **14**, 148–58.

Hockney, R.W., Warriner, R.A. and Reiser, M. (1974b). Two-dimensional models in semiconductor device analysis. *Electr Lett* **10**, 484–6.

Houston, P.A. and Evans, A.G.R. (1977). Electron drift velocity in n-GaAs at high electric fields. *Solid State Electronics* **20**, 197–204.

Hu, J., Tomizawa, K. and Pavlidis, D. (1989). Monte Carlo approach to transient analysis of HBT's with different collector designs. *IEEE Electron Device Letters* **10**, 55–7.

Hull, T.E. and Dobell, A.R. (1962). Random number generators. *Soc Indust Appl Math Rev* **4**, 230–54.

Ifrah, G. (1981). *Histoire Universelle des Chiffres*, Seghers, Paris.

Ikarashi, N., Sakai, A., Baba, T., Ishida, K., Motohisa, J. and Sakaki, H. (1990). High-resolution transmission electron microscopy of GaAs/AlAs heterointerfaces grown on the misoriented substrate in the ⟨110⟩ projection. *Appl Phys Lett* **57**, 1983–5.

Inoue, M. and Frey, J. (1980). Electron–electron interaction and screening effects in hot electron transport in GaAs. *J Appl Phys* **51**, 4234–9.

Inoue, M., Takenaka, N., Shirafuji, J. and Inuishi, Y. (1978). Hot carrier recombination radiation in GaAs diode. *Solid State Electronics* **21**, 1527–30.

Jacoboni, C. and Lugli, P. (1989). *The Monte Carlo Method for Semiconductor Device Simulation*, Springer Verlag, Vienna.

Jacoboni, C. and Reggiani, L. (1979). Bulk hot-electron properties of cubic semiconductors. *Advances in Physics* **28**, 493–553.

Jacoboni, C. and Reggiani, L. (1983). The Monte Carlo method for the solution of charge transport in semiconductors with application to covalent materials. *Rev Mod Phys* **55**, 645–705.

Jensen, K.L. and Buot, F.A. (1991). The methology of simulating particle trajectories through tunnelling structures using a Wigner distribution approach. *IEEE Trans Electron Devices* **38**, 2337–47.

Jørgensen, M.H. (1978). Electron–phonon scattering and high-field transport in n-type Si. *Phys Rev* **B18**, 5657–66.

Kaminska, M., Liliental-Weber, Z., Weber, E.R. and George, T. (1989). Structural properties of As-rich GaAs grown by molecular beam epitaxy at low temperatures. *Appl Phys Lett* **54**, 1881–3.

Kane, E.O. (1956). Energy band structure in p-type germanium and silicon. *J Phys Chem Sol* **1**, 82–99.

Kane, E.O. (1957). Band structure of indium antimonide. *J Phys Chem Sol* **1**, 249–61.

Kane, E.O. (1969). Basic concepts of tunnelling. In Burstein, E. and Lundqvist, S. (eds), *Tunnelling Phenomena in Solids*, Plenum Press, New York.

Kaneko, A., Kanno, T., Tomii, K., Kitagawa, M. and Hirao, T. (1991). Wedge-shaped field-emitter arrays for flat display. *IEEE Trans Electron Devices* **38**, 2395–7.

Kato, K. (1988). Hot-carrier simulation for MOSFET's using a high-speed Monte Carlo method. *IEEE Trans Electron Devices* **35**, 1344–50.

Katoh, R. and Kurata, M. (1989). Self-consistent particle simulation for (AlGa)As/GaAs HBT's under high-bias conditions. *IEEE Trans Electron Devices* **36**, 2122–8.

Katoh, R., Kurata, M. and Yoshida, J. (1989). Self-consistent particle simulation for (AlGa)As/GaAs HBT's with improved base-collector structures. *IEEE Trans Electron Devices* **36**, 846–53.

Katsnel'sson M.I. and Sadovskii, M.V. (1983). Electron–electron interaction in a self-consistent localisation theory. *Sov Phys Solid State* **25**, 1942–7.

Kaveh, M. and Wiser, N. (1984). Electron–electron scattering in conducting materials. *Adv Phys* **33**, 257–372.

Kay, L.E. and Tang, T.W. (1991a). An improved ionised-impurity scattering model for Monte Carlo simulations. *J Appl Phys* **70**, 1475–82.

Kay, L.E. and Tang, T.W. (1991b). Monte Carlo calculation of strained and unstrained electron mobilities in $Si_{1-x}Ge_x$ using an improved ionised-impurity model. *J Appl Phys* **70**, 1483–8.

Kayanuma, Y. and Fukuchi, S. (1982). Electron correlation in a pair of impurities in semiconductors with deformable lattice. *J Phys Soc Japan* **51**, 164–71.

Kelly, M.S. and Hanke, W. (1981). Electron-phonon interaction at a silicon surface. *Phys Rev* **B23**, 924–7.

Kim, K.J., Harmon, B.N., Chen L.Y. and Lynch, D.W. (1990). Optical properties and electronic structures of intermetallic components $AuGa_2$ and $PtGa_2$. *Phys Rev* **B42**, 8813–19.

Kim, H., Min, H.S., Tang, T.W. and Park, Y.J. (1991). An extended proof of the Ramo-Shockley theorem. *Solid State Electronics* **34**, 1251–3.

Kittel, C. (1961). *Introduction to Solid State Physics*, 2nd edn, John Wiley, New York.

Kittel, C. (1971). *Introduction to Solid State Physics*, 4th edn, John Wiley, New York.

Kizilyalli, I.C. and Hess, K. (1987). Ensemble Monte Carlo simulation of a velocity-modulation field-effect transistor (VMT). *Jap J Appl Phys* **26**, 1519–24.

Kizilyalli, I.C., Hess, K. and Iafrate, G.J. (1987). Electron transfer between adjacent channels simulated by ensemble Monte Carlo methods. *J Appl Phys* **61**, 2395–8.

Klingenstein, M., Kuhl, J., Rosenzweig, J., Moglestue, C. and Axmann, A. (1991). Transit time limited response of GaAs metal-semiconductor-metal photodiodes. *Appl Phys Lett* **58**, 2503–5.

Koscielniak, W.C., Pelouard, J.L. and Littlejohn, M.A. (1989). Dynamic behaviour of photocarriers in GaAs metal-semiconductor-metal photodetector with sub-half-micron electrode pattern. *Appl Phys Lett* **54**, 567–9.

Koscielniak, W.C., Pelouard, J.L. and Littlejohn, M.A. (1990). Intrinsic and extrinsic response of GaAs metal-semiconductor-metal photodetectors *IEEE Photonics Tech Lett* **2**, 125–7.

Kosina, H., Lindorfer, P. and Selberherr, S. (1991). Monte Carlo–Poisson coupling using transport coefficients. *Microelectron Eng* **15**, 53–6.

Kotel'nikov, Y.E. (1983). Plasmon-phonon oscillations in molecular crystals. *Sov Phys Solid State* **25**, 1930–2.

Kratzer, S. and Frey, J. (1978). Transient velocity characteristics of electrons in GaAs with Γ–L–X conduction band ordering. *J Appl Phys* **49**, 4064–8.

Krowne, C.M. (1983). Electron power loss in the (100) n channel of a Si metal-oxide-semiconductor field-effect transistor: I. intrasubband phonon scattering. *J Appl Phys* **54**, 2441–54.

Krowne, C.M. (1987). Analytical method for selecting random scattering angle for polar or acoustic phonon scattering of central valley conduction band electrons in GaAs. *IEE Proc Part I: Solid State and Electron Devices* **134**, 45–9.

Kuan, T.S., Batson, P.E., Jackson, T.N., Rupprecht, H. and Wilkie, E.L. (1983). Electron microscope studies of an alloyed Au/Ni/Au-Ge Ohmic contact to GaAs. *J Appl Phys* **54**, 6952–7.

Kubo, R. and Nagamiya, T. (eds). (1969). *Solid State Physics*, McGraw-Hill, New York.

Kurosawa, T. (1966). Monte Carlo calculation of hot electron problems. *J Phys Soc Japan* Suppl. **21**, 424–6.

Lambsdorff, M. (1990). Subpicosecond photoconductivity in ion irradiated Si and GaAs: theory and applications. Thesis, Max Planck Institute of Research into Solid State Matter (in German).

Lambsdorff, M., Klingenstein, M., Kuhl, J., Moglestue, C., Rosenzweig, J., Axmann, A., Schneider, J., Hülsmann, A., Leier, M. and Forchel, A. (1991). Subpicosecond characterization of carrier transport in GaAs-metal-semiconductor-metal photodiodes. *Appl Phys Lett* **58**, 1410–12.

Landau, L.D. and Lifshitz, E.M. (1959). *Statistical Physics: Course on Theoretical Physics*, Vol. 5, Pergamon Press, London.

Landolt and Börnstein, (1982). *Numerical Data and Functional Relationships in Science and Technology: Group III: Crystal and Solid State Physics. Vol. 17: Semiconductors*, (new series), Hellwege, K.H. (ed. in chief), Springer Verlag, Berlin.

Langdon, A.B. and Birdsall, C.K. (1970). Theory of plasma simulation using finite-size particles. *Phys of Fluids* **13**, 2115–22.

Lassnig, R. (1988). Remote ion scattering in GaAs–GaAlAs heterostructures. *Solid State Communications* **65**, 765–8.

Lax, M. and Birman, J.L. (1972). Intervalley scattering selection rules for Si and Ge. *Phys Stat Sol* **(b)49**, K153–K154.

Lax, M. and Hopfield, J.J. (1961). Selection rules connecting different points in the Brillouin zone. *Phys Rev* **124**, 115–23.

Leburton, J.P. (1984). Size effects on polar optical phonon scattering of 1-D and 2-D electron gas in synthetic semiconductors. *J Appl Phys* **56**, 2850–5.

Lebwohl, P.A. and Price, P.J. (1971a). Direct microscopic simulation of Gunn-domain phenomena. *Appl Phys Lett* **19**, 530–2.

Lebwohl, P.A. and Price, P.J. (1971b). Hybrid method for hot electron calculations. *Solid State Communications* **9**, 1221–4.

Lee, H.J. and Look, D.C. (1983). Hole transport in pure and doped GaAs. *J Appl Phys* **54**, 4446–52.

Lenard, A. (1960). On Bogoliubov's kinetic equation for spatially homogeneous plasma. *Annnals of Phys* **10**, 390–400.

Leslie, P.H. and Chitty, D. (1951). The estimation of population parameters from data obtained by means of the capture–recapture method: I. The maximum likelihood equations for estimating the death rate. *Biometrica* **38**, 269–92.

Littlejohn, M.A., Hauser, J.R., Glisson, T.H., Ferry, D.K. and Harrison J.W. (1978). Alloy scattering and high-field transport in ternary and quaternary III–V semiconductors. *Solid State Electronics* **21**, 107–14.

Long, A.P., Beton, P.H. and Kelly, M.J. (1987). Hot electron transport in $In_{0.53}Ga_{0.47}As$. *J Appl Phys* **62**, 1842–9.

Look, D.C., Stutz, C.E. and Evans, K.R. (1990). Unpinning of GaAs surface Fermi level by 200 °C molecular beam epitaxial layer. *Appl Phys Lett* **57**, 2570–2.

Löwdin, P.O. (1950). On the non-orthogonality problem connected with the use of atomic wave function in the theory of molecules and crystals. *J Chem Phys* **18**, 365–75.

Lowe, D. (1985). On the construction of transition rates in high-field quantum transport theories. *J Phys C: Solid State Phys* **18**, L209–L213.

Lugli, P. and Ferry, D.K. (1985). Electron–electron interaction and high-field transport in Si. *Appl Phys Lett* **46**, 594–6.

Lugli, P., Reggiani, L. and Jacoboni, C. (1986). Collision broadening and energy tails of hot-electron distribution function. *Superlattices and Microstructures* **2**, 143–6.

Lüthi, B. and Wyder, P. (1960) A Monte Carlo calculation for a size effect problem. *Helv Phys Acta* **33**, 667–74.

Madelung, O. (1972). *Festkörpertheorie*, Springer Verlag, Berlin.

Mahan, G.D. (1965). Temperature dependence of the band gap in CdTe. *J Phys Chem Sol* **26**, 751–6.

Mahan, G.D. (1972). *Polarons in Ionic Crystals and Polar Semiconductors*, Devreese, J.T. (ed.), North Holland, Amsterdam, 553–657.

Maksym, P.A. (1982). A theoretical study of hot photoexcited electrons in GaAs. *J Phys C: Solid State Phys* **15**, 3127–40.

Martin, G.M., Mitonneau, A. and Mircea, A. (1977). Electron traps in bulk and epitaxial GaAs crystals. *Electr Lett* **13**, 191–3.

Mašovič, D.R. and Vukajlović, F.R. (1983). Band structure calculations of cubic metals, elementary semiconductors and semiconductor compounds with spin-orbit interaction. *Comp Phys Comm* **30**, 207–17.

McLellan, A.G. (1991). The quantum theory of elastic constants. *J Phys: Condens Matter* **3**, 2247–62.

Merzbacher, E. (1961). *Quantum Mechanics*, John Wiley, New York.

Mickevicius, R. and Reklaitis, A. (1987). Monte Carlo study of nonequilibrium phonon effects in GaAs. *Solid State Communications* **64**, 1305–8.

Milman, V.Y., Antonov, V.N. and Nemoshkalenko, V.V. (1990). Relativistic pseudopotentials for transition metals: I. construction and application to atoms. *J Phys: Condens Matter* **2**, 7101–14.

Mirlin, D.N., Karlik, J.Y. and Sapega, V.F. (1988). Intervalley Γ–X scattering rate in gallium arsenide crystals. *Solid State Communications* **65**, 171–2.

Moglestue, C. (1981a). A Monte Carlo particle model study of the influence of the doping profiles on the characteristics of field-effect transistors. In Browne, B.T. and Miller, J.J.H. (eds), *Numerical Analysis of Semiconductor Devices and Integrated Circuits*, Boole Press, Dublin, 244–8.

Moglestue, C. (1981b). Monte Carlo particle modelling of local heating in n-type GaAs FET. *IEE Proc I: Solid State and Electron Devices* **128**, 131–3.

Moglestue, C. (1982). A Monte Carlo particle model and its applications to a study of the influence of the doping profiles on the characteristics of field-effect transistors. *COMPEL* **1**, 7–36.

Moglestue, C. (1983a). Carrier mobility at interfaces. In Miller, J.J.H. (ed.) Proc. NASECODE III 198–203.

Moglestue, C. (1983b). Negative differential resistivity in field-effect transistors. *IEE Proc I: Solid State and Electron Devices* **130**, 275–80.

Moglestue, C. (1983c). Monte Carlo particle modelling of noise in semiconductors. In Savelli, M., Lecoy, G. and Nougier, J.P. (eds), *Noise in Physical Systems and 1/f Noise*, Elsevier, Amsterdam, 23–5.

Moglestue, C. (1984a). Monte Carlo particle model study of the influence of gate metallisation and gate geometry on the AC characteristics of GaAs MESFETs. *IEE Proc I: Solid State and Electron Devices* **131**, 193–202.

Moglestue, C. (1984b). Monte Carlo particle model study of a microwave photodetector. *IEE Proc. I: Solid State and Electron Devices* **131**, 103–6.

Moglestue, C. (1984c). A Monte Carlo particle study of a semiconductor responding to a light pulse. In Board, K. and Oven, D.R.J. (eds), *Simulation of Semiconductor Devices and Processes*, Pineridge Press, Swansea, 153–63.

Moglestue, C. (1985). A Monte Carlo particle study of the intrinsic noise figure in GaAs MESFETs *IEEE Trans Computer Aided Design* **4**, 536–40. Also in *IEEE Trans Electron Devices* **32**, 2092–6.

Moglestue, C. (1986a). Hot, tepid and temperate electrons in bulk GaAs. *IEE Proc I: Solid State and Electron Devices* **133**, 35–46.

Moglestue, C. (1986b). Self-consistent calculation of electron and hole inversion charges at silicon–silicon dioxide interfaces. *J Appl Phys* **59**, 3175–83.

Moglestue, C. (1986c). Monte Carlo particle simulation of the hole–electron plasma formed in a p–n junction. *Electr Lett* **22**, 397–8.

Moglestue, C. (1990). Monte Carlo particle study of transport along the hetero-interface in a field-effect transistor. *Euro Microwave J* **1**, 439–46.

Moglestue, C. and Beard, S.J. (1979). A particle model simulation of field effect transistor. In Browne, B.T. and Miller, J.J.H. (eds), *Numerical Analysis of Semiconductor Devices*, Boole Press, Dublin, 232–6.

Moglestue, C., Rosenzweig, J., Kuhl J., Klingenstein, M., Lambsdorff, M., Axmann, A., Schneider, J. and Hülsmann, A. (1991a). Picosecond pulse response characteristics of GaAs metal–semiconductor–metal photodetectors. *J Appl Phys* **70**, 2435–48.

Moglestue, C., Buot, F.A. and Anderson, W.T. (1991b). Monte Carlo particle simulation of radiation-induced heating in GaAs field-effect transistors. *Appl Phys Lett* **59**, 192–4.

Moglestue, C., Rosenzweig, J., Axmann, A., Schneider, J., Lambsdorff, M., Kuhl, J., Klingenstein, M., Leier, H. and Forchel, A. (1990). Monte Carlo particle calculation and direct observation of the electron and hole contribution to the response of a GaAs metal–semiconductor–metal Schottky diode to a short light pulse. In Anastassakis, E.M. and Joannopoulos, J.D. (eds), *20th International Conference on the Physics of Semiconductors*, World Scientific, Singapore, 407–10.

Mon, K.K., Hess, K. and Dow, J.D. (1981). Deformation potentials of superlattices and interfaces. *J Vac Sci Technol* **19**, 564–6.

Moore, B.T. and Ferry, D.K. (1980a). Scattering of inversion layer electrons by oxide polar mode generated interface phonons. *J Vac Sci Technol* **17**, 1037–40.

Moore, B.T. and Ferry, D.K. (1980b). Remote polar phonon scattering in Si inversion layers. *J Appl Phys* **51**, 2603–5.

Nedjalkov, M. and Vitanov, P. (1989). Iteration approach for solving the Boltzmann equation with the Monte Carlo method. *Solid State Electronics* **10**, 893–6.

Nedjalkov, M. and Vitanov, P. (1990). Application of the iteration approach to the ensemble Monte Carlo technique. *Solid State Electronics* **33**, 407–10.

Nedjalkov, M. and Vitanov, P. (1991). Monte Carlo technique for simulation of high energy electrons. In Miller, J.J.H. (ed.), *NASECODE VII: Transactions*, Boole Press and James & James Science Publisher's, Dublin and London, 525–31.

Neidert, E.R., Philips, P.M., Smith, S.T. and Spindt, C.A. (1991). Field-emission triodes. *IEEE Trans Electron Devices* **38**, 661–5.

von Neumann, J. (1951) Various techniques used in connection with random digits. *National Bureau of Standards, Applied Mathematics Series* **12**, 36–8.

Nougier, J.P., Vaissière, J.C. and Gasquet, D. (1981). Duration of collision in semiconductors. *J de Phys Supp* **42**, C7-283–C7-292.

Ong T.C., Terrill, K.W., Tam, S. and Hu, C. (1983). Photon generation in forward-biased silicon p–n junctions. *IEEE Electr Dev Lett* **4**, 460–2.

Onga, S., Yoshio, T., Hatanaka, K. and Yasuda, Y. (1976). Effects of crystalline defects on electrical properties in silicon films on sapphire. *Jap Appl Phys Supp* **V15**, 225–31.

Osman, M.A. and Ferry, D.K. (1987). Electron–hole interaction and high-field transport of photo-excited electrons in GaAs. J Appl Phys **61**, 5330–6.

Ottaviani, G., Reggiani, L., Canali, C., Nava, F. and Alberigi-Quaranta, A. (1975). Hole drift in silicon. *Phys Rev* **B**, 3318–29.

Park, D.H. and Brennan, K. (1990). Monte Carlo simulation of 0.35-μm gate-length GaAs and InGaAs HEMT's. *IEEE Trans Electron Devices* **37**, 618–28.

Park, Y.J., Navon, D.H. and Tang, T.W. (1984). Monte Carlo simulation of bipolar transistors. *IEEE Trans Electron Devices* **31**, 1724–30.

Pelouard, J.L., Hesto, P. and Castagné, R. (1988). Monte Carlo study of the double heterojunction bipolar transistor. *Solid State Electronics* **31**, 333–6.

Pfeiffer, L., West, K.W., Stormer, H.L. and Baldwin, K.W. (1989). Electron mobilities exceeding 10^7 cm^2/Vs in modulation-doped GaAs. *Appl Phys Lett* **55**, 1888–90.

Phillips, A. and Price, P.J. (1977). Monte Carlo calculations on hot electron energy tails. *Appl Phys Lett* **30**, 528–30.

Phillips, J.C. and Kleinman, L. (1959). New method for calculating wave functions in crystals and molecules. *Phys Rev* **116**, 287–94.

Poli, P., Rota, L. and Jacoboni, C. (1989) Weighted Monte Carlo for electron transport in semiconductors. *Appl Phys Lett* **55**, 1026–8.

Pollak, E. and Miller, W.H. (1984). New physical interpretation for time in scattering theory. *Phys Rev Lett* **2**, 115–18.

Požela, J. and Reklaitis, A. (1980). Electron transport properties in GaAs at high electric fields. *Solid State Electronics* **23**, 927–33.

Price, P.J. (1973). Transport properties of the semiconductor superlattice. *IBM J Res and Dev* **Jan. 1973**, 39–46.

Price, P.J. (1981). Monte Carlo calculation of electron transport in solids. *Semiconductors and Semimetals* **14**, 249–308.

Pucel, R.A., Hous, H.A. and Statz H. (1975). Signal and noise properties of gallium arsenide microwave field-effect transistors. *Adv Electr Electron Phys* **38**, 195–265.

Rees, H.D. (1968). Calculation of steady state distribution function by exploiting stability. *Phys Lett* **26A**, 416–7.

Reggiani, L., Vaissière, J.C., Nougier, J.P. and Gasquet, D. (1981). *J de Phys Supp* **42**, C7-357–C7-367.

Reggiani, L., Kuhn, T. and Varani, L. (1990). A Monte Carlo simulator including generation recombination processes. In Eccleston, W. and Rosser, P.J. (eds), *Proc 20th European Solid State Device Research Conf (ESSDERC 90)*, 489–92.

Resta, R. (1977). Thomas–Fermi dielectric screening in semiconductors. *Phys Rev* **B16**, 2717–22.

Rhoderick, E.H. (1978). *Metal–Semiconductor Contacts*, Clarendon Press, Oxford.

Riddoch, F.A. and Ridley, B.K. (1983). On the scattering of electrons by polar optical phonons in quasi-2D quantum wells. *J Phys C: Solid State Physics* **16**, 6971–82.

Ridley, B.K. (1982). The electron–phonon interaction in quasi-two-dimensional semiconductor quantum well structures. *J Phys C: Solid State Physics* **15**, 5899–917.

Rieger, M. and Vogl, P. (1989). New lattice gas method for semiconductor transport simulations. *Solid State Electronics* **32**, 1399–403.

Rigand, A., Nicolet, A. and Savelli, M. (1973). Noise calculation by the impedance-field method: application to single injection. *Phys Stat Sol* **18a**, 531–43.

Rockett, P.I. (1987). Simple technique to improve the computational efficiency of Monte Carlo carrier transport simulations in semiconductors. *IEE Proc I: Solid State and Electron Devices* **134**, 101–4.

Rockett, P.I. (1988). Monte Carlo study of the influence of collector region velocity overshoot on the high-frequency performance of AlGaAs/GaAs heterojunction bipolar transistors. *IEEE Trans Electron Devices* **35**, 1573–9.

van de Roer, T.G. and Widdershoven, F.P. (1986). Ionised impurity scattering in Monte Carlo calculations. *J Appl Phys* **59**, 813–15.

Röpke, G. and Höhne, F.E. (1981). Quantum transport theory of impurity and electron–electron scattering resistivity in n-type semiconductors using the correlation function method. *Phys Stat Sol* **(b)107**, 603–15.

Rosenzweig, J., Moglestue, C., Axmann, A., Schneider, J., Hülsmann, A., Lambsdorff, M., Kuhl, J., Klingenstein, M., Leier, H. and Forchel, A. (1991). Characterisation of picosecond GaAs metal–semiconductor–metal photodetectors. *Physical Concept of Materials for Novel Optoelectronic Device Applications II: Device Physics and Applications, SPIE Proceedings*, **1362**, 168–78.

Rössler, U. (1984). Nonparabolicity and warping in the conduction band of GaAs. *Solid State Communications* **49**, 943–7.

Rota L., Jacoboni, C. and Poli, P. (1989). Weighted ensemble Monte Carlo. *Solid State Electronics* **32**, 1417–21.

Ruch, J.G. and Kino, G.S. (1968). Transport properties of GaAs. *Phys Rev* **174**, 921–31.

Sanghera, G.S., Chryssafis, A. and Moglestue, C. (1980). Monte Carlo particle simulation of n-type GaAs field-effect transistors with p-type buffer layer. *IEE Proc I: Solid State and Electron Devices* **127**, 203–6.

Sano, N., Aoki, T., Tomizawa, M. and Yoshii, A. (1990). Electron transport and impact ionisation in Si. *Phys Rev* **B41**, 12122-8.

Seeger, K. (1982). *Semiconductor Physics. An Introduction*. Springer Series in Solid State Sciences, Vol. 40, Springer Verlag, Berlin.

Selberherr, S. (1984). Process and device modelling for VLSI. *Microelectron Reliab* **24**, 225-57.

Shichijo, H. and Hess, K. (1981). Band-structure-dependent transport and impact ionisation in GaAs. *Phys Rev* **B23**, 4197-207.

Shur, M. (1990). *Physics of Semiconductor Devices*, Prentice Hall, Englewood Cliffs, New Jersey.

Siggia, E.D. and Kwok, P.C. (1970). Properties of electrons in semiconductor inversion layers with many occupied electric subbands. I. Screening and impurity scattering. *Phys Rev* **B2**, 1024-36.

Snowden, C.M. (1989). Classical and semiclassical models. In Snowden, C.M. (ed.) *Semiconductor Device Modelling*, Springer Verlag, Berlin, 16-32.

Stern, F. (1972). Self-consistent results for n-type Si inversion layers. *Phys Rev* **B5**, 4891-9.

Strauch, D. and Dorner, B. (1990). Phonon dispersion in GaAs. *J Phys Condensed Matter* **2**, 1457-74.

Streitwolf, H.W. (1969). On space group selection rules. *Phys Stat Sol* **33**, 217-23.

Streitwolf, H.W. (1970). Intervalley scattering selection rules for Si and Ge. *Phys Stat Sol* **37**, K47-K49.

Sze, S.M. (1969). *Physics of Semiconductor Devices*, Wiley Interscience, New York.

Tajima, Y. and Shibata, K. (1978). Noise analysis of GaAs FETs with negative mobility. *Trans IECE Japan* **E61**, 585-92.

Takahashi, T. (1990). High-T_c superconductors as heavily doped semiconductors studied by photoemission and X-ray absorption. In Anastassakis, E.M. and Joannopoulos, J.D. (eds), *20th Int Conf on the Physics of Semiconductors*, World Scientific, Singapore, 2189-96.

Takenaka, N., Inoue, M. and Inuishi, Y. (1979). Influence of inter carrier scattering on hot-electron distribution function in GaAs. *J Phys Soc Japan* **47**, 861-8.

Takenaka, N., Inoue, M. and Inuishi, Y. (1980). Analysis at hot electron transport using experimentally obtained distribution functions. *J Phys Soc Japan, Suppl* **A49**, 325-8.

Tanaka, S., Kawata, T., Hakimoto, T., Kobayashi, H. and Saito, H. (1983). Picosecond dynamics of optical gain due to electron-hole plasma in GaAs under near band-gap excitation. *J Phys Soc Japan* **52**, 677-85.

Terashima, K., Hamaguchi, C. and Taniguchi, K. (1985). Monte Carlo simulation of two-dimensional hot electrons in n-type Si inversion layers. *Superlattices and Microstructures* **1**, 15-19.

Theodorou, D.E. and Queisser, H.J. (1979). Carrier scattering by impurity potentials with spatially variable dielectric functions. *Phys Rev* **B19**, 2092-8.

Thomas, D.G. (1961). Excitons and band splitting produced by uniaxial stress in CdTe. *J Appl Phys Supp* **32**, 2298-304.

Througnumchai, K, Asada, K. and Sugano, T. (1986). Modelling of a 0.1-μm MOSFET on SOI structure using Monte Carlo simulation technique. *IEEE Trans Electron Devices* **33**, 1005-11.

Till, S.J. and Herbert, D.C. (1983) The intra-collisional field effect in semiconductors: II. Numerical results. *J Phys C: Solid State Physics* **16**, 5849-66.

Ting, C.S. and Nee, T.W. (1986). Quantum theory of transient transport in an interacting system of electrons, impurities, and phonons. *Phys Rev* **B10**, 7056-68.

Toennies, J.P. (1990). Surface phonons, past, present and future *Superlattices and Microstructures* **7**, 193-200.

Tomisawa, K., Yokoyama, K. and Yoshii, A. (1988). Non-stationary carrier dynamics in quarter-micron Si MOSFET's *IEEE Trans Computer Aided Design* **7**, 254-8.

Trakhtenberg, L.I. and Flerov, V.N. (1983). Multiphonon theory of kinetic processes in amorphous dielectrics. *Sov Phys JETP* **58**, 146-54.

Tsironis, C. and Dekkers, J.J.M. (1980). Gunn oscillations in thin GaAs epilayers and MESFETs. *IEEE Proc I: Solid State and Electron Devices* **127**, 241-9.

Valdmanis, J.A. (1987). 1 TH_z-bandwidth prober for high-speed devices and integrated circuits. *Electr Lett* **23**, 1308-10.

Vasconcellos, A.R. and Luzzi, R. (1980). Coupled electron-hole plasma-phonon system in far-from-equilibrium semiconductors. *Phys Rev* **B22**, 6355-63.

Vogl, P., Hjalmarson, H.P. and Dow, J.D. (1983). A semi-empirical tight-binding theory of the electronic structure of semiconductors. *J Phys Chem Sol* **44**, 365-78.

Walukiewicz, W., Lagowski, L., Jastrzebski, L., Lichtensteiger, M. and Gatos, H.C. (1979). Electron mobility and free-carrier absorption in GaAs: determination of the compensation ratio. *J Appl Phys* **50**, 899-908.

Wang, T. and Hess, K. (1985). Calculation of the electron velocity distribution in high electron mobility transistors using an ensemble Monte Carlo method. *J Appl Phys* **57**, 5336-9.

Wang, W.B., Ockman, N., Yan, M. and Alfano, R.R. (1989). The intervalley $X_6 \rightarrow \Gamma_6$, L_6 scattering time in GaAs measured by ultrafast pump-probe infra-red absorption spectroscopy. *Solid State Electronics* **12**, 1337–45.

Warriner, R.A. (1977). Distribution function relaxation times in gallium arsenide. *IEE Proc I: Solid State and Electron Devices* **1**, 92–6.

Wiley, J.D. (1970). Valence-band deformation potential for the III–V compounds. *Solid State Communications* **8**, 1865–8.

Wiley, J.D. and di Domenico, M. (1970). Lattice mobility of holes in III–V compounds. *Phys Rev* **B2**, 427–33.

Willig, H.A. and de Santis, P. (1977). Modelling of Gunn domain effects in GaAs MESFETs. *IEE Proc I: Solid State and Electron Devices* **13**, 537–9.

Yamaguchi, E. (1984). Theory of defect scattering in two-dimensional multisubband electronic systems on III–V compound semiconductors. *J Appl Phys* **56**, 1722–7.

Zappe, H.P. and Moglestue, C. (1990). Electroluminescence from Gunn domains in GaAs/AlGaAs heterostructure field-effect transistors. *J Appl Phys* **68**, 2501–3.

Zener, C. (1934). A theory of the electrical breakdown of solid dielectrics. *Proc Royal Soc* **A145**, 523–9.

van der Ziel, A. and Chenette, E.R. (1978). Noise in solid state devices. *Adv Electron Phys* **46**, 313–83.

Ziman, J.M. (1960). *Electrons and Phonons*, Clarendon Press, Oxford.

Zimmermann, J. and Constant, E. (1980). Application of Monte Carlo technique to hot carrier diffusion noise calculation in unipolar semiconducting components. *Solid State Electronics* **23**, 915–25.

Zollner, S., Kircher, J., Cardona, M. and Gopalan, S. (1989). Are transverse phonons important for Γ–X-intervalley scattering? *Solid State Electronics* **32**, 1585–9.

Zollner, S., Schmid, U., Christensen, N.E., Grein, C.H., Cardona, M. and Ley, L. (1990). LMTO and EPM calculations of strained valence bands in GaAs and InAs. In Anastassakis, E.M. and Joannopoulos, J.D. (eds) *20th International Conference on The Physics of Semiconductors*, World Scientific, Singapore, 1735–8.

Zook, J.D. (1964). Piezoelectric scattering in semiconductors. *Phys Rev* **136**, A869–A878.

List of Symbols

SUBSCRIPT LABELS

1,2,3	Components
$2o$	Second order optical phonon
a	Anion site
ac	Acoustic phonon
Ai	Acceptor type i
al	Alloy
b	Site in Bravais cell
C	Conduction band
c	Cation site
cr	Crystal
Di	Donor type i
e	Electron
ep	Electron-plasmon
ex	Exchange
h	Hole
hh	Heavy hole
i	Components, electron index or counting index
ii	Ionised impurity scattering
iv	Intervalley
j	Components, electron index or counting index
$\mathbf{k}$	Wave vector of electron or hole, of fermion
L	L-minimum of conduction band
l	Longitudinal component or orbital quantum number
$\mathbf{l}$	Position of Bravais cell in crystal
lh	Light hole
Lt	Lattice
m	Magnetic quantum number
n	Principal quantum number or energy level
o	Optical phonon
po	Polar optical phonon
$\mathbf{q}$	Wave vector of phonon or boson
s	Spin
sf	Surface or interface
so	Split-off hole
T	Total
t	Transversal component
V	Valance band
X	X-minimum of conduction band
x, y, z	Cartesian components
α	Valence or conduction band extremum
Γ	Γ-minimum of conduction band
θ	Polar angle component

$\boldsymbol{\kappa}$	Set of quantum numbers
ϕ	Azimuth angle component
λ	Angular quantum number
ξ, η, ζ	Cartesian components of Fourier transforms

SUPERSCRIPTS

A	Phonon absorption
E	Phonon emission
i, j	Components
0	Unperturbed state
p	Polarisation
T	Transpose
x, y, z	Cartesian components
$+$, $-$	Ionisation charge

SYMBOLS

Note: The sub or superscript ι (iota) refers to one of the symbols given above.

$\lvert 0\rangle_k \lvert 1\rangle_k$	Empty and occupied, respectively, state of wave vector **k**
A	Cross-sectional area
A'	Parameter relating to the band structure of the valance band
$\mathbf{A}$	Electromagnetic vector potential
a	Lattice constant,
$\mathbf{a}_1$, $\mathbf{a}_2$, $\mathbf{a}_3$	Basis vectors of cubic crystallographic cell
a_B	Bohr radius
a_g	Parameter entering G ($\mathbf{k}$, $\mathbf{k}'$)
a_i	Lattice constant along a crystallographic cell edge for non-cubic crystals
$\mathbf{a}_i$	Basis vector of crystallographic unit cell
a_k	Bloch function overlap parameter
a_q^p, a_q^{+p}	Annihilation and creation operators for bosons of polarisation p and wave vector $\mathbf{q}$
a_r	Auxiliary parameter for generating random numbers
$\mathbf{B}$	Magnetic flux
B'	Parameter relating to the band structure of the valence band
$\mathbf{B}_H$	Brook–Herring transformation matrix
b_g	Parameter entering G ($\mathbf{k}$, $\mathbf{k}'$)
$\mathbf{b}_i$, $\mathbf{b}_1$, $\mathbf{b}_2$, $\mathbf{b}_3$	Basis vector of primitive crystallographic unit cell
$\mathbf{b}_i^*$, $\mathbf{b}_1^*$, $\mathbf{b}_2^*$, $\mathbf{b}_3^*$	Reciprocal basis vector of primitive crystallographic unit cell
B_f	Bulk factor
b_{rp}	$3/d_{rp}$
B_z	z component of magnetic flux
C	Various constants
c	Velocity of light
C'	Parameter relating to the band structure of the valence band

C_A	Stoichiometric ratio of atoms sharing a crystallographic site
C_{gd}	Gate to drain capacitance
C_{gs}	Gate to source capacitance
c_g	Parameter entering $G(\mathbf{k}, \mathbf{k}')$
$\mathrm{c}_{\mathbf{k}}$	Bloch function overlap parameter
$c_{\mathbf{k}}$, $c_{\mathbf{k}}^{+}$	Annihilation and creation operator for fermions of wave vector **k**
c_k, c_n	Expansion coefficient of perturbed wave function in terms of unperturbed ones
c_r	Coefficient in formula for random number generator
c_{tr}	Trapping rate parameter
C_{sd}	Source to drain capacitance
C_{sg}	Source to gate capacitance
$C(T)$	Noise correlation function
$C(\omega)$	Fourier transform of noise correlation function
D	Denominator
$\mathbf{D}$	Electric displacement
d	Depletion depth
D_n	Electron diffusivity
D_p	Hole diffusivity
d_{rp}	Average distance from heterojunction
$D_S(\mathbf{q})$, $D_S(\mathbf{k})$	Density of states for phonons and electric carriers, respectively
D_S^a	Density of available or free states
$D_S^a(E)$	Density of states per unit interval of energy
E	Energy
E'	Energy after scattering
e	Elementary electronic charge
E_0	Unperturbed energy
E_n^0, E_k^0	Unperturbed energy level n, k
E_a^p	Energy of acoustic phonons of polarisation p
E_{al}	Potential inside a hard sphere
E_C	Valence band energy gap
E_{cc}	Binding energy of a Cooper pair
E_F	Fermi energy
E_G	Energy gap between the conduction and the valence band
E_{imp}	Threshold energy for impact ionisation
E_{iv}^p	Energy of intervalley phonons of polarisation p
E_k^j	Kinetic energy associated with motion in the j direction
E_M	Deformation potential energy
$E_{\min}$, $E_{\max}$	Minimum and maximum energy, respectively
E_{nt}	Energy level n associated with energy ladder t
E_o^p	Energy of optical phonons of polarisation p
e_s	Superparticle charge
E_{so}	Energy of split-off hole
E_T	Total energy of the entire crystal
E_{tr}	Trap energy level
E_V	Valence band edge energy
E_ι	Energy
$\mathbf{F}$	Electric field

$f(E)$	Fermion or boson distribution function
F_n^C	Correlation function for n-particles
F_{int}	Electric field perpendicular to an interface
$f_h(E)$	Hole distribution function
$F_j(\eta)$	Fermi integral of order j
f_j	Particle distribution within a narrow range of energies
F_k	Cartesian component of electric field
F_L	Lorentz force
F_{min}	Intrinsic minimum noise figure
F_{ox}	Electric field in oxide
$f(t)$	Function of t, statistical distribution
F_x, F_y, F_z	Cartesian components of electric field
F_ι	Component of electric field
$\mathbf{G}$	$2\pi\mathbf{g}$: reciprocal lattice vector
$\mathbf{G}$	Electromagnetic tensor
g	Degeneration factor
$\mathbf{g}$, $\mathbf{g}_3$, $\mathbf{g}_4$	Contravariant lattice vectors
$\mathbf{G}_{bb'}(\mathbf{h})$	Interatomic force tensor
$\mathbf{G}_{bb'\mathbf{q}}$	Fourier component of interatomic force tensor
$\mathbf{G}_{bb,\mathbf{q}ij}$	Matrix element of $\mathbf{G}_{bb,\mathbf{q}}$
$\mathbf{G}_i$, $\mathbf{G}_j$	Reciprocal lattice vectors
g_i	Degeneration factor for donor or acceptor atom type i
G_j	Number of quantum states within a narrow energy range
$G(\mathbf{k}, \mathbf{k}')$	Overlap integral between Bloch functions
$\mathbf{G}_{1b;1'b'}$	Interatomic force tensor
g_m	Transconductance
G_p	Power gain
$G(\mathbf{r}, \mathbf{r})$	Green's function
Gu	Unilateral gain
G_v	Voltage gain
$\mathbf{H}$	Magnetic field
$\mathbf{h}$	Distance between atoms
$\hbar$	Plancks' constant divided by 2π
$\mathcal{H}_0$	Unperturbed Hamiltonian
$\mathcal{H}_1, \mathcal{H}_2, \ldots$	Part Hamiltonians
h_c	Size of mesh cell
$\mathcal{H}_{\text{el-ph}}$	Electron–phonon interaction Hamiltonian
h_i	Size of mesh cell
$\mathcal{H}^{ij}$	Pseudopotential Hamiltonian
$\mathcal{H}_{kp}$	$\mathbf{k}\bullet\mathbf{p}$ perturbation Hamiltonian
$H_n(q)$	Hermitian polynomial of degree n
$\mathcal{H}_{ph}$	Photon Hamiltonian
$\mathcal{H}_{tr}$	Trapping Hamiltonian
h_x, h_y, h_z	Size of mesh cell along the three Cartesian directions
$\mathcal{H}_\iota$	Hamiltonian
i	Imaginary unit $[\sqrt{-1}]$
$\mathbf{I}$	Identity tensor
I, I_1, I_2	Electric current
I_d	Drain current
I_g	Gate current
I_i	Current through terminal i of an N-port

I_i'	Extrapolated current through terminal *i* of an *N*-port
$I_i(\omega)$	Fourier transform of I_i
$\langle I_{nc} \rangle$	Fluctuation current
I_{ov}	Wave function overlap integral
I_s	Source current
I_y	*y* component of current
j	As label: position of electrons or holes, Energy interval for arrangement of particles
$J(T)$	Johnson noise function
$\mathbf{J}$	Current density
$\mathbf{k}$	Fermion wave vector
$\mathbf{k}'$	Wave vector just after scattering
$\|\mathbf{k}\rangle$	Wave function of electron or hole of wave vector **k**
$\mathbf{K}$, $\mathbf{K}_i$	Difference wave vector
$\mathbf{k}_0$	Wave vector at beginning of field adjusting time step
k_B	Boltzmanns' constant
k_{Br}	Brillouin radius or distance to a surface of the Brillouin zone
k_j	Wave vector component
$\mathbf{k}_m$	Position of conduction band minimum
k_{ph}	Photon wave number
$\mathbf{k}_s$	Wave vector of state **k** prior to the action of a perturbation
k_x, k_y, k_z	Cartesian components of wave vector
$\mathbf{k}_\iota$	Wave vector band extremum or component
l	Orbital quantum number
$\mathbf{l}$	Triplet of three integers
l_1, l_2, l_3	Integers
L_g	Gate length
L_{gd}	Gate to drain distance
L_j	Dimension of crystal along j^{th} component
l_j	Sundry integer
$L_j^i(\rho)$	Associated Legendre polynomial
ln	Natural logarithm
$\log_{10}$	Brigg's logarithm
L_{sd}	Source to drain distance
L_{sg}	Source to gate distance
L_x, L_y, L_z	Width of crystal along Cartesian components
m	Magnetic quantum number
m^*	Effective mass
$\mathbf{m}^*$	Effective mass tensor
m_0	Rest mass of free electron
M_a	Mass of atom occupying an anion site
M_b	Mass of atom labelled *b* in Bravais cell
M_c	Mass of atom occupying a cation site
m_D^*	Density of states effective mass
M_{fs}	Matrix element for transition from state *k* to *s* and *vice versa*
M_i	Mass of ion at $\mathbf{r}_1$
m_{ij}^*	Component of effective mass tensor
M_{ijk}	Components of piezoelectric tensor

$M_{kk'}$	Matrix element for transition from **k** to **k**′
$M_{kk,\iota}$, $M_{kk',\iota}{}^{i}$	Matrix element for scattering type ι
m_r^*	Reduced effective mass
m_ι^*	Effective mass
N	Number of real particles
N	Triplet of the integers N_i, N_j, N_k
n	Miscellaneous integers, principal quantum number
$\|n\rangle$	Dirac notation for Ψ_n
N_1, N_2, N_3, N_i, N_j	Number of Bravais cells in the crystal along the directions of three basis vectors of the Bravais cell
N_A	Number of different acceptors
N_C	Density of cation states
n_c'	Electron density factor
N_{cr}	Number of Bravais cells in a crystal
n_{depl}	Sheet number density of depleted charges
N_D	Number of different donors
n_e	Number density of holes
n_i, n_j, n_k	Integer variables
n_{im}^+	Number density of ionised atoms
n_{inv}	Inversion sheet charge number density
N_j	Number of occupied quantum states within a narrow range of energies
n_o^p	Occupation number of optical phonons of polarisation p
N_{ps}	Measure for how far a pseudopotential calculation is to be carried out
$n_{\mathbf{q}}^p$	Occupation number of phonons of wave vector **q** and polarisation p
N_r	Length of sequence of flat random numbers
n_r	Refractive index
$n_{iv,\mathbf{q}}^p$	Occupation number of intervalley phonons of wave vector **q** and polarisation p
$\|n_{\mathbf{q}}^p\rangle$	Wave function of n phonons of polarisation p and wave vector **q**
N_s	Number of superparticles
N_t	Duration of simulation in field-adjusting time steps
N_u	Number of atoms in Bravais cell
N_v	Number of valence electrons used in a pseudopotential calculation
N_x, N_y, N_z	Number of mesh cells along the three Cartesian directions
n_x, n_y, n_z	Integer variables
N_Γ	Number of different scattering events
N_ι	Quantity, number
$n\lambda\iota$	Number density
$\mathcal{O}$	General quantum operator
P	Electric polarisation vector
P	Momentum
$\mathbf{P}_i$	Momentum of electron or hole labelled i
p_h	Number density of holes
P_{imp}	Impact ionisation parameter
$\mathbf{P}_{\mathrm{lb}}$	Momentum of nucleus labelled b in Bravais cell at $\mathbf{r}_1$
$p_l^m(\xi)$	Associated Legendre function

$\mathbf{P}_{\mathbf{q}b}$	Fourier component of nuclear momentum
$P(t)$	Probability distribution of distribution of scattering times
p_x, p_y, p_z	Cartesian components of **p**
P_z	Expectation value of z component of electron momentum
$\mathbf{q}$	Phonon or boson wave vector
q, q_1, q_2, q_3	Magnitude and components of **q**
Q_c	Doping charge of mesh cell
q_c	Magnitude of wave vector below which plasmon scattering can take place
Q_i	Charge flowing through terminal i of an N-port
$\mathbf{q}_{\text{iv}}^{p}$	Phonon wave vector coinciding with a conduction band minimum
$\mathbf{q}_k$	Cartesian component of the phonon wave vector
Q_n	Probability distribution of nth scattering event
q_x, q_y, q_z	Cartesian components of phonon or boson wave vector
R	Generation and recombination rate, Radius of electronic cloud
$\mathbf{r}$	Position vector
r	Magnitude of position vector
$\mathbf{r}_a, \mathbf{r}_c$	Position of cation and anion site, respectively, in the Bravais cell
r_{al}	Radius of hard sphere
R_d	Drain resistance
r_d	Displacement from equilibrium in deformed lattice
R_g	Gate resistance
R_i	Internal resistance
r_i, r_k	Random numbers
$\mathbf{r}_i, \mathbf{r}_j$	Position of electron or hole
$\mathbf{r}_\mathbf{l}$	Position of reference corner of Bravais cell
$\mathbf{r}_{\mathbf{l}b}$	Position of nucleus labelled b in unit crystallographic cell labelled **l**
$\mathbf{r}_{\mathbf{l}b}^{0}$	Equlibrium position of nucleus labelled b in Bravais cell at $\mathbf{r}_\mathbf{l}$
$\mathbf{R}_\mathbf{N}$	Displacement across crystal
R_s	Source resistance
$\mathbf{S}$	Unit vector perpendicular to a surface
s	Spin quantum number
$\mathbf{s}$	Spin matrix (vector)
$\|s\rangle$	Quantum state prior to perturbation
$\mathbf{S}_1$	Transformation matrix
$\|S\rangle$	s-orbital
S_e	Entropy
S_{ij}	Stress component
s_{ij}	s-parameter
$\mathbf{S}_{\mathbf{l}b}$	Displacement of nucleus labelled b in Bravais cell at $\mathbf{r}_\mathbf{l}$
s_p	Free flight path
$\mathbf{S}_{\mathbf{q}b}$	Displacement of nucleus labelled b in Bravais cell at $\mathbf{r}_\mathbf{l}$
$\mathbf{S}_{\mathbf{q}b}$	Fourier component of atomic position
$\mathbf{S}_x, \mathbf{S}_y, \mathbf{S}_z$	Rotation matrices connecting directions of symmetry in the Brillouin zone
s_x, s_y, s_z	Cartesian components of the spin operator **s**

Symbol	Meaning
$S(\omega)$	Noise power spectrum
t	Time or independent variable
$t_1, t_2, \ldots$	Scattering times
T	Duration of a simulation
$\mathfrak{T}$	Probability of transmission through a barrier
t_D	Thickness of absorbing layer of a photodetector
T_e	Temperature of electron or hole distribution
δt	Field-adjusting time-step
T_f	Temperature of a distribution
t_f	Final value
t_k	Time variable
t_{kf}, t_{ks}	Final and start value, respectively
T_{ks}, T'_{ks}	Probability of transfer from state k to s
T_{Lt}	Lattice temperature
T_{1b}	Kinetic energy of nucleus labelled b in Bravais cell at $\mathbf{r}_1$
t_s	Start value
T_ι	Temperature
U	Perturbation Hamiltonian
U_{al}	Alloy scattering potential
U_d	Dislocation potential
U_e	Interaction between an electron and all nuclei of the crystal
$U_{e,1b}$	Interaction between an electron and one nucleus
U_G	Overlap integral, transition matrix between two quantum states
$u_k(\mathbf{r})$	Bloch function
U_{q1b}	Interaction between an electron and a nucleus labelled b in unit crystallographic cell labelled $\mathbf{l}$
u_k	Bloch function
U_{1b}	Potential energy of nucleus labelled b of Bravais cell at $\mathbf{r}_1$
$U_{1b;1'b'}$	Energy of interaction between atoms at $\mathbf{r}_1$ and $\mathbf{r}'_1$
$\mathbf{u}^p$	Velocity of acoustic phonons of polarisation p
$\mathbf{U}^p_{iv}$	Velocity of intervalley phonons of polarisation p
$\mathbf{v}$	Group velocity of charge carriers
V_1, V_2	Input and output bias
v_1, v_2	Pseudopotential form factors
V_A	Antisymmetric part of pseudopotential
V_a	Volume of crystallographic unit cell
V_B	Heterojunction barrier
V_C	Symmetric part of pseudopotential
V_{cr}	Volume of crystal
V_d	Drain potential of bias
V_{dr}	Drift velocity
V_g	Gate potential
V_{gs}	Gate bias
V_j	Voltage of terminal j of an N-port
$V_j(\omega)$	Fourier transform of V_j
$V(\mathbf{r})$	Electrostatic potential of an ionised impurity at $\mathbf{r}$
V_s	Symmetric part of pseudopotential
V_S	Schottky contact potential

v_s	Saturation drift velocity
V_{sf}	Surface potential
V_u	Volume of Bravais cell
V_ι	Volume
W	Width of transistor
w	Width of quantum well
W_1, W_2	Input and output power
$W(\mathbf{k},\mathbf{k}')$	Rate of transfer from state **k** to **k**′
$\|X\rangle$	*p*-orbital with lobes along the *x* axis
x_0	Seed of random number generator
x_i	*x* component of $\mathbf{r}_i$
X_i'	Random integer number
x_{1b}	*x* component of $\mathbf{S}_{1b}$
y	Admittance matrix
$\|Y\rangle$	*p*-orbital with lobes along the *y* axis
y_i	*y* component of $\mathbf{r}_i$
y_{ij}, y_{11}, y_{12}, y_{21}, y_{22}	*y* parameters
y'_{kj}	*y* parameter for a device of width different from the unit one
y_{1b}	*y* component of $\mathbf{S}_{1b}$
$Y_1^m(\theta,\phi)$	Spherical harmonic function
Z	Number of elementary charges of nucleus or ionised atom
$\|Z\rangle$	*p*-orbital with lobes along the *z* axis
Z_1, Z_2	Input and output impedance
Z_b	Number of elementary charges of a nucleus labelled *b*
z_i	*z* component of $\mathbf{r}_i$
Z_l	Load impedance
z_{1b}	*z* component of $\mathbf{S}_{1b}$
α	Non-parabolicity factor
α_{1b}, α_{2b}, α_{3b}, α_{ib}	Non-negative numbers smaller than 1
α'	Lagrange multiplier
α_λ	Absorption coefficient for light
α_s	Electron cloud parameter
β	Angle as dummy integration variable
β'	Lagrange multiplier
β_{ph}	Phonon screening length
β_s	Screening length for carriers
Γ	Sum of all real scattering rates
Γ_0	Sum of all scattering rates including self-scattering
γ_0	Factor in connection with Fröhlich scattering
Γ_j	Number of arrangements of particles within a small range of energies
$\Gamma(j)$	Gamma function
γ_r	Factor in connection with ionised impurity scattering
Δ	Small, but not infinitesimal, increment
$\Delta(E)$	Thermal disequilibrium function
$\Delta\mathbf{I}$	Incremental current
$\delta(x)$	Delta function of x
δ_{nm}, δ^{nm}	Kronecker delta

δT	Field adjusting time step
$\Delta \mathbf{V}$	Incremental bias
ε	Dielectric constant
$\boldsymbol{\varepsilon}$	Dielectric tensor
ε_0	Permittivity of vacuum
ε_{ox}	Dielectric constant of oxide
ζ	Wavefunction perpendicular to a barrier
η	Reduced energy
η'	Position of conduction band minimum along $\langle 111 \rangle$
η_F	Reduced Fermi energy
θ	Polar angle
λ	Wavelength
λ'	Fourier parameter
λ_0	Self-scattering rate
λ_D	Debye length
λ_{ee}	Intercarrier scattering rate
λ_i	Scattering rate from mechanism i
λ_{imp}	Impact ionisation rate
λ_{ni}	Neutral impurity scattering rate
λ_{pe}	Piezoelectric scattering rate
λ_{po}	Polar optical phonon scattering rate
λ_{rl}	Release rate
λ_{sp}	Spontaneous photon generation
λ_{st}, λ_{st}^{ι}	Stimulated photon emission or absorption rate
λ_{tr}	trapping rate
λ_{ι}	Scattering rate for scattering type ι
μ	Relative permeability
$\boldsymbol{\mu}$	Relative magnetic permeability tensor
μ'	Fourier parameter
μ_0	Permeability of vacuum
μ_n, μ_p	Mobility of electrons and holes, respectively
ν	Frequency
ν_{11}, ν_{12},. . .	Force constant parameters to calculate the phonon spectrum
ν_s	Number of particles or superparticles required to obtain charge neutrality in a mesh cell
ν_{ι}	Degeneracy factor
ν_c	Number of equivalent energy minima
ξ	Distance from centre of Brillouin zone
ξ'	Position of conduction band minimum along $\langle 100 \rangle$
$\xi_{\mathbf{q}b}^{p}$	Phonon polarisation vector for atom labelled b with polarisation mode of a phonon, p
$\xi_{\mathbf{q}bj}^{p}$	Component of ξ_{qbp}
Ξ_{ι}	Deformation potential
ρ	Mass density
ρ_{depl}	Depletion charge sheet density
ρ_e	Charge density
ρ_{sf}	Surface sheet charge density
$\rho_{i,j,k}$,$\rho_{i,j}$	Discretised charge density in 3 and 2 dimensions, respectively

σ	Product of numbers
σ_e	Specific conductivity
$\sigma_x, \sigma_y, \sigma_z, \boldsymbol{\sigma}$	Pauli spin matrices
τ	Time of free flight
$\boldsymbol{\tau}$	$\frac{1}{2}\mathbf{r}_c$
τ_E	Energy relaxation time
τ_p	Momentum relaxation time
τ_t	Transit time of a signal through a device
τ_p	Period time
$\upsilon, \upsilon_1, \ldots$	Various constants used to calculate the phonon spectra
ϕ	Azimuth
Φ	Electrostatic potential
Φ_{im}	Electrostatic potential surrounding an ionised impurity
$\Phi_{k,j,i}, \Phi_{j,i}$	Discrete value for electrostatic potential in three or two dimensions, respectively
ϕ_m	Work function
$\Phi^p_{\mathbf{q}}$	Periodic potential due to oscillatory motion of polarisation p
$\Phi_{\mathbf{rr}'}$	Electrostatic potential at $\mathbf{r}$ from a mobile carrier at $\mathbf{r}'$
Ξ	Susceptibility tensor
Ψ_A, Ψ_B	Electronic wave function in area A and B, respectively
Ψ	Perturbed wave function
Ψ^0	Unperturbed wave function
Ψ_k	Electronic wave function $\vert\mathbf{k}\rangle$
Ψ^i_k	Electronic wave function associated with the ith Cartesian component
$\Psi^0_n(0)$	Unperturbed wave function at time 0
$\psi_{\text{nlms}}, \psi_{\text{nlms},\mathbf{k}}$	Atomic orbital
Ψ_ι	Wave function
Ψ_{ik}	Envelope wave function perpendicular to a quantum well
Ψ_{nk}	Envelope wave function perpendicular to a quantum well for the nth energy level of energy ladder t
Ψ_r	Radial part of electronic wave function
ω	Angular frequency
$\omega^p_{iv,\mathbf{q}}$	Angular frequency of intervalley phonons of wave vector $\mathbf{q}$ and polarisation p
ω_{ks}, ω_{kn}	Energy difference between levels divided by $\hbar$
Ω_n	Eigenvalue of any general quantum operator
ω^p_o	Angular frequency of optical phonon of polarisation p
ω_p	Plasma frequency
$\omega^p_{\mathbf{q}}$	Angular frequency of a phonon of wave vector $\mathbf{q}$ and polarisation p
$\uparrow$	Spin up
$\downarrow$	Spin down

Index